BIOMEMBRANES

Volume 9

BIOMEMBRANES

A series edited by Lionel A. Manson
The Wistar Institute, Philadelphia, Pennsylvania

Recent Volumes in this Series

1972 . Biomembranes . Volume 3
Passive Permeability of Cell Membranes
Edited by F. Kreuzer and J. F. G. Slegers

1974 . Biomembranes . Volume 4A
Intestinal Absorption
Edited by D. H. Smyth

1974 . Biomembranes . Volume 4B
Intestinal Absorption
Edited by D. H. Smyth

1974 . Biomembranes . Volume 5
Articles by Richard W. Hendler, Stuart A. Kauffman, Dale L. Oxender,
Henry C. Pitot, David L. Rosenstreich, Alan S. Rosenthal, Thomas K. Shires,
and Donald F. Hoelzl Wallach

1975 . Biomembranes . Volume 6
Bacterial Membranes and the Respiratory Chain
By N. S. Gel'man, M. A. Lukoyanova, and D. N. Ostrovskii

1975 . Biomembranes . Volume 7
Aharon Katzir Memorial Volume
Edited by Henryk Eisenberg, Ephraim Katchalski-Katzir, and Lionel A. Manson

1976 . Biomembranes . Volume 8
Articles by Robert W. Baldwin, William C. Davis, Paul H. DeFoor,
Carl G. Gahmberg, Sen-itiroh Hakomori, Reinhard Kurth, Lionel A. Manson,
Michael R. Price, and Howard E. Sandberg

1977 . Biomembranes . Volume 9
Membrane Transport—An Interdisciplinary Approach
By Arnošt Kotyk and Karel Janáček

A Continuation Order Plan is available for this series. A continuation order will bring delivery of each new volume immediately upon publication. Volumes are billed only upon actual shipment. For further information please contact the publisher.

BIOMEMBRANES, Volume 9

MEMBRANE TRANSPORT
An Interdisciplinary Approach

Arnošt Kotyk and Karel Janáček

Institute of Microbiology
Czechoslovak Academy of Sciences, Prague

PLENUM PRESS • NEW YORK AND LONDON

ISBN-13: 978-1-4684-3335-7   e-ISBN-13: 978-1-4684-3333-3
DOI: 10.1007/978-1-4684-3333-3
Library of Congress Catalog Card Number: 75-18942

Scientific Editor: Prof. Dr. J. Koštíř, DrSc.
Scientific Advisers: Prof. Dr. J. Koryta, DrSc.
Doc. Dr. Ľ. Drobnica, CSc.

Published in coedition by ACADEMIA
Publishing House of the Czechoslovak
Academy of Sciences, Prague

and outside of the Socialist countries by
PLENUM PRESS
A Division of Plenum Publishing Corporation
227 West 17th Street, New York, N.Y. 10011

PREFACE

Not many years ago, problems of membranes and transport attracted the attention of but a few dozen enthusiasts, mainly physiologists who recognized the significance of membranes for the stabilization of the general steady state of organisms. The first symposium organized some fifteen years ago could boast of the attendance of perhaps fifty scientists (the remaining fifty were not yet sure that membranes was the topic of their choice), ranging in specialization from physical chemistry to bacterial genetics, who clairvoyantly decided to study what now has become the number one subject at most congresses of biophysics, physiology, and even biochemistry and microbiology.

As is the case with many rapidly developing fields, the interest in membranes and transport seems to be growing out of bounds and the whole field of membranology, interdisciplinary as it is, has penetrated into the realms of a number of branches of physics, chemistry, and biology. Its subject is primarily biological and, although much has been done in the world to increase the "exactness" of biology over the past thirty years, one cannot strive for a rigorous mathematical description of biological phenomena since, as M. H. Jacobs wrote appropriately back in 1935, "In the first place, it is utterly hopeless for the biologist with the means at present at his disposal, to reduce the variables that enter into his problem to the small number usually encountered in physical investigations. He is compelled, therefore, regretfully but of necessity, to be content with a lesser degree of precision in his results than that attainable in the so-called 'exact sciences'. It follows that in dealing with most biological problems it is not only useless, but actually unscientific, to carry mathematical refinements beyond a certain point, just as it would be both useless and unscientific to employ an analytical balance of the highest precision for obtaining the growth-curve of a rat."

The present book is rather an attempt at bringing together data on membranes and their primary function, viz. transport of substances, in a relatively small volume. Recent literature abounds with compendia, highly specialized monographs and conference proceedings dealing with various aspects of membranes, but it is our feeling that there is a need for an introductory text that would integrate at least most of the available information about membranes and transport.

Several years ago, it was possible to include in a single volume devoted exclusively to transport studies (Kotyk and Janáček, Plenum Press, New York, 1970; 2nd edition 1975) a large section of comparative aspects, including various microorganisms, plants and animal tissues. At present, there are so many pieces of information available from various biological objects that they deserve specialized reviews. For this reason, only generally valid and most important observations are included in the present book while, on the other hand, due attention is being paid to membrane functions other than transport, including their physical chemistry and biochemistry.

We are aware of the fact that some chapters of the present book are better than others because some of the authors' personal experience and perhaps even hobbies are reflected in them. We hope that the reader will not fail to peruse the more erudite monographic works that we refer to at the appropriate places in the text if he wishes to obtain more enlightenment.

In preparing the manuscript we were greatly aided, both through consultations and through technical assistance, by some of our co-workers, in particular Dr. K. Sigler, Dr. J. Horák, and Dr. R. Rybová of this laboratory. The most competent photography of Dr. A. Wolf of this Institute is gratefully acknowledged. When mentioning these collaborators, we do not overlook the general stimulating atmosphere created by the whole staff of the department of membrane transport. Without it, this book could not have been written.

Arnošt Kotyk
Karel Janáček

Laboratory for Cell Membrane Transport
Institute of Microbiology
Czechoslovak Academy of Sciences

CONTENTS

Preface

8

1. INTRODUCTION

"Whence is it that Nature doth nothing
in vain; and whence arises all that Order
and Beauty which we see in the World?"

Isaac Newton, Opticks

The principal aim of the present book is to show how the inter-disciplinary scientific branch of steadily growing importance, which may be called biological membranology, explains phenomena observed on biological membranes in terms of physics, chemistry and biology. This introduction follows a similar purpose, limiting itself, however, to considerations of the most general nature; its second section re-presents a speculative attempt to evaluate the biological survival value of membrane transport and the third considers the transport as a physical phenomenon taking place in space and time, whereby certain general characteristics of its physical description by mathe-matical formulae are predetermined. Still, although most diverse developments in the whole realm called by our ancestors natural history and natural philosophy are thus seen to contribute directly or indirectly to membranological theories, some of the scientific exploits of the past are especially closely related to them. The first section of this introduction mentions briefly these milestones in the early history of membranology.

1.1. IMPORTANT EVENTS
IN THE HISTORY OF MEMBRANOLOGY

It may be a relatively fair estimate to state that the science of membranology began with German physiologists and botanists at about the middle of the last century. In the forties, one of the celebrated pupils of the great physiologist Johannes Peter Müller (1801 to 1858), Emil Heinrich Du Bois - Reymond (1818 – 1896) describes the electrical potential difference across surviving frog skin (Untersuchungen über tierische Elektricität, 1. Band, Berlin, 1848) and in 1851 the German physiologist Hugo von Mohl (1805 – 1872) describes plasmolysis of plant cells, assuming that their cell wall functions as a membrane (Grundzüge der Anatomie und Physiologie der vegetabilischen Zelle, Braunschweig). In 1855, the Swiss-born botanist Karl Wilhelm von Nägeli (1817 – 1891) explains the osmotic behavior of cells by the presence of a semipermeable cell membrane (Nägeli, K. und Cramer, K.: Pflanzenphysiologische Untersuchungen, Zürich) and in the same year the physiologist Adolf Eugen Fick (1829 – 1901) derives the phenomenological laws of diffusion (Über Diffusion, *Poggendorffs Annalen*, **94**, 59). In 1877, the botanist Wilhelm Pfeffer (1845 – 1920) publishes his "Osmotische Untersuchungen" (Leipzig) in which he postulates the existence of the cell membrane on the basis of similarities in the behavior of cells and osmometers with artificial semipermeable membranes (from deposited copper ferrocyanide, prepared shortly before by Moritz Traube).

In the eighties, the Dutch botanist Hugo Maria de Vries (1848 to 1935) continued with osmotic studies on plant cells, believing that the whole protoplasmic layer between the plasmalemma and the tonoplast functions as a membrane (e.g., Plasmolytische Studien über die Wand der Vacuolen, *Jahrbuch wiss. Botanik*, **16**, 465). His research served as a basis of the physicochemical theories of osmotic pressure and electrolyte dissociation by the Dutchman, Jacobus Hendricus van't Hoff (1852 – 1911), and the Swede, Svante August Arrhenius (1859 – 1927). In 1888, an equation for the liquid junction potential was derived by the German physicist and chemist Walther Hermann Nernst (1864 – 1941); in 1890, the German physical chemist and philosopher Wilhelm Ostwald (1853 – 1932) stressed the probable role of membranes in the generation of bioelectric phenomena. Between the years 1895 and 1902, Ernst Overton (1865 – 1933) measured the permeability of cell surfaces to a great number of substances and demonstrated the relation between their permeability and lipid solubility (e.g., Über die osmotischen Eigenschaften der lebenden Pflanzen- und Tierzelle, *Vierteljahrsschr. Naturforsch. Ges. Zürich*, **40**, 159, 1895; Beiträge zur allgemeinen Muskel- und Nervenphysiologie, *Pflügers Arch.* **92**, 115, 1902.

Edward Waymouth Reid (1862 − 1948) appears to have been the first to introduce a split chamber to study transepithelial transport (Transport of fluid by certain epithelia, *J. Physiol.* 26, 436, 1901). In 1902, Julius Bernstein (1839 − 1917) explains the electrical phenomena in living cells by a membrane hypothesis (Untersuchungen zur Thermodynamik der bioelektrischen Ströme, *Pflügers Arch.* **92**, 521). The existence of the cell membrane was later proved beyond any doubt by experiments of Rudolf Höber (1873 − 1953), who demonstrated a high electrical resistance of whole erythrocytes and a low one of their interior (e.g., Physikalische Chemie der Zelle und der Gewebe, 6th edition, Leipzig, 1926) and by experiments of Robert Chambers (1881 − 1957), who demonstrated by microinjections that substances not permeating easily across the cell surface diffuse freely in the cell interior (A microinjection study on the permeability of the starfish egg, *J. Gen. Physiol.* **5**, 189, 1922). Studies on artificial membranes date back to the pioneering work of Leonor Michaelis (1875 − 1949) who used collodion membranes during his stay in Nagoya, Japan, in the early twenties (see, e.g., *Kolloid Z.*, **10**, 575, 1933) to demonstrate their selective permeability. Likewise, the work of Karl Sollner on membranes started as early as 1930 (*Z. Elektrochem.* **36**, 36, 234, 1930).

With P. J. Boyle and Edward J. Conway, who observed the similarity between distribution of potassium ions in cells and the Gibbs − Donnan equilibrium (Potassium accumulation in muscle and associated changes, *J. Physiol.* **100**, 1, 1941), with H. B. Steinbach who showed that the cell membrane is permeable to sodium ions (Sodium and potassium in frog muscle, *J. Biol. Chem.* **133**, 695, 1940), and with the concept of a sodium pump extruding sodium ions from muscle fibers, as proposed by R. B. Dean (Theories of electrolyte equilibrium in muscle, *Biol. Symp.* **3**, 331, 1941) we are approaching the discovery of active transport and the stage of biological membranology, which forms the subject matter of the present book.

1.2. EVOLUTION AND TRANSPORT

There is little doubt that the transport faculties of living cells and organisms developed in the same way as any other of their useful properties, i.e. as a result of spontaneous random mutations which have afterwards proved their survival (adaptive) value in the process of natural selection. A rather powerful argument in favor of this proposition can be recognized in the fact that in a number of cases elaborate transport mechanisms capable of concentrating nutrients in cells can be seen to disappear when they are of little or no survival value, a satisfactory inflow of these substrates being ensured for long evolutionary periods by some other means. This is obviously the case with a number of organ cells and with anucleate erythrocytes as well as with some species of yeast cells cultivated for millenia.

In this connection we would like to draw the attention to two rather common features of the evolutionary process.

(*1*) The first of these principles, even if its occurrence may not be a logical necessity, is obviously often encountered in the process of evolution: *a single problem of adaptation is often solved in a number of different ways*. We can cite, e.g., the diversity of devices used by various organisms to solve the problem of locomotion or of direct as well as indirect utilization of the sun radiation energy.

(*2*) The second principle appears to be an obvious corollary of the combination of random mutations with natural selection: *There may be a number of adaptive values of a given single mutation and not all of them may be apparent at the same time*. Thus a mutation which was seemingly useless at the time of its appearance may (provided that it was not lethal) prove advantageous in a number of ways when the organism itself and/or its environment is changed.

The lack of uniformity in the mechanisms responsible for transport of nonelectrolyte nutrients across cell membranes may well serve as an illustration of the first principle — simple diffusion, mediated diffusion as well as active transport may be encountered in this field (see, e.g., Kotyk, 1973, and section 4.5. of the present book). Already the simple diffusion across a membrane of lipoid character seems to be an important factor rendering the basic biochemical processes much more efficient. Most important intermediate metabolites (like the organic acids of Krebs' cycle) are very polar compounds and hence they diffuse only slowly across lipoid membranes and do not readily escape from cells, so that sequences of biochemical reaction steps inside the cells are possible (Kubišta, 1974). Concerning the more sophisticated mechanisms, it has been proposed (Holden, 1968) that evolution of specific and even active (i.e., uphill-accumulating) transport mechanisms possibly preceded the evolution of biosynthetic pathways, "or occurred at the same time as an alternative solution to the stress of nutrient depletion". Indeed, an increase in the rate of passive transport of the required nutrient, achieved by the mechanism of a specific equilibrating transport, or even the capability of an active transport system to maintain the intracellular concentration of the nutrient at a higher level than that in the environment, certainly helps to sustain a higher rate of growth than in cells which are less endowed. The abundant growth of organism so developed then might result in a still more pronounced depletion of the environment of the given substrate, with the result that mutants possessing more efficient transport mechanisms (e.g., those with a higher affinity for the substrate) or, alternatively, capable of converting a more abundant precursor into the growth-limiting nutrient, will have a selective advantage. If, on the other hand, the transport mechanisms did not develop at this early stage of evolution, they were likely to develop only much later. For, it is argued, if the biosynthetic enzymes have a lower level of structural complexity than the transporting mechanisms and hence appeared earlier in evolution, a selective advantage may be obtained only by sparing the energy required by a complex biosynthetic pathway under conditions where the biosynthesis is already highly developed and the environment partly re-enriched in organic substrates (Holden, 1968). Be it as it may, "there are more advanced cell types where most unnecessary properties have been weeded out (like the plurality of carrier systems for a single compound in many unicellular organisms), and there are the less advanced cell types which do not yet possess the obviously efficient coupled transports of higher organisms and particularly of their specialized tissues" (Kotyk, 1973).

The second principle, that of a single mechanism of mutational origin serving a multiple biological purpose is perhaps best apparent in the case of the most wide-spread active ion-transporting mechanism, the so-called sodium pump. The hypothesis that the active sodium transport is primarily a volume regulating device appears to be ill-founded now. Not only are sodium pumps operative in plant cells in which the problem of volume control is entirely solved by the presence of a rigid cell wall, but also in animal cells a sodium-independent regulation of the cell volume was demonstrated (Rorive *et al.*, 1972). There are, however, a number of other ways in which sodium pumps are useful for cells and organisms. The gradients of the sodium electrochemical potential serve as energy reservoirs in excitability phenomena as well as in some of the so-called secondary active transports of sugars and amino acids (see section 5.4.). Several intracellular enzymes are activated by potassium ions; sodium ion activation of such enzymes is much rarer or even negative (see, e.g., Ussing, 1960). Although it may not be easy to decide which of the two properties developed first, whether the sodium extrusion from cells or the potassium activation of intracellular enzymes, their combination could represent a useful device of metabolic control. May it not be that the basal metabolism of some cells in the cold is reduced not only by the cold itself but also due to a cold-induced sodium inflow into cells? Finally, homeostasis of whole organisms with respect to salt and water is maintained by specialized tissues in which full use is made of the ability of cells to transport sodium actively — without an active sodium transport the migration of life from seas to fresh water and on dry land would be hardly possible. Thus not only are the transport mechanisms a product of evolution, they are also its integral part; without membrane transport the whole evolution as we know it would be unthinkable.

1.3. TRANSPORT IN SPACE AND TIME

Physical, chemical and biological phenomena of membrane transport take place in space and time. A description of these phenomena is hence bound to include the fundamental physical concepts of space and time explicitly or implicitly. The aim of the present section of the introduction is to characterize the mathematical means used to describe transport phenomena and to point out how they reflect the properties of space and time variables.

As in other branches of natural science, statements of qualitative or semiquantitative nature (like "the substance does or does not permeate a given membrane" or "the substance is accumulated on one side of the membrane") are only an initial step toward a quantitative description of membrane transport phenomena. From the time of Newton and Leibniz, scientists have attempted to derive quantitative laws by establishing, from fundamental principles, such relations between states as are only infinitesimally distant in space and time. Thus, instead of trying to comprehend in one step the laws governing various processes as a whole, they are looking first for microscopic laws of evolution known as differential equations. By establishing relations between states neighbouring in time, the differential equations are an expression of the deterministic character of natural phenomena, of causality in nature.

There are two different types of differential equations, partial differential equations, containing more independent variables, and ordinary differential equations, containing only one independent variable. It might seem that the behavior of quantities which are of interest in transport studies, e.g., of concentrations, is properly described only by partial differential equations, quantities of this type having a definite value at each point of space (given by three independent variables corresponding to the coordinates of Euclidean space) and being, moreover, dependent on a fourth independent variable, on time. Still, in the description of membrane transport we often encounter ordinary differential equations which are genetically related to the description of movement of indestructible particles, the coordinates of which are solely functions of time. Actually, they will appear as plausible approximations of the appropriate partial differential equations, the dependence of, say, concentration, on space variables being replaced with the qualitative notions of "inside" and "outside" of the membrane-surrounded compartment. The plausibility of such an approach rests in this case on the assumption that equilibration in the media adjacent to the membrane proceeds at a much faster rate than the transport across the membrane itself.

A new and possibly very promising approach, alternative to the straightforward description of the membrane transport by ordinary differential equations, has emerged quite recently in the formalism of network thermodynamics (Kedem, 1972; Oster et al., 1973). Using a suitable algorithm, differential equations describing a system can be obtained directly from the network graphs; the graphs, at the same time, reveal the topology of the system (the way in which individual parts of the system are interconnected) and thus they contain more information than the differential equations themselves (Oster et al., 1973).

It will be seen that in a number of cases the differential equations need not be actually integrated to obtain valuable mathematical formulae governing membrane phenomena. At least in two important instances are the differential equations solved by other means.

(*1*) The time derivative is approximated by the quotient of a finite change of the quantity and of the elapsed time. Typically, the so-called initial velocities of transport are measured under various conditions and then tested graphically in order to see whether they satisfy this or that differential equation.

(*2*) In biological systems, the time-independent steady states play as important a role as thermodynamic equilibria do in other physical systems. For both these stationary states, the time derivatives are zero and the differential equations become algebraic equations.

Ordinary differential equations commonly encountered in the field of membrane transport show the property of being autonomous, i.e., the time derivatives occurring in them are not, or need not be, explicit functions of time. This property is both convenient and interesting. Convenient, for it allows the nature of the solutions to be studied even when they are not accessible in an analytical form, using the qualitative theory of differential equations. The time differential is easily excluded from the sets of such equations and relations between the remaining variables may be studied in a phase plane or phase space. It is interesting, too, since it points to an important characteristic of time, its homogeneity. Any process described by an autonomous differential equation can be reproduced at any time, the only prerequisite being that all other conditions be reproduced – the law expressed by an autonomous differential equation is eternally valid and applicable. The existence of laws of this kind is somewhat less obvious in biology than in physics; it may be argued that, due to the evolution of the organic world and to the development of individual organisms, the biological and physiological time is less homogeneous than the physical one. However, the need to account for such phenomena by nonautonomous equations is probably quite remote.

Whereas the time used in the description of membrane phenomena may be homogeneous to a very good approximation, it certainly lacks another of the symmetry properties, the isotropy, i.e. the property of being equivalent in all directions. Time flows in one direction only; transport phenomena are accompanied by dissipation of energy and once they have taken place they can never be returned completely, without leaving some changes on the surroundings of the system studied. It is for this reason that thermodynamics is of great value in considerations of membrane transport.

Space, too, is homogeneous and the same observations on membrane transport are found to be reproducible all over the world (or, at least, one wishes them to be). As for the other symmetry property, empty space is obviously isotropic but space filled with matter (e.g., a membrane) need not be isotropic and may thus show different properties in different directions. This consideration is of importance when the so-called active transport, a flow of substance across a membrane coupled to a chemical reaction in the membrane or in its proximity, is considered. According to the principle called the Curie−Prigogine principle (see, e.g., Katchalsky and Curran, 1965), in an isotropic medium there cannot be any direct coupling between vectorial and scalar quantities so that a vectorial flow of a substance cannot be directly driven by a scalar affinity of a chemical reaction In an anisotropic membrane, however, such coupling is permissible.

2. MEMBRANES

"And where it is of a less thickness the Attraction may be proportionally greater, and continue to increase, until the thickness do not exceed that of a single Particle of the Oil. There are therefore Agents in Nature able to make the Particles of Bodies stick together by very strong Attractions. And it is the Business of experimental Philosophy to find them out."

    Isaac Newton, Opticks

    The view that all phenomena of life are more or less intimately associated with the existence of cell membranes has become a powerful candidate for superseding the often quoted dictum of F. Engels that life is a form of existence of proteins. Indeed, various types of membranes serve as the locale of most enzyme activities of a cell, provide the physical basis of locomotion, intercellular contacts, secretion and uptake, are involved in protein synthesis and cell division. To explore the manifold functions of cell membranes would mean to write a book on biochemistry, physiology and biophysics combined, something that no serious scholar could attempt at this age of specialization. Still, while bearing in mind the title of the book and concentrating on one particular function of cell membranes, *viz.* the transport of ions and molecules, we shall first briefly review the existing knowledge of the structure and function of biological membranes in general.

    The word "membrane" has been used for over a century (*cf.* von

Mohl, 1851; Nägeli, 1855) to designate the boundary of a cell serving both as a mechanical barrier between the cell interior and its surroundings and as a semipermeable partition permitting the passage of water and of some but not all solutes. While the last-named function is at the core of most investigations into the transport of substances, the first, mechanical attribute is now recognized to belong rather to structures that lie externally to the cell membrane proper, namely the various types of wall-like structures to be discussed later (*cf.* p. 133). On the other hand, membranes of one kind or another have been discovered to form the structural framework of the whole cell and give its architecture a unique common denominator.

There are perhaps two principal features that permit to distinguish a membrane from other morphological cell constituents: (*1*) The typical trilaminar, "railroad-track" appearance seen after negative staining in an electron microscope, (*2*) the high lipid (particularly phospholipid) content, accompanied by varying amounts of protein and only occasionally by rather minute amounts of other types of compounds, especially polysaccharides and, in some special membranes, possibly by nucleic acids.

2.1. SURFACES AND INTERFACES

The striking stability of membranes in aqueous solutions and their spontaneous formation from suitable mixtures of lipids with water has been at one time considered analogous to the spreading of monomolecular films of an immiscible substance lighter than water (typically a lipid) on the surface of water. It is well known (e.g. Langmuir, 1933) that such films can exist in several different states of molecular freedom, that can be likened to gaseous, liquid and solid states of matter. If the surface area of the solvent (usually water) is large enough, the second-component molecules tend to disperse randomly, with little or no interaction between them. Using a larger-size fatty acid, the mean area at the disposal of a "floating" molecule will be above 0.5 nm^2. These monolayers correspond to low surface pressures, such as exerted in a Langmuir trough, a device employed to spread or compress surface films and measure the force exerted by the film in a given state to oppose compression. The pressures in this type of spreading are generally less than 0.001 N m^{-1} and the situation may be considered as the "gaseous" state. It is properly described by a "two-dimensional" gas law which may be written as

$$\pi(A_m - A_0/N) = 10^5\, kT \tag{2.1}$$

where π is the surface pressure in $N\,m^{-1}$, A_m is the molecular area in nm^2, A_0 is the limiting molar area, N is the Avogadro constant $(6.023 \cdot 10^{23}\, mol^{-1})$, k is the Boltzmann constant $(1.38 \cdot 10^{-23}\, J\,K^{-1})$ and T is the absolute temperature. Hence a plot of π versus A_m will yield an equilateral hyperbola (part **A** of Fig. 2.1.)

It the surface pressure is increased, interactions between the floating molecules will occur and a coherent film will be formed on the surface, the area taken up by a molecule corresponding roughly to $0.5\ nm^2$ at the highest compression, i.e. several times the smallest cross section of the molecule. This situation, corresponding to the liquid three-dimensional state, is termed expanded monolayer and exists at surface pressures of $0.001 - 0.01\ N\,m^{-1}$. This state is properly described by

$$(\pi - \pi_0)(A_m - A_0'/N) = 10^5\, kT \tag{2.2}$$

where π_0 is the surface pressure at the transition from the gaseous state and A_0' the limiting molar area under the new pressure conditions (part **B** of Fig. 2.1).

If the pressure is increased further (beyond $0.01\ N\,m^{-1}$) the molecules on the surface will tend to arrange themselves in a pattern of closest packing, in the case of fatty acids, aldehydes, alcohols, phospholipids, etc., in such a way that the polar heads of the molecules will be immersed in water while the hydrocarbon chains will stick out into the surrounding air. This packing pattern, termed condensed monolayer, still has some degree of translational motion since the heads are hydrated and have a larger cross-sectional area than the hydrophilic tails. The equation describing the situation is then

$$\pi = ask(A_0''/N - A_m) \tag{2.3}$$

where a is the activity of the solute ($a = f_a c$, c being concentration, f_a the activity coefficient), s the surface area of the solution and k a constant (part **C** of Fig. 2.1).

A further increase in pressure will result in stripping the polar heads of their solvation water molecules and in the formation of a condensed monolayer which is solid and practically noncompressible. The equation describing this situation is like (2.3) with a different constant replacing k (part **D** of Fig. 2.1). That this particular type of

24

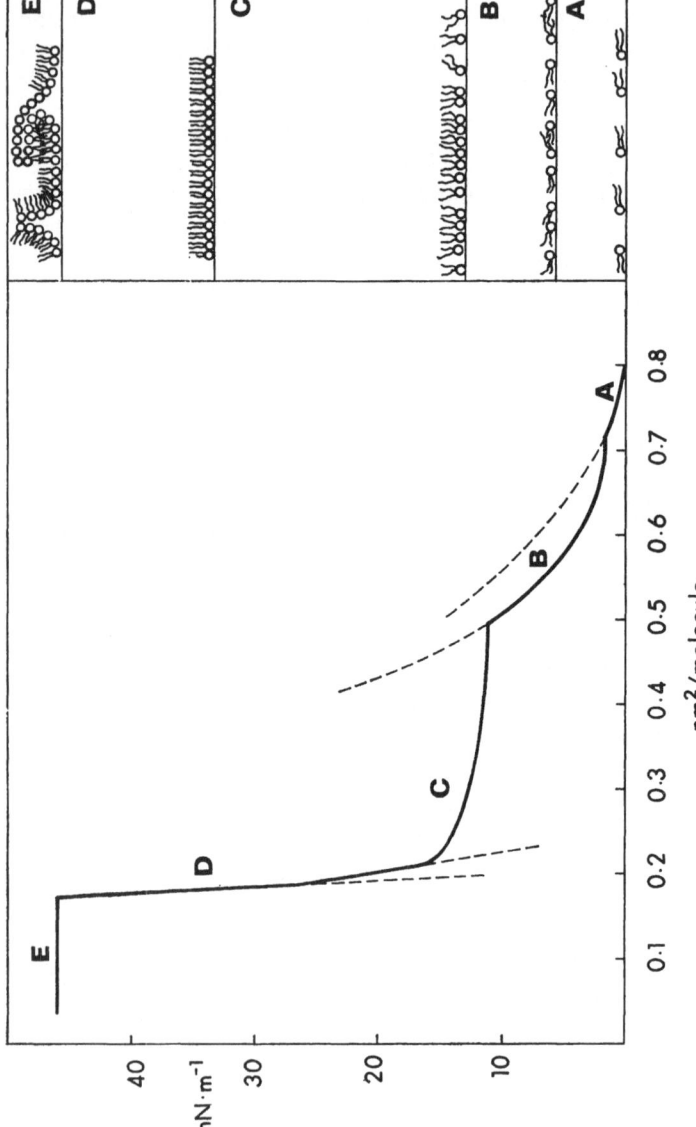

Fig. 2.1. A generalized force-area curve for a long-chain fatty acid. The right-hand panel shows schematically the position of the lipid molecules at the air—water interface. Explanation of the curve segments is given in the text.

packing occurs is supported by the fact that in a series of homologues of amphiphiles the cross-sectional area per molecule (about 0.2 nm^2) is practically independent of the hydrocarbon chain length.

If the limiting pressure is exceeded, the monolayer will gradually collapse, giving finally rise to a multilayered structure as in part **E** of Fig. 2.1. The equation describing the situation is

$$\pi = ask''. \tag{2.4}$$

The ability to form surface monolayers is not limited to lipids, proteins being known to spread in films on the surface of solutions (especially ammonium sulfate). However, the spreading of protein (achieved artificially) often results in a collapse of the tertiary structure or at least in partial reversible unfolding. In fact, mixed monolayers, both of miscible and immiscible components, can be prepared, the latter forming larger or smaller patches on the liquid surface, corresponding to separate phases.

It is of practical interest that surface monolayers have been used successfully for the determination of molecular weight both of proteins and lipids. If πA is plotted against π (A being the total area of a Langmuir trough), one obtains for the "gaseous" phase, i.e., at low solute concentrations, a straight line with a slope of nA_0, intersecting the vertical axis at nRT; this yields the number of moles n, as well as the limiting molar area A_0.

Monolayers need not form only on the surface of a liquid. If a suitable proportion of lipid, organic solvent and water (e.g., 50 mg dipalmitoyllecithin, 2 mg steroid, 0.5 ml benzene-chloroform of density equal to one, 50 ml 2 mM CsCl) is sonicated (e.g., 10 min at 45 °C with 70 W power) spherical vesicles about 50 nm in diameter are formed which enclose the organic solvent (Sackmann and Träuble, 1972).

The tendency to form monolayers on the surface is associated with the degree of immiscibility of the solute in the solvent and the ability to alter the surface tension of the solution. The surface pressure exerted by the bar in a Langmuir through is not to be confused with surface tension which is a qualitative property of liquids at the boundary with a gaseous phase and is due to the tendency of liquids toward cohesion. An analogous, interfacial, tension exists at the boundary between two immiscible liquids or a liquid and a solid. Every liquid tends to take up the smallest possible volume and hence form spheres but when dealing with larger amounts of liquids this trend is overcome by the pull of gravity which causes the liquid to distribute itself flat. The two quantities are related by $\pi = \sigma_0 - \sigma$, so that the surface pressure π is equal to the difference in the surface tension of the pure solvent σ_0 and of the solution σ. The surface tension, like the surface pressure, is expressed in dynes per cm or preferably in newtons per m (Table 2.1).

TABLE 2.1.   Values of interfacial tension σ of some liquids at 20 °C (in N m^{-1})

Liquid: air			
Liquid	σ	Liquid	σ
Mercury	0.484	Toluene	0.028
Water	0.073	Carbon tetrachloride	0.027
Glycerol	0.064	Acetone	0.023
Aniline	0.043	Ethanol	0.022
Carbon disulfide	0.034	Hexane	0.018
Benzene	0.029	Diethyl ether	0.017

Liquid: water			
Liquid	σ	Liquid	σ
n-Octane	0.051	Benzene	0.035
Carbon tetrachloride	0.045	Aniline	0.006

The surface tension decreases linearly with increasing temperature to a point lying close to the critical temperature of the liquid to disappear completely at the critical temperature (the temperature above which a given compound cannot exist as liquid no matter what pressure be applied—e.g., it is 647 K for water and 304 K for carbon dioxide). What is more important in the context of film formation is the consequence of the Gibbs surface isotherm derived for the relationship between surface concentration and changes in surface tension with the bulk concentration of solute. The internal energy change dU is expressed in terms of change of entropy dS, change of volume dV at a given pressure p, surface tension σ, surface area A

and chemical potentials μ of the two components, n'_1 and n'_2, thus:

$$dU = T \, dS - p \, dV + \sigma \, dA + \mu_1 \, dn'_1 + \mu_2 \, dn'_2. \quad (2.5)$$

The primed n's refer to concentration in a column of solution including the surface layer and are related to concentration in the same volume of solution excluding the surface layer in the following manner:

$$\Gamma = \frac{n' - n}{A} \quad (2.6)$$

Γ being called the surface excess of moles of solute per unit surface area.

Equation (2.5) can be integrated and then differentiated completely to yield

$$dU = T \, dS + S \, dT - p \, dV - V \, dp + \sigma \, dA + A \, d\sigma +$$
$$+ \mu_1 \, dn'_1 + n'_1 \, d\mu_1 + \mu_2 \, dn'_2 + n'_2 \, d\mu_2$$

whence (by comparison with eq. 2.5)

$$S \, dT - V \, dp + A \, d\sigma + n'_1 \, d\mu_1 + n'_2 \, d\mu_2 = 0$$

so that, at constant temperature and pressure,

$$A \, d\sigma + n'_1 \, d\mu_1 + n'_2 \, d\mu_2 = 0. \quad (2.7)$$

Inside the solution, the Gibbs–Duhem equation will hold:

$$n_1 \, d\mu_1 + n_2 \, d\mu_2 = 0. \quad (2.8)$$

The equilibrium condition is obtained by subtracting eq. (2.8) after multiplying it with a suitable constant such that $n'_1 = kn_1$, from eq. (2.7):

$$A \, d\sigma + (n'_2 - kn_2) \, d\mu_2 = 0. \quad (2.9)$$

Rearrangement of this equation gives

$$-\frac{d\sigma}{d\mu_2} = \frac{n'_2 - kn_2}{A} = \Gamma_2. \quad (2.10)$$

Since n_1 does not enter the equation, the surface concentration of component 2 $(= \Gamma_2)$ is independent of volume. The chemical potential change $d\mu_2$ being equal to $RT \, da/a$, eq. (2.10) becomes

$$\Gamma = -\frac{a \, d\sigma}{RT \, da} \quad (2.11)$$

where a is the activity of solute in the given solution.

Equation (2.11) is the Gibbs surface isotherm (or adsorption equation) and shows that compounds decreasing the surface tension will concentrate at the surface since if $d\sigma/da$ (sometimes called the surface activity of a substance) is negative, $\Gamma(= \Delta a)$ will be positive. Thus, surface-active agents (soaps, detergents, etc.) will decrease the surface tension by accumulating on the surface of a liquid. The actual value of Γ can be arrived at from an empirical equation due to von Szyszkowski (1908)

$$\sigma_0 - \sigma = \alpha \ln (1 + \beta a) \tag{2.12}$$

where σ_0 is the surface tension of pure solvent, σ that of the solution, α and β are constants relating to the solute. Derivation of eq. (2.12) yields

$$-d\sigma/da = \alpha\beta/(1 + \beta a)$$

which may be substituted in (2.11) to give

$$\Gamma = \alpha\beta a/RT(1 + \beta a). \tag{2.13}$$

(The similarity with a Langmuir adsorption isotherm is obvious on rearranging to $\Gamma = (\alpha/RT) . a/(1/\beta + a)$.) The value of α is identical for a given homologous series (e.g., 12.95 for monocarboxylic fatty acids) while the value of β depends strikingly on the molecular species (e.g., it is 6.1 for propionic acid but 2 327 for caproic acid). It may be seen that Γ tends toward a limiting value equal to α/RT as concentration is raised and that it is reached the sooner the greater the value of β which may be compared with the surface activity of the given compound.

It should be realized that the surface monolayer film of a lipid on water, although superficially resembling a biological "half-membrane" and although being identical with one half of an artificial black membrane (see p. 95), is a far cry from a true biological membrane. One arrives somewhat closer at what appears like a typical railroad-track membrane if one increases vastly the concentration of lipid in the lipid-water system, particularly when working again with amphipathic lipids, i.e., those that contain both hydrophilic and hydrophobic portions. As the "concentration" of lipid in the system rises, there will be not only a superficial film formed corresponding to eq. (2.13) but the lipid molecules will begin to form aggregates within the solution, the so-called micelles. These clusters may contain tens or hundreds of lipid molecules arranged as shown in Fig. 2.2. The solution may still appear transparent but light scattering and particularly X-ray scattering will reveal the presence and size of the aggregates formed. Their shape may range from spheres to ellipsoids to rods and generally the higher the concentration the greater the participation of rods (but there are examples where spheres persist at all concentrations, e.g., with palmityltrimethylammonium chloride; or where rods are found at all concentrations, e.g. with sodium oleate).

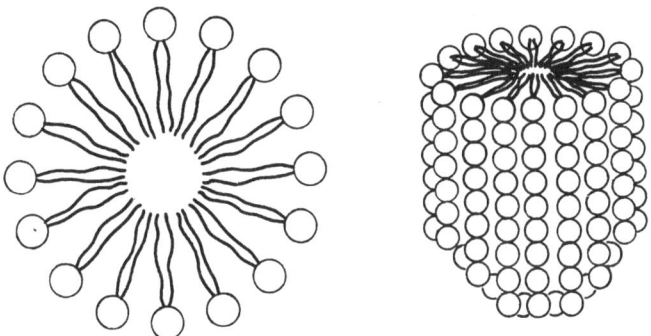

Fig. 2.2. A schematic drawing of micelles of amphiphilic molecules in an aqueous solution.

If the concentration of lipid is increased further (generally above 50% v/v) an abrupt change in viscosity occurs and the system becomes anisotropic. The organization of lipids is now one of several patterns, containing periodic repeats in one or two dimensions and is called mesomorphic or liquid-crystalline (in contrast with crystalline structure where the periodic repeats occur in three dimensions). The word smectic is occasionally used to describe this semicrystalline state (from Greek smektikos = cleansing, purifying, since the structure was first described in detergent mixtures); the smectic phase contains parallel planes or sheets of aggregated molecules (in contrast with the nematic state, rather rare indeed, where parallel chains or strands are formed). The sequence of semicrystalline phases, in the order of increasing lipid concentration, is hexagonal – deformed hexagonal – rectangular – complex hexagonal – cubic – lamellar but rarely do all the patterns occur. Thus, sodium stearate will pass from hexagonal to deformed hexagonal, complex hexagonal and lamellar, while sodium oleate from hexagonal to rectangular, complex hexagonal and lamellar. The arrangement of lipid molecules in the various patterns is shown in Fig. 2.3. The structure designated as deformed hexagonal is surmised from a particular set of reflections in X-ray diffraction, the one called cubic is postulated from analogy with the structure of anhydrous soaps. Distinction between hexagonal type I and hexagonal type II is possible on the basis of intensity of X-ray reflections and also from the fact that in type I the area taken up by a given molecule is almost independent of the hydrocarbon chain length and lipid concentration

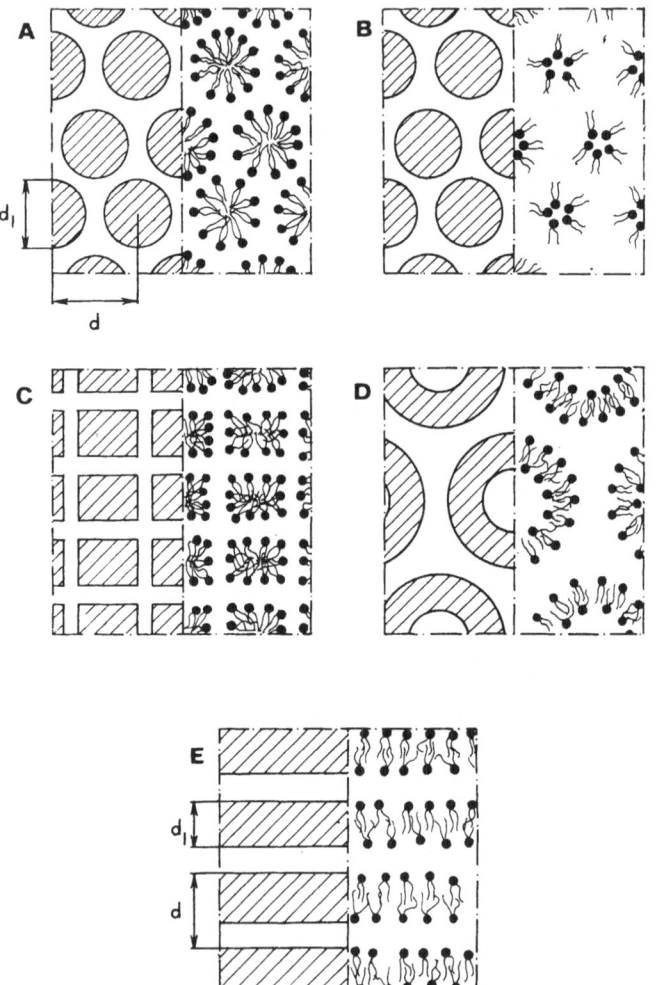

Fig. 2.3. Schematic structure of some liquid-crystalline phases of lipid—water systems. **A** Hexagonal I, **B** hexagonal II, **C** rectangular, **D** complex hexagonal, **E** lamellar. d Repeat distance, d_1 diameter of lipid cylinder or lamella. (Adapted from Luzzati, V. and Husson, F. (1962). *J. Cell Biol.* **12**, 210.)

whereas in type II the area is a complex function of both. Type I is found in water mixtures of lysolecithin or soaps, type II in water mixtures of phosphatidyl ethanolamine, phosphatidyl choline, mitochondrial lipids and other more biological substances.

At very low water contents (about $10-15\%$) lipids tend to form gels or coagels which may also have a lamellar structure, the coagel containing two phases, liquid water and crystalline lipid, unlike the gel which is homogeneous but usually metastable (Vincent and Skoulios, 1966).

It is to be noted that transitions similar to those described for lipid-water mixtures occur in anhydrous lipids, particularly soaps, as temperature is decreased and an isotropic (possibly micellar) liquid passes over to a liquid crystalline state and finally to a solid crystalline state (Luzzati, 1968).

The phase diagrams of a number of lipid-water mixtures have been obtained, some of them rather complicated. As an example, based on the work of Abrahamsson and co-workers (1972), the system of sodium sulfatide and water is shown in Fig. 2.4.

In many cases, the existence of a given repeating structure is detectable without the aid of X-ray diffraction or other sophisticated techniques. Electron microscopy of a phosphotungstate-fixed mixture

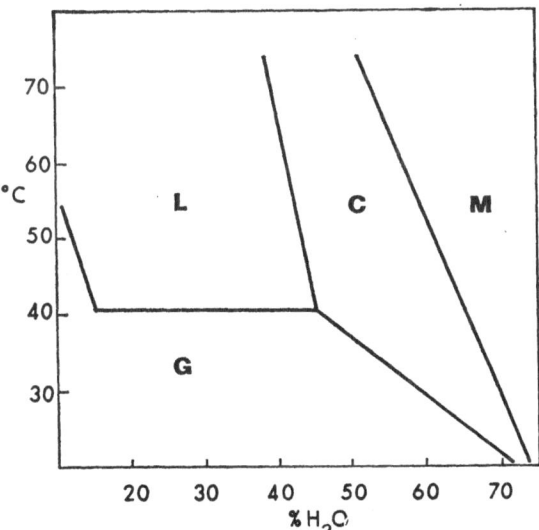

Fig. 2.4. Phase diagram of sodium sulfatide in water. **L** Lamellar phase, **C** cubic phase, **M** micellar phase, **G** gel phase. (Adapted from Abrahamsson *et al.*, 1972.)

of lecithin with water or an alkylphosphoric acid will reveal stacked or concentric lamellae, their repeat distance being 5 – 10 nm, depending on the amount of aqueous phase present. These "membranes" have been called myelin figures because they resemble the multilayered myelin sheath of a nerve axon (Fig. 2.5; *cf.* Fig. 2.35E). In other cases, hexagonal arrays are formed (Fig. 2.6).

In all the patterns assumed by lipids in water the obvious tendency is to ensure the maximum polar-to-polar and nonpolar-to-nonpolar interactions, this being dictated by the necessity of forming patterns with the lowest free energy content. The principal factor active in steering the molecules into their positions are the hydro-

Fig. 2.5. Myelin figures—a negatively stained preparation of lecithin and cholesterol. (Taken with kind permission from Glauert, A. M. and Lucy, J. A. (1968). In: *The Membranes* (ed. by Dalton, A. J. and Haguenau, F.), p. 1. Academic Press, New York—London.)

50 nm

Fig. 2.6. **A** Hexagonal phospholipid structure in 3 % water. The dark spots are probably aqueous cylinders surrounded by lipid. (Taken with kind permission from Stoeckenius, W. (1962). *J. Cell Biol.* **12**, 221.)
B The hexagonal phase of a lecithin-cholesterol-saponin complex. The light rings appear to be composed of subunits (arrow). (Taken with permission from Glauert and Lucy, as in Fig. 2.5.)

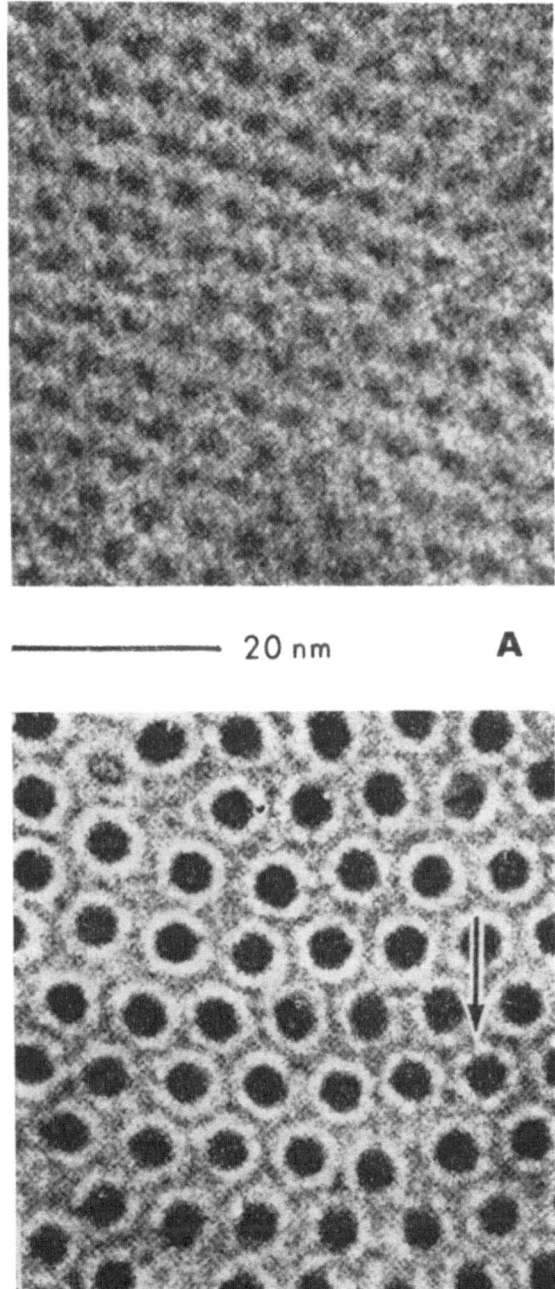

—————— 20 nm   A

—— 20 nm   B

phobic interactions (Kauzmann, 1959). To demonstrate the nature of these interactions, let us take the simple example of methane distributed between benzene and water. The change in free energy (properly, the unitary free energy, i.e. the total free energy minus translational entropy, the so-called cratic entropy) on going from benzene to water is 10.9 kJ mol^{-1}, reflecting the higher solubility of methane in benzene than in water. However, the change in enthalpy is negative, -11.7 kJ mol^{-1}, so that it is the negative entropy change, -75 J K^{-1} mol^{-1} at 25 °C, which makes the methane reside rather in benzene than in water. The entropy decrease is apparently associated with an ordered arrangement of water molecules about the methane, either in a cage form (clathrate) or not. The same consideration holds for the "dissolving" of lipids (or proteins or any kind of high-molecular compound).

With phospholipids, carrying both positive and negative charges at different parts of their polar heads, electrostatic interaction acts as an additional factor in the orientation of their molecules in an aqueous medium. In fact, of naturally occurring lipids, it is only the phospholipids that can form extensive bilayer structures oriented with their polar heads toward water and their hydrocarbon tails toward each other.

Van der Waals forces make a relatively minor contribution to the structural patterns of lipids in water, perhaps most importantly by ensuring the closest possible packing of the hydrocarbon chains of lipids with no gaps or spaces between them.

We have thus arrived at an artificial system containing only lipid and water but resembling the biological membrane in its appearance under the electron microscope as well as in its dimensions. However, it was pointed out at the beginning of this chapter that perhaps a half of the weight of biological membranes is due to proteins. How would then a protein—lipid—water mixture behave as compared with a lipid—water mixture? Especially X-ray diffraction studies have shown that two principal structural arrangements appear, just like in the protein-free mixtures: a lamellar one and a hexagonal one (Shipley *et al.*, 1969; Gulik-Krzywicki *et al.*, 1969).

Depending on the lipid-protein ratio, the spacing between repeating patterns will change, the assumption being that when more protein is present, there will be an increasing number of protein layers intercalated between layers of lipids. Analogously, the situation will develop in the hexagonal array (Fig. 2.7). Pertinent studies were

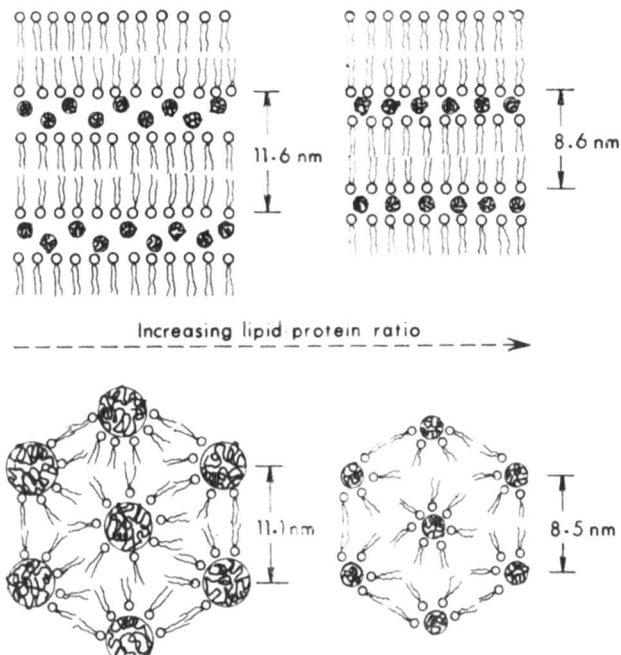

Fig. 2.7. Schematic representation of the lamellar and hexagonal phases formed in a mixture of cytochrome c, phospholipids and water. The repeat distance is shown to decrease as the relative amount of phospholipids increases. (Redrawn with permission from Shipley, G. G. (1973). In: *Biological Membranes* (ed. by Chapman, D. and Wallach, D. F. H.), p. 1. Academic Press, London—New York.)

undertaken with oxidized cytochrome c, lysozyme and serum albumin, and a variety of phospholipids. As demonstrated by the example shown in Table 2.2, both pH and temperature influence the lattice spacing but the range of lattice dimensions tallies with the dimensions of biological membranes estimated both by X-ray diffraction and by electron microscopy.

In the next chapter it will be seen that, although the range of biological membrane functions precludes a simple reconstitution from dispersed components, many of the characteristics described so far for purely artificial membrane-like structures are retained in cell membranes and may be vital for the maintenance of membrane structure as a whole. However, before making such comparisons, we should know more about the chemical composition and supramoecular organization of actual biological membranes.

TABLE 2.2.   Characteristics of artificial mixtures of
lysozyme—phosphatidyl inositol—water (from Gulik-Krzywicki *et al.*,
1969)

pH	°C	Lattice type	Spacing nm
4	0—35	two-dimensional square	6.35
5.5—7	0—35	two-dimensional hexagonal	8.0—8.25
7	0	two-dimensional hexagonal	9.6
7	30	lamellar	10.6
7	40	lamellar	8.3
8	25	lamellar	10.6
8	40	lamellar	8.3

2.2.   CHEMICAL COMPOSITION

The variety of chemical species occurring in biological membranes is impressive but more than 95% of the bulk in practically all membranes falls into two groups, lipids* and proteins. We shall now describe in some detail the constituents of the two groups and proceed from the static chemical composition to the structural arrangement in biological membranes.

The percent content of lipids and proteins in various membranes is shown in Table 2.3.

2.2.1.   Lipids

2.2.1.1.   Chemistry

Although not necessarily all lipid compounds found in nature occur in membranes (leaf waxes and fatty alcohols, insect cuticle waxes; reserve triglycerides in adipose tissue; steroids used as means of defense or as hormones, pheromones, isoprenoids in latex) the

---

* The term lipids is used here in the more physical sense of the word, designating hydrophobic, oil-soluble compounds. From the chemist's point of view, lipids should be probably restricted to esters of glycerol and sphingosine.

spectrum of lipids actually found in membranes is rather broad and includes practically all groups of lipids plus a number of complex substances of lipid-carbohydrate-protein nature.

There exist different types of classification of lipids: here we shall use a simple system, recognizing glycerol-substituted lipids, sphingosine-substituted lipids, sterols and other lipids.

TABLE 2.3. Protein and lipid content of various membrane preparations (from Dewey and Barr, 1970; Suomalainen et al., 1973; Wallach, 1972; Wolfe, 1964)

Membrane	% Protein	% Lipid
Bovine myelin	22	78
Human erythrocyte[a]	49	44
Ehrlich ascites cell membrane	70	30
Liver plasma membrane	60	40
Intestinal villi	85[b]	15
Inner mitochondrial membrane	78	22
Outer mitochondrial membrane	55	45
Retinal rod outer segment	59	41
Brain synaptic vesicles	66	34
Rat liver microsomes[c]	62	32
Rabbit muscle microsomes[c]	54	22
Chloroplast lamellae	45	55
Micrococcus lysodeikticus[d]	50	28
Bacillus megaterium[e]	68	20
Pseudomonas aeruginosa	60	35
Halophilic bacteria	65	35
Acholeplasma	70	30
Baker's yeast plasma membrane[f]	46 (27, 35)	38 (46, 38)
Candida utilis plasma membrane[f]	39	40
Amoeba[f]	43	32
Myxoviruses[g]	73	20
Leukoviruses[h]	64	31

[a] Also about 7 % carbohydrate; [b] a great part of this may be nonmembrane protein; [c] the remaining material is apparently RNA from attached ribosomes; [d] contains 15—20 % carbohydrate; [e] contains 6—10 % carbohydrate; [f] the remaining material is mainly carbohydrate from adhering contamination by the cell wall; [g] also some 6 % carbohydrate and 1 % RNA; [h] also some 6 % carbohydrate and 2 % RNA.

A. Based on substituted glycerol

The backbone of these lipids is formed by the substituted tri-hydroxy-alcohol glycerol

$$
\begin{array}{l}
CH_2OR^1 \\
|\ \\
R^2OCH \\
|\ \\
CH_2OR^3
\end{array}
$$

the three substituents R^1, R^2, and R^3 being generally different.

a. *Containing only* C, H, O

If R^1 is a fatty ester, $CH_3(CH_2)_nC{=}O$ ($n = 10-20$), R^2 and $R^3 =$ H, the compound is a monoglyceride; if R^1 and R^3 are fatty esters, the compound is a diglyceride; if all three R's are fatty esters, it is a triglyceride.

If R^1 or R^2 is a fatty ether, $CH_3(CH_2)_nCH_2-$ ($n > 10$), the compound is a glycerol ether.

If R^1 and R^2 are fatty esters and R^3 is a galactoside, the compound

is a monogalactosyl diglyceride. If R^3 is a digalactoside,

the compound is a digalactosyl diglyceride (both are neutral glyco-lipids). Analogously, a monoglucosyl and a diglucosyl diglycerides have been described, as well as other, rather rare, compounds, such as 6-sulfo-6-deoxyglucosyl diglyceride (Okaya, 1964).

b. *Containing* C, H, O, P

R^1 and R^2 are fatty esters. If R^3 is a phosphate ester, $-\overset{\overset{\displaystyle O}{\|}}{\underset{\underset{\displaystyle OH}{|}}{P}}-O^-$,

the compound is phosphatidic acid. If R^3 is an inositol phosphate ester,

the compound is phosphatidyl inositol. The inositol moiety may be further phosphorylated in position 4 (or 4 and 5).

If R^3 is a glycerol phosphate ester,

the compound is called phosphatidyl glycerol. If it is a phosphatidylglycerol phosphate ester,

the compound is diphosphatidyl glycerol or cardiolipin.

c. *Containing* C, H, O, P, N

R^1 and R^2 are fatty esters. If R^3 is ethanolamine phosphate ester,

the lipid is phosphatidyl ethanolamine (one of the cephalins). If R^3 is serine phosphate ester, the lipid is phosphatidyl serine (another of the cephalins). If R^3 is choline phosphate ester,

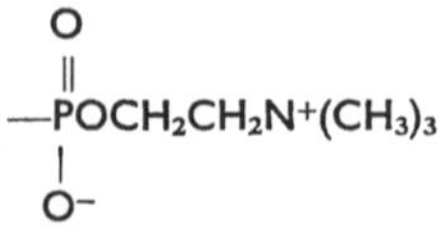

$$
\begin{array}{c}
O \\
\parallel \\
-POCH_2CH_2N^+(CH_3)_3 \\
\mid \\
O^-
\end{array}
$$

the lipid is phosphatidyl choline or lecithin.

Occasionally, the occurrence of lysolecithin or lysophosphatidyl ethanolamine has been reported, these compounds being formed from the parent ones by splitting off the fatty acyl residue R^2.

If R^3 is aminoethyl phosphonate ester,

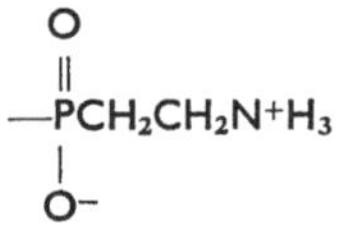

$$
\begin{array}{c}
O \\
\parallel \\
-PCH_2CH_2N^+H_3 \\
\mid \\
O^-
\end{array}
$$

the compound is a phosphonolipid. Some protozoans contain also phosphatidyl-N-(2-hydroxyethyl)alanine, the R^3 being here

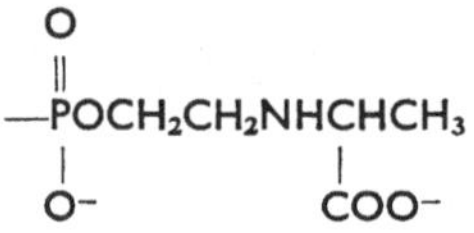

$$
\begin{array}{c}
O \\
\parallel \\
-POCH_2CH_2NHCHCH_3 \\
\mid \qquad\qquad \mid \\
O^- \qquad\quad COO^-
\end{array}
$$

Many bacteria contain the so-called lipoamino acids in which the R^3 substituent is

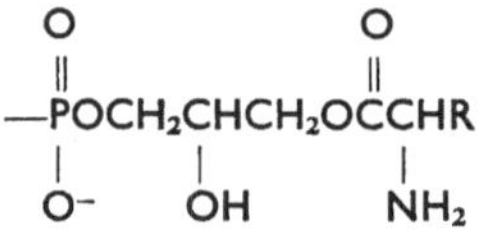

$$
\begin{array}{c}
O \qquad\qquad\quad O \\
\parallel \qquad\qquad\quad \parallel \\
-POCH_2CHCH_2OCCHR \\
\mid \qquad \mid \qquad\quad \mid \\
O^- \quad OH \qquad NH_2
\end{array}
$$

the R representing an amino acid residue (Arg, Phe, Thr, Leu, Orn, Ala, Lys, Glu, Asp, His having been found so far).

R^1 is a vinyl ether $CH_3(CH_2)_nCH=CH-$ ($n > 9$). The compounds, usually substituted at R^3 with ethanolamine phosphate, are called plasmalogens. A phosphonoplasmalogen is also known.

B. Based on substituted sphingosine

The backbone of the following lipids is the substituted amino-alcohol sphingosine

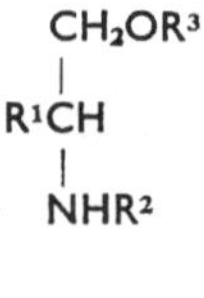

$$
\begin{array}{c}
CH_2OR^3 \\
\mid \\
R^1CH \\
\mid \\
NHR^2
\end{array}
$$

in which substituent R^1 is known to occur in three different forms: with $CH_3(CH_2)_{14}CH(OH)-$ it is dihydrosphingosine; with $CH_3(CH_2)_{12}CH=CHCH(OH)-$ it is sphingosine proper; with $CH_3(CH_2)_{13}CH(OH)CH(OH)-$ it is phytosphingosine (in all cases $R^2 = R^3 = H$). Substituent R^2 is either an acyl group $CH_3(CH_2)_nC=O$

with $n = 14-24$, or an α-hydroxy acyl group $CH_3(CH_2)_nCHC=O$.

$$\begin{array}{cc} & | \quad | \\ & OH \end{array}$$

The main distinctions arise from the type of substituent R^3. If $R^3 = H$, the resulting amide is called ceramide. If it is a choline phosphate ester, the lipid is sphingomyelin (R^3 may also be an ethanolamine phosphate ester or even an aminoethyl phosphate ester).

If R^3 is a glucoside

(or infrequently galactoside), the lipid belongs to the cerebrosides (a class of *N*-containing glycolipids). An important group of compounds, the gangliosides, contain an oligosaccharide as R^3 and are comprised of hematosides, globosides and blood-group substances. The common components of the oligosaccharide substituents are glucose, galactose and particularly *N*-acetylgalactosamine and *N*-acetylneuraminyl residues. *N*-Acetylneuraminic or sialic acid is a pyranose derivative of the following formula:

(*cf.* with *N*-acetylmuramic acid of bacterial walls; p. 135).

Some of the gangliosides are listed below.

Monosialoganglioside (GM₁)	Galactosyl-*N*-acetylgalactosaminyl--(*N*-acetylneuraminyl)-galactosyl-glucosyl ceramide
Disialoganglioside	*N*-Acetylneuraminyl-galactosyl--*N*-acetylgalactosaminyl-(*N*-acetyl-neuraminyl)-galactosyl-glucosyl ceramide

CTH	α-Galactosyl-galactosyl-glucosyl ceramide
Hematoside	N-Acetylneuraminyl-galactosyl-glucosyl ceramide
Disialohematoside	N-Acetylneuraminyl-N-acetylneuraminyl-galactosyl-glucosyl ceramide
Globoside 1 (cytolipin)	N-Acetylgalactosaminyl-galactosyl-galactosyl-glucosyl ceramide (if second to third galactose is linked 1 → 4, it is cytolipin K of humans; if the link is 1 → 3, it is cytolipin R of the rat)
Forssman glycolipid	N-Acetylgalactosaminyl (α1→3)-N-acetyl--acetylgalactosaminyl (β1→3)-galactosyl(α1→4)-galactosyl(β1→4)-glucosyl ceramide

The blood group substances are glycolipids of the ganglioside type present on the surface of red cells as well as in tissue fluids and secretions. The sugar terminal units defined in these substances are the following: O-α-N-acetyl-D-galactosaminyl(1→3)-O-β-D-galactosyl-(1→4)-N-acetyl-D-glucosaminyl (group A), 3-O-α-D-galactosyl(1→3)-D-galactosyl (group B), O-α-L-fucosyl(1→2) and O-β-N-acetyl-D-glucosaminyl (group O or H), N-acetylneuraminyl (group M and N), 3-O-β-D-galactosyl-N-acetylglucosaminyl or 4-O-α-L-fucosyl-N-acetyl-glucosaminyl (Le^a).

A special group of glycolipids are the so-called phytoglycolipids where the ceramide is substituted with an oligosaccharide phosphate ester (= R^3 of the sphingosine backbone).

Quite recently (Lester *et al.*, 1974), three sphingolipids containing a sugar residue have been discovered in the fungus *Neurospora crassa* and in baker's yeast. The major one in yeast is an inositol phosphoryl-ceramide with R^1 being a dihydroxyaliphatic chain with 16 or 18 carbon atoms, R^2 an α-hydroxy acyl group with 26 carbon atoms.

The last group of sphingosine (or ceramide) derived lipids are the sulfatides where R^3 is a glucosyl sulfate.

C. Sterols

Only a few sterols have been found in major amounts in the membranes of eukaryotic (but not prokaryotic) cells:

cholesterol

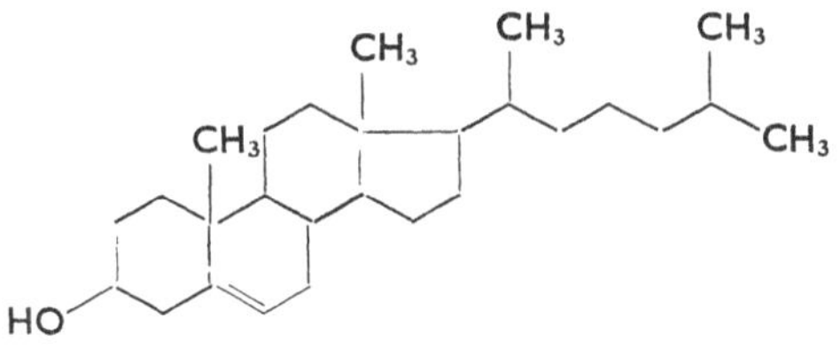

ergosterol

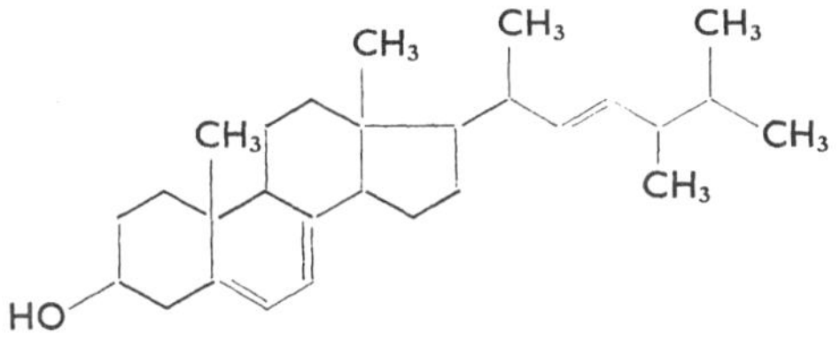

β-sitosterol

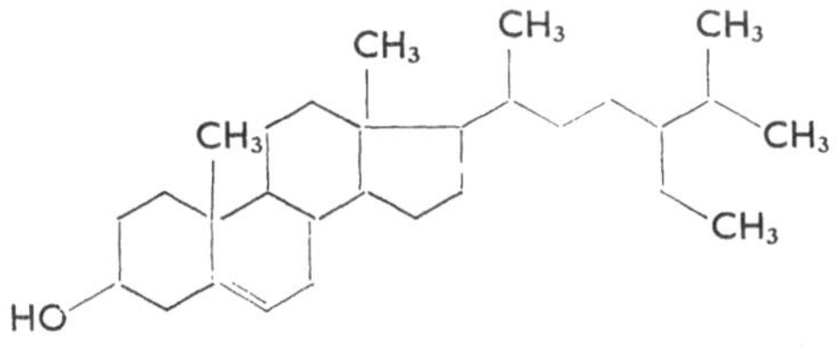

zymosterol

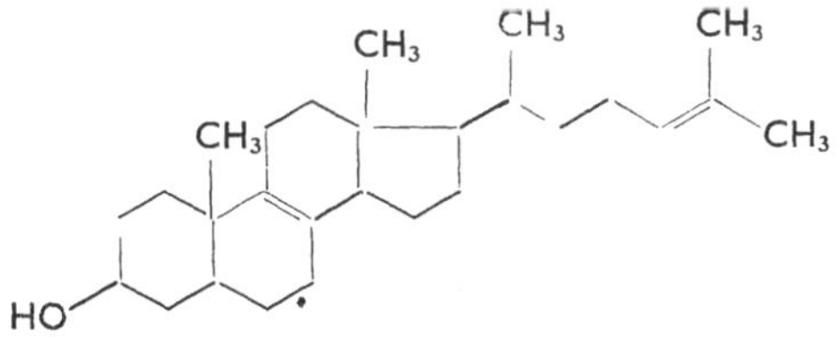

Other sterols have been reported from various sources, particularly from lower plants (Goodwin, 1973). They include campestrol, brassicasterol, cycloartenol, desmosterol, fucosterol, spinasterol, ascosterol, lanosterol and a number of others, as well as several sterol glycosides.

D. Minor lipids

Although of restricted occurrence and insignificant abundance, some of these lipids may play very important roles in special metabolic reactions (derivatives of β-carotene in the vision process; vitamin K in the terminal respiration shunt and blood clotting initiation, etc.). They may be classified as follows (basically according to Law and Snyder, 1972).

a. *Terpenoid compounds, e.g.* β-carotene

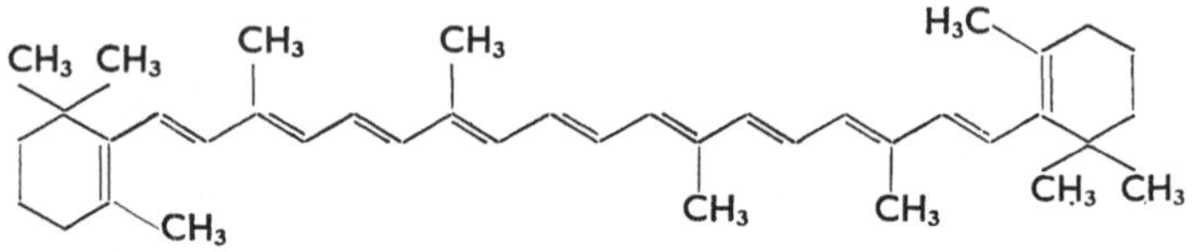
CH3 CH3
CH3
CH3
H3C
CH3
CH3
CH3
CH3 CH3

vitamin K_1

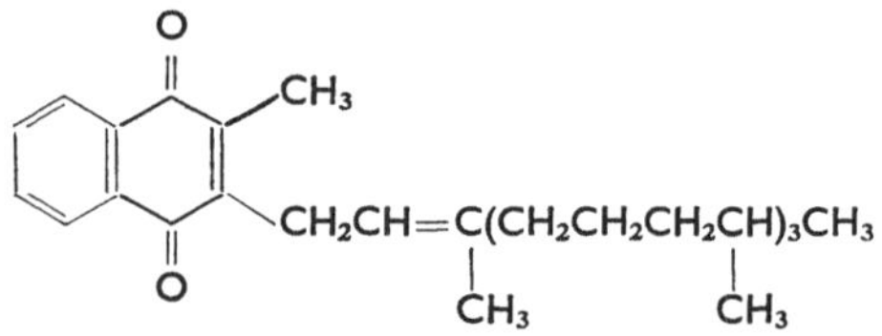
O
CH3
CH2CH=C(CH2CH2CH2CH)3CH3
CH3
CH3
O

polyisoprenoid alcohols

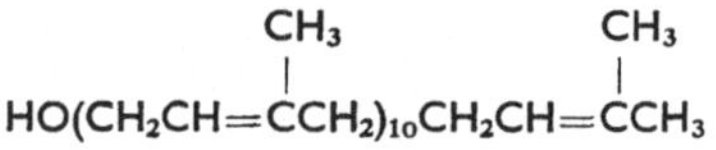
CH3
CH3
HO(CH2CH=CCH2)10CH2CH=CCH3

squalene

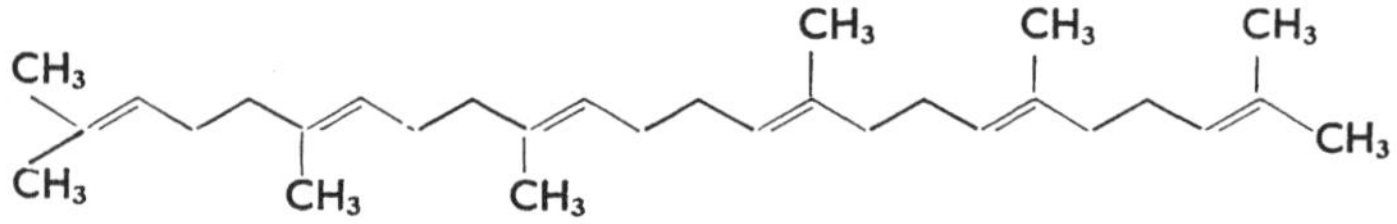
CH3
CH3
CH3
CH3
CH3
CH3
CH3
CH3

b. *Hydrocarbons*, particularly odd-numbered ones, such as $C_{31}H_{64}$.

c. *Wax esters* of the type

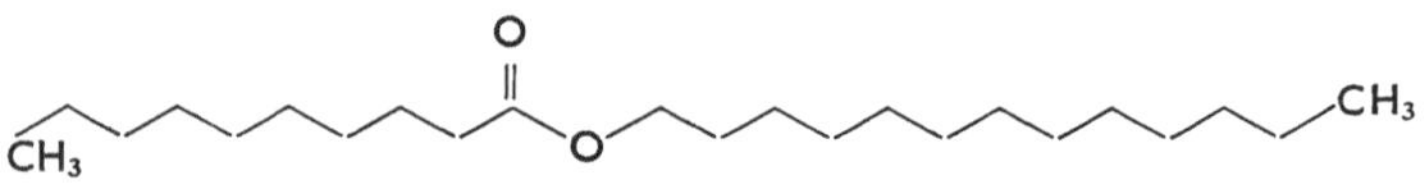
O
CH3
O
CH3

d. *Sugar esters* of the type

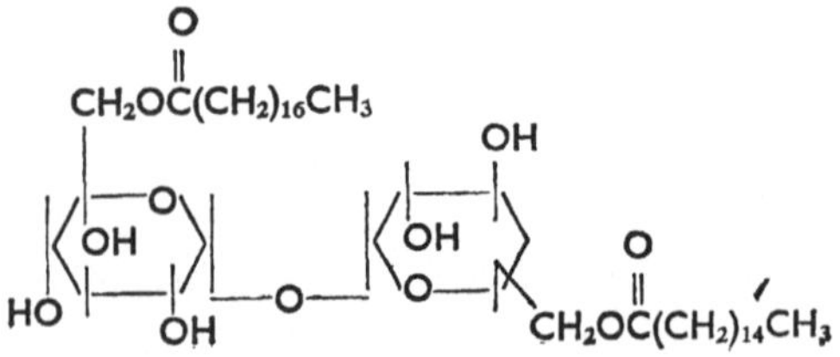
O
CH2OC(CH2)16CH3
OH
O
OH
HO
OH
O
OH
O
O
CH2OC(CH2)14CH3

e. *Esters of diols* of the type

$$
\begin{array}{c}
\text{O} \\
\parallel \\
CH_2OC(CH_2)_{14}CH_3 \\
\mid \\
CH_2OC(CH_2)_{16}CH_3 \\
\parallel \\
\text{O}
\end{array}
$$

f. *Esters of monohydroxyalcohols* of the type

$$
\begin{array}{c}
\text{O} \\
\parallel \\
CH_3OC(CH_2)_{14}CH_3
\end{array}
$$

g. *Sulfolipids*, such as

$$
\begin{array}{c}
CH_3(CH_2)_7CH(CH_2)_{12}CH_2OSO_3^- \\
\mid \\
OSO_3^-
\end{array}
$$

as well as halosulfolipids, such as

$$
\begin{array}{c}
\text{Cl} \quad\; OSO_3^- \qquad\qquad \text{Cl} \\
\mid \qquad \mid \qquad\qquad\quad \mid \\
CH_3(CH_2)_5CHCHCHCHCH_2CH(CH_2)_8CCH_2OSO_3^- \\
\mid \quad\; \mid \quad\; \mid \quad\;\; \mid \\
\text{Cl} \quad \text{Cl} \quad \text{Cl} \quad\;\; \text{Cl}
\end{array}
$$

h. *Lipopolysaccharides.* These represent large molecules of the outer layer of Gram-negative bacterial cell wall and will be discussed on p. 139.

i. *Other components* extractable with organic solvents, although generally not considered as lipids, include quinones and chlorophyll, occurring almost exclusively in plant and algal chloroplasts.

2.2.1.2. Distribution

Law and Snyder (1972) compiled an informative table on the occurrence of various lipid groups in nature. In a modified and extended form, it is reproduced here as Table 2.4. The table includes compounds that may not occur in membranes in major amounts (wax esters, hydrocarbons). It shows some significant differences between prokaryotic (bacterial) and eukaryotic (other) cells, most striking being the occurrence of sterols and sphingomyelins.

The relative representation of the various lipid groups varies from organ to organ but there is a considerable lack of variability for a given membrane among related species as well as taxonomically

TABLE 2.4.   Distribution of lipid groups in nature

Lipid group	Viruses	Eubacteria	Myco-bacteria	Fungi	Plants	Protozoa	Inverteb-rates	Vertebrates
Triglycerides	?	traces	rare	+	+	+	+	+
Phospholipids	+	+	+	+	+	+	+	+
Phosphonolipids	?	—	rare	—	—	+	+	+
Plasmalogens	?	—[a]	?	?	?	?	+	+
Sphingomyelins	+	—[a]	—	+	+	+	+	+
Glycolipids	+	+	+	+	+	+	+	+
Sterols	+	traces	—	+	+	+	+	+
Hydrocarbons	—	—	?	?	+	?	+	+
Wax esters	—	?	?	?	+	?	+	+
Sulfolipids	—	?	?	?	+	+	?	+

[a] Only in some anaerobes.

distant groups. An exception to this rule is represented by erythrocyte membranes where the variety is overwhelming. Some examples of this variability are shown in Tables 2.5, 2.6 and 2.7.

TABLE 2.5.   Percent content of phospholipids in the cell membranes of various organs of the rat (adapted from Rouser et al., 1968)

Phospholipid[a]	Liver	Spleen	Heart	Lung	Kidney	Erythrocyte
PC	52.3	47.2	41.2	48.2	36.6	47.5
PE	25.2	27.2	33.8	23.3	28.8	21.5
PS	3.9	9.0	3.2	9.7	8.1	10.8
PI	9.0	6.3	4.0	4.2	6.1	3.5
PG	—	—	1.1	2.3	0.2	3.8[b]
DPG	4.4	2.4	13.0	1.1	7.1	
PA	0.4	0.6	0.2	0.3	0.3	0.3
SP	4.5	7.3	3.5	10.8	12.9	12.8

[a] Abbreviations used: PC, phosphatidyl choline; PE, phosphatidyl ethanolamine; PS, phosphatidyl serine; PI, phosphatidyl inositol; PG, phosphatidyl glycerol; DPG, cardiolipin; PA, phosphatidic acid; SP, sphingomyelin. [b] Lysophosphatidyl choline.

TABLE 2.6. Weight percent composition of phospholipids from various sources (adapted from Galliard, 1973; Jakovcic et al., 1971; Suomalainen et al., 1973; Wolfe, 1964)

Phospholipid	Brain								Kidney				Saccharomyces cerevisiae	Potato microsomes	Sugar beet protoplasts	Apple mitochondria	Euglena gracilis	Micrococcus lysodeikticus
	Man	Cow	Rattlesnake	Frog	Goldfish	Octopus	Fly	Lobster	Man	Rat	Mouse	Frog						
PC	29.2	29.6	39.0	43.2	49.6	38.8	18.6	58.4	37.9	36.6	38.5	36.7	45	45	38	45	47	—
PE	35.0	35.7	36.7	39.0	33.6	37.7	54.5	25.7	30.8	28.8	28.4	32.1	15	33	15	35	15	—
PS	17.6	16.9	9.1	10.2	8.8	11.8	5.3	5.7	7.0	8.1	7.6	6.5	7.3[a]	1	—	3	4	—
PI	2.0	2.7	3.9	2.4	3.6	2.6	2.7	1.2	6.1	6.1	6.3	6.1	20.3	16	9	5	9	5
PG	—	—	—	—	—	—	—	—	0.6	0.2	0.5	0.3	0.3	1	38	7	10	—
DPG	0.4	0.8	0.5	1.1	0.6	2.3	1.8	0.6	4.2	7.1	7.5	5.2	3.3	1	—	trace	14	68
PA	0.5	1.0	0.6	1.5	2.0	?	—	1.7	0.6	0.3	0.3	0.3	0.8	3	—	5	—	—
SP	13.6	13.2	10.1	2.7	1.7	1.0	?	6.7	12.8	12.9	11.0	13.0	2[b]	—	—	—	—	24[c]

Abbreviations as in Table 2.5. [a] According to Baraud et al. (1970) baker's yeast contains no phosphatidyl serine; [b] unidentified; [c] glycolipid.

TABLE 2.7. Percent content of lipids in erythrocyte membranes of various animals (adapted from Dawson et al., 1960; Dewey and Barr, 1970; Rouser et al., 1968)

Lipid	Cat	Cow	Dog	Guinea pig	Horse	Pig	Rabbit	Rat	Sheep	Man
Phospholipids	61.3	64.8	52.6	55.6	52.0	59.8	65.8	67.0	63.2	65.9
PC[a]	30.5	0	46.9	41.1	42.4	23.3	33.9	47.5	0	28.9
PE	22.2	25.1	22.4	24.6	24.3	29.7	31.9	21.5	26.2	27.2
PS	13.2	15.3	15.4	16.8	18.0	17.8	12.2	10.8	14.1	13.0
PI	7.4	3.7	2.2	2.4	0.3	1.8	1.6	3.5	2.9	1.3
PA	0.8	0.3	0.5	4.2	0.3	0.3	1.6	0.3	0.3	2.2
SP	26.1	46.2	10.8	11.1	13.5	26.5	19.0	12.8	51.0	26.9
Others	0.3	1.7	1.8	0.3	1.7	0.9	0.3	3.8	4.8	—
Cholesterol	26.8	27.5	24.7	27.0	24.5	26.8	28.9	24.7	26.5	22.2
Gangliosides	8.8	5.5	11.8	2.2	15.5	3.3	4.5	6.3	7.8	
Other glycolipids	3.1	2.2	10.9	15.2	8.0	10.1	0.8	2.0	2.5	10

[a] The various phospholipid fractions are expressed as percent of total phospholipids. Abbreviations as in Table 2.5.

Unfortunately, there are no comparative data regarding the occurrence in membranes of various other lipids although some regularities have already emerged, among them:

1. Animal membranes contain cholesterol, plant membranes contain β-sitosterol, yeast and fungal membranes ergosterol and zymosterol as major sterols.

2. Glycolipids in the broad sense do not occur in mitochondria and probably are restricted to plasma membranes.

3. Only mitochondria contain greater amounts of cardiolipin.

4. The composition of phospholipids in bacterial membranes is somewhat peculiar; the major component appears to be phosphatidyl glycerol and its amino acyl derivative. Phosphatidyl ethanolamine and phosphatidyl serine occur sporadically while sphingolipids and phosphatidyl choline are very rare.

Various glycolipids are of importance, e.g., monoglucosyl and diglucosyl diglycerides.

5. *Acholeplasma* cells, although prokaryotic like bacteria and although containing a great of phosphatidyl glycerol (36%) and

glucosyl diglycerides (46%) among its polar lipids, as bacteria do, show a striking content of cholesterol ($10-30\%$ of total lipid).

6. Virus lipids are apparently derived from their host cells and are thus of corresponding composition.

One may consider the distribution of various lipids from the aspect of occurrence of polar groups in their molecules. Classifying lipids as uncharged (cholesterol, triglycerides, polyhydroxy compounds), zwitterionic (phosphatidyl ethanolamine, phosphatidyl choline), weak acids (carboxylic acids and phosphatidyl serine) and strong acids (phosphatidic, sulfo acids) we arrive at a predominance of the uncharged class in myelin, erythrocytes and plant tissues and a prevalence of zwitterionic and weakly acidic substances in mitochondrial membranes.

2.2.1.3. Fatty acids

In all lipids based on glycerol and sphingosine one or more hydrocarbon chains derived from a fatty acid are present. It may be of interest to compare the pattern of fatty acid representation (1) in different lipids and (2) in different species. The variability is shown in Tables 2.8 and 2.9. The predominance of 16 : 0, 18 : 0, 18 : 1,

TABLE 2.8.   Percent fatty acid content in various phospholipids of human erythrocytes (adapted from Dewey and Barr, 1970; Rouser et al., 1968)

Source of fatty acids	Carbon atoms:double bonds										
	16:0	16:1	18:0	18:1	18:2	20:4	22:0	22:5	22:6	24:0	24:1
PC	33.0	1.0	11.7	20.6	18.2	5.0	—	5.4	1.1	—	—
PE	18.9	0.6	8.0	25.2	7.0	21.9	4.7	3.1	2.9	—	—
PS	7.1	0.4	41.6	13.0	2.8	19.7	—	2.9	4.2	—	—
SP	41.3	0.1	9.1	5.2	3.7	0.1	8.0	—	—	15.0	15.5
Whole membrane	28.2	0.7	15.1	18.3	10.6	10.8	—	4.0	2.1	?	?

Some authors have detected the following additional fatty acids; PC 14:0, 20:0, 20:2; PE 12:0, 14:0, 20:3; PS 20:3; SP 12:0, 14:0, 20:0, 22:4, 22:5.

TABLE 2.9. Percent fatty acid content in the total erythrocyte lipids from various species (compiled from several sources)

Animal	Carbon atoms: double bonds													
	12:0	14:0	16:0	16:1	18:0	18:1	18:2	18:3	20:0	20:3	20:4	22:5	22:6	24:1
Rat	—	0.3	31.1	2.2	13.7	18.5	7.2	—	—	2.3	24.0	—	—	—
Rabbit	—	—	20.1	2.6	17.8	11.0	21.5	—	—	—	18.5	—	—	—
Guinea pig	—	—	16.1	2.6	26.6	12.3	15.7	—	—	—	16.1	—	—	—
Sheep	0.6	1.6	16.0	2.3	9.7	49.6	11.5	0.7	0.9	1.9	1.4	—	—	—
Cow	—	—	12.1	2.7	14.1	34.5	21.1	—	—	—	4.8	—	—	—
Pig	—	—	21.4	2.4	10.4	32.1	23.2	—	—	—	6.4	—	—	—
Dog	—	—	16.9	1.7	19.0	14.2	12.9	—	—	—	30.8	—	—	—
Cat	—	—	20.1	2.6	17.8	11.0	21.5	—	—	—	18.5	—	—	—
Monkey	—	0.3	15.6	0.5	15.2	15.7	11.0	—	—	1.0	12.3	4.2	10.8	7.5?
Chicken	—	0.2	28.7	0.8	8.8	21.7	31.8	0.9	—	0.6	4.6	1.2	3.2	—

It should be noted that the analyses vary somewhat depending on the technique and degree of refinement applied by the various authors.

18 : 2 and 20 : 4 acids is obvious in most of the species tested, palmitic and oleic acids leading the field. However, in microorganisms and plants there appears to be a much greater variety in fatty acid patterns.

To begin with, in addition to saturated and unsaturated fatty acids, there exist cyclopropane fatty acids with one or more of the three-membered rings along the hydrocarbon chain in various bacteria and both cyclopropane and cyclopropene fatty acids in some plants (*cf.* Gunstone, 1970). Another rarity found practically only in bacteria are the branched-chain fatty acids (mainly with 15 and 17 carbon atoms). The importance of all these acids for membrane fluidity will be discussed later in the chapter. The differences between microbial and plant species and the animal kingdom may be grasped from a comparison of Tables 2.8 and 2.9 with Table 2.10.

2.2.1.4. Solubilization

Extraction of lipid material from membranes presents no particular problem in cases when it is of no concern whether lipid protein interactions are disrupted in the process. Lipids can be extracted from cells or tissues with a number of organic solvents or their mixtures, such as aqueous acetone, tertiary amyl alcohol, isobutyl alcohol, *n*-butanol, ether, etc. The usual procedure is homogenization with a mixture of chloroform and methanol (2 : 1, v/v), using a 20-fold excess of solvent over tissue with subsequent washing of the extract with water (one-fifth of the volume is recommended) to remove the water-soluble components (some sugars and most amino acids are extracted by the chloroform-methanol treatment). The only lipid that remains unextracted is phosphatidyl inositol (particularly in the polyphosphorylated form) and this must be removed by using chloroform-methanol plus hydrochloric acid or, as suggested by Folch-Pi and Stoffyn (1972) for brain proteolipids, by applying five volumes of chloroform-methanol (1 : 1, v/v) and 0.5 volume 2*M* KCl in water. The lower phase now contains all phospholipids, cholesterol, polyphosphoinositol, as well as lipoprotein, the upper phase contains gangliosides (plus protein and low-molecular weight components).

Since the extraction of lipids occurs almost as well at subzero as at room temperature it is advisable to work at the lowest practicable temperature, there being less risk of chemical changes of the lipids. Likewise, some lipid components being rapidly degraded by oxygen,

TABLE 2.10. Percent fatty acid content in the cell envelopes or external membranes of some plants and microorganisms (adapted from Cho and Salton, 1966; Galliard, 1973; Nurminen and Suomalainen, 1971; Tourtellotte, 1972)

Source of fatty acid	Carbon atoms: double bonds								
	10—13:0	14:0	14:1	16:0	16:1	17:0	18:0	18:1	18:2, 3
Saccharomyces cerevisiae[a]	7	6	6	6	51	1	4	18	<1
Acholeplasma laidlawii[b]	7.5	24.8	—	53.5	—	—	3.0	6.6	4.4
Bacillus licheniformis[c]	—	4.2	—	3.9	12	—	0.2	11.1	—
Micrococcus lysodeikticus[d]	0.4	4.4	—	0.2	—	—	—	—	—
Aerobacter aerogenes	—	5.7	—	56.4	5.7	3.8	—	9.6	18.8
Escherichia coli B	—	6.8	2.1	42.0	2.3	—	1.7	25.7	—
Serratia marcescens[e]	—	10.5	—	55.5	0.9	—	6.1	3.1	2.6
Apple fruit	—	—	—	25	—	—	4	7	65
Moss (*Hypnum*)[f]	—	1	—	14	5	—	2	7	44
Chlorella pyrenoidosa[g]	—	+	—	+	3	trace	+	14	43

[a] There are a number of fatty acids with more than 20 carbon atoms present, albeit in minute amounts, the order of decreasing occurrence being 26 : OH, 26 : 0, 20 : 0, 22 : 0, etc. Differences depending on strain character and type of cultivation may be substantial (*cf.* Longley *et al.*, 1968). According to Baraud *et al.* (1970) the plasma membrane phospholipids contain appreciable amounts of 3,7,11-trimethyldodecanoic acid. [b] Like in many bacteria, the fatty acid content of *A. laidlawii* membranes can be influenced by growth on a fatty acid. Thus growth on penta-decanoic acid will produce membranes containing 83% of this acid, growth on isopalmitic acid will yield 79% of this acid in membranes. [c] Contains also 50.4 % branched 15 : 0 acid and 28.2 % branched 17 : 0 acid. [d] Contains 85.4 % branched 15 : 0 acid. [e] Contains nearly 20 % 9, 10-methylenehexadecanoic acid (a cyclopropane fatty acid). [f] Also 12 % of 20 : 4, 7 % of 20 : 5, and 5 % of 22 : 0. [g] Total saturated acids represent 17 %; also some 18 : 4, 16 : 2 and 16 : 3 acids are present.

the extraction procedure should be done in the absence of air (under argon or nitrogen).

Once extracted, the lipid mixture can be resolved into classes and subclasses by a number of chromatographic techniques, among them chromatography on silicic acid, on Sephadex LH (a lipophilic variety), thin layers of alumina or other adsorbents. For further analysis, various techniques are available, including the application of phospholipases.

Fatty acyl esters can be hydrolyzed either in acid or in alkaline media and the fatty acids liberated then esterified and as esters analyzed elegantly by gas-liquid chromatography.

There is a wealth of methods applicable to the definitive analysis and much can be learned from compendia such as those of Johnson and Davenport (1971) or Lowenstein (1969).

2.2.2. Proteins

2.2.2.1. Types and occurrence

Unlike the lipids which (1) have a single major role in membranes, viz. that of maintaining their mechanical stability and overall hydrophobicity and (2) are easily extracted and analyzed as chemical individuals, membrane proteins have proved to be singularly difficult both to extract and to analyze structurally as well as functionally.

It will be recalled from Table 2.3 that proteins usually represent more than one-half of the membrane dry weight; still, no universal structural protein or even a subunit has been established. In fact, in cases where the protein content is relatively high (mitochondrial inner membrane, bacterial cytoplasmic membranes) the membranes are endowed with numerous enzymic functions, the inference being that membrane proteins are largely enzymes. Although this is apparently not so, most membrane proteins do play a more or less specific role: in enzyme catalysis, as receptors of hormonal or antigenic signals, as recognition elements in membrane transport as well as pinocytosis and chemotaxis, as transmembrane carriers for low-molecular substances, etc. There are thus few proteins with no apparent physiological function left.

The rather crude methods of extraction and separation of membrane proteins generally do not permit to analyze the whole spectrum of proteins present, the highest number extracted from a given membrane being about 30. However, it must be obvious that an average cell, both a eukaryotic organ cell or a unicellular organism, contains probably 30−50 different proteins whose sole role is to select solutes for transport, plus a variety of proteins with other functions.

Specific proteins, mostly of enzyme character, of various membranes will be described in the appropriate section. Here we shall consider some characteristic protein types occurring in membranes and examine whether any of the protein may be considered as a structural backbone of a membrane.

The molecular weights of polypeptides extracted from membranes vary from about 10 000 to 240 000. (The "miniproteins" of mol. wt. of 6000 reported by Laico and associates (1970) were later shown by the same group headed by Dreyer (1972) to be artifactual.)

Erythrocytes. Depending on the solubilization procedure, 8−20 polypeptides have been resolved, the major ones being (according to Bretscher, 1973) peptide *a* (100 000), a glycoprotein (30 000) and tektin A (220 000 + 240 000), this last component being probably identical with the spectrin isolated by Clarke (1971). Steck (1974) reports 8 major polypeptides and 2 glycoproteins in the same material. After treatment with 2-chloroethanol, the molecular weights of the peptides all lie between 10 000 and 50 000 (Zahler, 1969), the inference being that the large molecular weights reported above may be due to association of subunits. Apparently, none of the components is a structural protein.

Mitochondria. Apart from a multitude of enzymes (*cf.* Table 2.26) these organelles were reported to contain a structural protein (Richardson *et al.*, 1963) with mol. wt. of about 23 000, the protein showing an affinity for phospholipids and ATP. Analogous proteins were detected in liver microsomes, bovine erythrocytes and spinach protoplasts and assumed to form a part of the integral membrane structure. However, no universal occurrence of these proteins could be established and the concept was eventually abandoned (Green *et al.*, 1968). In fact, even the inner and outer mitochondrial membranes differ completely in their individual polypeptides (Schnaitman, 1969).

Bacteria. In *Escherichia coli*, the outer envelope (see p. 134) appears to contain six polypeptides while the plasma membrane proper (with its array of enzymes) has 27 major polypeptides plus many minor ones, none of them in comparable amounts in the two layers (Schnaitman, 1970). Recently, many more were detected by using two-dimensional gel electrophoresis (Johnson *et al.*, 1975). In *Acholeplasma laidlawii*, disc-gel electrophoresis revealed 20−30 polypeptides with molecular weights from 15 000 to 100 000, none of them predominating in any way (Morowitz and Terry, 1969).

Interesting observations exist on membrane protein mutants of various bacteria (for a review see Machtiger and Fox, 1973), the deletions affecting either the function of a membrane-located enzyme or the number of proteins eluted from the membrane. An extreme case in this connection is the strain *Bacillus megaterium* PP where more than 90% of membrane proteins is formed by a single type of polypeptide (Patterson and Lennarz, 1970).

Other membranes. In several instances, the distribution of extracted proteins was rather asymmetric, there being a single predominant polypeptide present, although by no means a "structural" protein. Thus, rhodopsin is the single important polypeptide in the rod outer segments, brain myelin contains three major polypeptides, similarly to sarcoplasmic reticulum. The purple membrane of *Halobacterium halobium* was found to contain a single protein. In these cases it is rather likely that the protein or few proteins present in the membrane do in fact play a structure-preserving role but it is not their primary function. Still, the existence of a functional structural protein is supported by experiments where thorough washing of liver microsomes with sodium chloride, sodium carbonate and sodium bicarbonate removes practically all proteins but a few with a predominant 52 000 polypeptide with no apparent function (Hinman and Phillips, 1970).

In their amino acid composition, membrane proteins do not exhibit any peculiarities in comparison with the composition of soluble enzymes. The hydrophobic amino acids represent about one-third of the total; acidic amino acids amount to about 20%, basic ones to about 12% but the carboxylic groups may be more strongly amidated so that no net charge arises from the apparent preponderance of the acidic over the basic amino acids (Table 2.11). The only difference distinguishing the membrane proteins from other cell proteins is the low amount of cysteine (apparently no disulfide bridges are formed in membrane proteins). Although in their average the membrane proteins are no more hydrophobic than soluble proteins, an important distinction arises when extrinsic and intrinsic membrane proteins are considered. Dividing amino acids into polar (Asp, Asn,

TABLE 2.11. Amino acid composition (in molar %) of proteins of various cell membranes (from Engelman and Morowitz, 1968; Haurowitz, 1963; Longley et al., 1968; Steck and Fox, 1972; Wallach, 1972)

Amino acid	Human erythrocyte	Ehrlich ascites tumour plasma membrane	Ehrlich ascites endoplasmic reticulum	Rat liver bile fronts	Baker's yeast[a] plasma membrane	Acholeplasma laidlawii[b]	Human serum albumin	Insulin
Lys	5.2	6.3	6.5	7.2	7.4	6.37	10.7	2.0
His	2.4	2.6	2.1	2.6	2.5	1.46	2.9	3.6
Arg	4.5	4.7	5.2	5.2	4.2	2.95	4.5	2.3
amide NH$_3$	6.9	14.7	10.8	12.4	?	?	8.0	14.2
Asp	8.5	8.8	8.7	9.3	10.8	11.4	9.9	5.7
Glu	12.2	10.1	10.6	12.0	10.2	8.23	15.0	15.5
Thr	5.9	5.5	5.4	5.3	10.9	6.77	5.4	3.1
Ser	6.3	6.6	6.2	6.0	10.9	6.40	4.4	6.2
Pro	4.3	5.2	5.4	4.9	5.1	3.60	5.6	2.8
Cys	1.1	trace	trace	0.9	1.0	0.19	6.6	11.0
Met	2.0	2.7	2.5	2.3	1.5	2.33	1.1	—
Gly	6.7	8.5	7.7	7.8	8.3	6.95	2.7	6.9
Ala	8.2	7.8	7.6	8.0	7.7	8.23	8.9	3.7
Val	7.1	6.6	6.7	6.6	6.1	7.55	8.4	9.5
Ile	5.3	6.1	5.1	5.1	5.5	7.39	1.7	14.2
Leu	11.3	10.1	10.3	9.6	9.6	9.79	11.6	14.2
Tyr	2.4	3.1	3.4	2.7	7.8	4.81	3.3	7.7
Phe	4.2	4.8	4.8	4.5	7.8	5.40	6.0	5.4
Trp	2.5	1.5	1.5	—	1.2	?	0.1	—

[a] Ornithine was also detected. [b] Also 1.73% glucosamine and 5.75% galactosamine.

Glu, Gln, Lys, Arg), intermediate (Ser, Thr, Tyr, His, Gly) and nonpolar (Ala, Val, Leu, Ile, Cys, Met, Pro, Phe, Trp) ones and assigning to the first group the polarity of unity, to the second of one-half and to the third of zero, and expressing them in molar percent, Vanderkooi and Capaldi (1972) arrived at an interesting distribution of polarities among proteins in general. While 206 nonmembrane proteins had polarities of 46 ± 6%, intrinsic proteins had polarities ranging from 30 to 40%, extrinsic ones from 41 to 53%, the difference being apparently associated with the localization of the former within the membrane, of the latter on the membrane surface.

2.2.2.2. Solubilization

The task of solubilizing proteins in a native form is incomparably more difficult than analogous work with lipids. The ease of denaturation and of possible subtle changes in the tertiary and particularly quaternary (subunit) structure have been frequently used by critics as arguments against the validity of deductions drawn from analyses of membrane proteins made soluble by one means or another; still, the natural way for an investigator at this stage of experimental capabilities is to start from a thorough analysis of constituents before proceeding to a synthetic approach that might give a more definitive picture of what a membrane is like. Hence, proteins must be released from membranes and sorted out. The applicable techniques are numerous, practically all of them based on disruption of weak bonds between individual proteins as well as between proteins and lipids. It is an obvious disadvantage of the methods that, depending on the potency of the solubilizing agent, proteins may be broken down to smaller units in the process, these units either being able to reaggregate or not if the solubilizing agent is removed. Most estimates of molecular weights of membrane proteins are thus hampered and one may encounter values that refer obviously to a single polypeptide chain, as well as others that would embrace most of the membrane protein in a few dozen supermolecules.

The solubilizing procedures may be divided as follows (*cf.* Steck and Fox, 1972; Coleman, 1973; Jakoby, 1971).

1. Proteins that are merely adsorbed on the membrane surface are readily washed out at a low ionic strength, even with distilled water or with e.g. $10^{-5} M$ NH$_4$OH.

2. Periplasmic proteins are immediately released by removing the external obstacle of the cell wall by treatment with lysozyme or penicillin (in bacteria), or with snail-gut or analogous enzymes (in yeasts and fungi). Occasionally, it is difficult to distinguish between a true periplasmic protein and an extracellular enzyme which can cross the wall without any special processing. A procedure frequently employed for the release of periplasmic proteins has been the osmotic shock, pioneered by Heppel (e.g., Nossal and Heppel, 1966) in bacteria.

In a frequently used modification, intact cells are suspended in cold 0.5 M sucrose (buffered with Tris-HCl and with some EDTA added) and after a few minutes centrifuged and rapidly dispersed in ice-cold 10^{-4} M MgCl$_2$; the supernatant liquid is collected and used for further work. A partial list of proteins thus released from *Escherichia coli* is given in Table 2.12. Sucrose has sometimes been

TABLE 2.12. **Proteins known to be released from *Escherichia coli* by an osmotic shock (adapted from Heppel, 1971)**

Alkaline phosphatase	Asparaginase (II)
5′-Nucleotidase	Penicillinase
2′ : 3′-Cyclic-nucleoside monophosphate phosphodiesterase	Deoxyribose-phosphate aldolase
Acid phosphatases (of three kinds)	Purine-deoxyribonucleoside phosphorylase
Ribonuclease I	
Deoxyribonuclease I	Phosphopentomutase
Glucose-1-phosphate adenylyltransferase	Various binding proteins

replaced with saturated mannitol (e.g., Schwencke *et al.*, 1971) or even concentrated salt solutions. There are indications that some periplasmic proteins may be released by simple transfer of cells to distilled water without previous exposure to high osmolarities (Opekarová *et al.*, 1975).

3. Removal of cations can occasionally release a membrane protein if it is held by ionic bonds or if ions are essential for its integrity. The case in point is the application of ethylenediamine tetraacetic acid for solubilization of adenosinetriphosphatase from *Streptococcus faecalis* protoplasts, from the mitochondrial inner membrane and from thylakoid membranes.

4. High ionic strength may be effective in removing electrostatically bound proteins, $0.5-1.2$ M NaCl being employed to this end.

5. Similarly to the preceding case, various denaturing agents in high concentrations are occasionally used under the name of chaotropic agents. These substances disturb the ionic milieu in the membrane, break the weak bonds and, moreover, decrease substantially the actual concentration of water at the membrane. They include particularly salts of thiocyanate perchlorate, as well as guanidine and urea. The concentrations of these last-named compounds

are usually 6 M. A drawback of the technique is the fact that the solubilized peptide chains are frequently devoid of activity and, in the case of urea application at neutral pH, may be heavily carbamylated.

6. Treatment with hydrolytic enzymes is frequently used as a preliminary step in membrane breakdown. Papain and trypsin have been employed widely both for the release of enzymes (e.g., β-fructosidase, L-β-naphthylamidase) and for disruption of larger complexes, even of tissue. Phospholipases have been used for releasing numerous lipid-associated proteins (e.g., 3-hydroxybutyrate dehydrogenase, mitochondrial NADH dehydrogenase).

7. Sonication can be used to break membranes by mechanical shearing into small fragments but there is little hope that proteins separated from lipids will be obtained by this technique.

8. Organic solvents, by extracting lipids from membranes, may in favourable cases release some of the proteins even in a native (or rather active) state. An effective solvent in this context is 2-chloroethanol in water (9 : 1) at pH below 2 or, alternatively, N,N-dimethylformamide in water. Some authors prefer mixtures of phenol – formic acid – water (e.g. 14 : 3 : 3). n-Butanol or n-pentanol, or even a mixture of n-butanol with urea, will release proteins in an association with lipids, in some cases as the active complexes.

9. Detergents, by being amphiphilic or amphipathic, will bind hydrophobically to membrane proteins while interacting through their polar moieties with the aqueous medium, thus pulling the proteins out of the membrane matrix. Like in the preceding case, lipid-protein complexes are likely to be extracted in this way. The nonionic detergents of the Triton type, Lubrols and various Tweens have been used most widely. The cationic detergents include in particular cetyltrimethylammonium bromide (used successfully for solubilization of rhodopsin); the anionic ones are sodium cholate and deoxycholate and, most important, sodium dodecyl sulfate which dissociates proteins into monomeric units by forming regular complexes with them. Such complexes are readily separated by gel electrophoresis and their molecular weight can be estimated.

10. Various other compounds, such as the organic mercurial mersalyl, have been used for special purposes (Cantrell, 1973).

Further processing and analysis of released proteins can be effected in a number of ways, various chromatographic and electrophoretic techniques being used (for detailed information see, e.g., Jakoby, 1971).

2.2.3. Carbohydrates

In many reports on membrane composition, the content of carbohydrates looms relatively large side by side with proteins and lipids. However, there is probably very little free carbohydrate in cell membranes, most of the sugar residues being constituents of glycolipids and glycoproteins. This is in sharp contrast with the composition of cell walls, particularly of plant or fungal origin, where true carbohydrates, such as cellulose, glucan, mannan, chitin and the like, are found as the major structural components (*cf.* p. 141).

For the sake of completeness we shall only briefly enumerate the carbohydrate parts of the complex membrane molecules.

The sugar moieties of glycolipids are described on p. 41. The structure of bacterial lipopolysaccharides and of peptidoglycans will be mentioned on p. 139 and p. 134, respectively.

The principal monosaccharides of glycoproteins are similar to those of glycolipids, *viz.* D-galactose, D-glucose, *N*-acetylglucosamine, *N*-acetylgalactosamine, sialic acid, D-fucose, but also D-mannose and D-xylose, plus a spattering of less usual sugars, such as are found in lipopolysaccharides (the question remains whether this is not a contamination caused by inadequate separation of membranes).

In glycoproteins, the monosaccharides usually form heterologous sequences of several units, the oligosaccharide being linked either *O*-glycosidically to the hydroxyl group of serine and threonine or *N*-glycosidically to asparagine.

The glycoproteins are found in many membranes but may be completely absent in some. An important glycoprotein of human erythrocyte membranes was already mentioned previously; it can be resolved into three sialoglycopeptides, their composition being characterized by tetrasaccharide groupings, but larger oligosaccharides are also present. There may be over 60% carbohydrate in this material (*cf.* Winzler, 1969).

An outstanding representative of the glycoprotein family is the retinal rod component rhodopsin — its species from bovine retina contains about 4% carbohydrate, arranged as an oligosaccharide chain attached to a 28 000 polypeptide (Heller and Lawrence, 1970).

The role of glycoproteins as cell surface receptors of signals (hormones, antigens, phytohemagglutinins, etc.) has been recognized in many instances (e.g., Kornfeld and Kornfeld, 1970).

Glycoproteins represent major constituents of viral envelopes (as much as 40% of total virus proteins) their molecular weights being rather alike. Thus, that of the glycoprotein of group A arboviruses is 52 000, that of rhabdoviruses 67−80 000, those of paramyxoviruses 53−56000 and 65−74000, those of orthomyxoviruses then 24−32000, 45−51 000, 49−58 000 and 74−78 000 (for more details see Klenk, 1973).

2.3. STRUCTURE OF MEMBRANES

In this section, the organization of membrane components, including their interactions, will be considered. It was during the last few years that considerable insight into the membrane structure has been gained, particularly through the use of previously unavailable techniques of physical chemistry. Before presenting what has been deduced about membrane structure, let us first survey the methodological approaches that assisted in these deductions.

2.3.1. Physico-chemical techniques

2.3.1.1. X-Ray diffraction

Crystalline solids can act as a three-dimensional diffraction grating for X-rays if the interatomic or intermolecular distances of the lattice are of the same order of magnitude as the wavelength of the X-rays, i.e. about 10^{-10} m (0.1 nm). Ordered constructive and destructive interference occurs as the wavefronts are diffracted, obeying the Bragg equation

$$2d \sin \theta = n\lambda \qquad (2.14)$$

d being the distance between repeating lattice planes, 2θ the angle between the incident and the diffracted beams, n the sequential number of a constructive maximum and λ the wavelength of the incident X-ray. With greater values of d (as in macromolecular systems), the angle will be rather small−hence we often speak of low-angle diffraction spectroscopy in the context of lipid or membrane studies.

Every crystalline sample can be envisaged as an array of the smallest repeating units, the so-called unit cells. It is the properties of this unit cell that determine the features of the diffracted waves. Thus, the positions of intensity maxima are determined by the unit cell dimensions; the intensity of the maximum depends on the distribution of electrons in the unit cell; the angular length of the maximum depends on the degree of relative orientation between the unit cells; the angular width of the maximum depends on the number of unit cells in the crystal examined.

The intensity of the diffraction pattern obtained is related to the amplitude F of a given wavefront by

$$I = F^2 \qquad (2.15)$$

so that the absolute amplitude but not its phase can be computed. Using the Fourier relationship for the vectorial electron density in real space (in the x direction in this particular case) it is defined by

$$\varrho(x) = \int F(r)\, e^{-2\pi\, irx}\, dr \qquad (2.16)$$

where $F(r)$ is the so-called structure factor, r being the reciprocal spacing (in nm^{-1}), i then $(-1)^{0.5}$. To obtain the correct phase of each $F(r)$, several ingenious techniques are in use. The complete solution

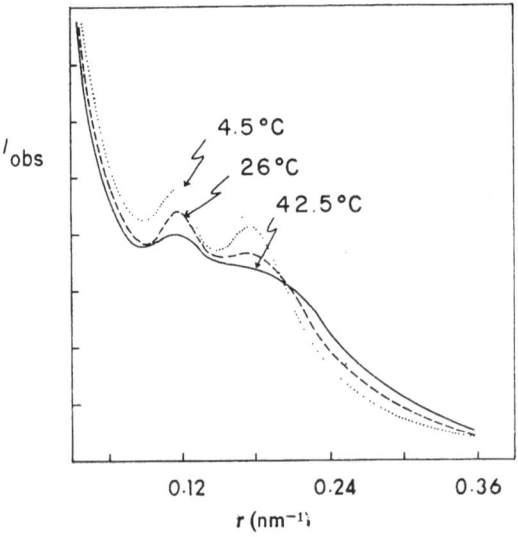

Fig. 2.8. Microdensitometer tracing of X-ray diffraction patterns from frog retinal rod membranes at different temperatures. The high intensity recorded near zero corresponds to the incident beam. I_{obs}, relative observed intensity; r, reciprocal space distance. (Taken with kind permission from Blasie and Worthington, 1969).

for $\varrho(x)$ yields the number of electrons at different intervals of the repeat spacing and, with the use of computers, can be interpreted to reveal the mutual positions of the component atoms or molecules.

In membrane studies, the usual procedure is to read the diffraction pattern intensity with a microdensitometer (a plot of observed intensity reciprocal spacing, such as in Fig. 2.8) from which an electron density profile (density *vs.* radial distance d in nm) is calculated.

X-ray diffraction studies have been applied both to artificial and native phospholipid membranes where, for a hexagonal packing, a sharp peak at $d = 0.42$ nm is obtained. At higher temperatures (when the hydrocarbon chains lose their rigidity) the peak becomes diffuse and is shifted to 0.46 nm. Refined techniques have shown that (at least in myelin) the packing distance of hydrocarbon chains is 0.46 to 0.48 nm, that there is a lipid bilayer $5-7$ nm thick (depending on hydration) and that the bilayer is coated with protein. The membranes are apparently symmetrical with respect to dimensions but asymmetrical with respect to electron density, probably due to different participation of cholesterol at the two membrane faces (*cf.* Caspar and Kirschner, 1971).

2.3.1.2. *Infrared spectroscopy*

This technique is widely used in structural analysis particularly in organic chemistry and has found only limited application in the analysis of membrane lipid and protein structure. The absorption spectrum of a sample in the $2-15$ μm region is due to absorption by vibrating atoms in molecules, various bands being ascribed empirically to certain functional groups. Since water obscures some of the important sections of the spectrum, organic chemists dissolve their samples in organic solvents or disperse then in KBr pellets; for membrane studies, ordinary water may be replaced with D_2O to achieve comparable results. Alternatively, air-dried films of lipids in AgCl can be analyzed. Examples of infrared spectra of membrane material are shown in Fig. 2.9.

In protein analysis, the amide bands are of particular importance, there being absorption maxima due to the amide grouping at 1652, 1630 and 1535 cm^{-1}, the first of these supporting the α-helical structure, the second the pleated sheet (β-conformation) structure. With the exception of some mitochondrial "structural-type" protein, all membranes were shown to possess a predominance of the α-helix.

Only recently spectroscopy based on Raman scattering of IR laser light has been applied to the study of membrane constituents and even whole membranes, the technique showing promise in the definition of various interactions (e.g., Lutz and Breton, 1973).

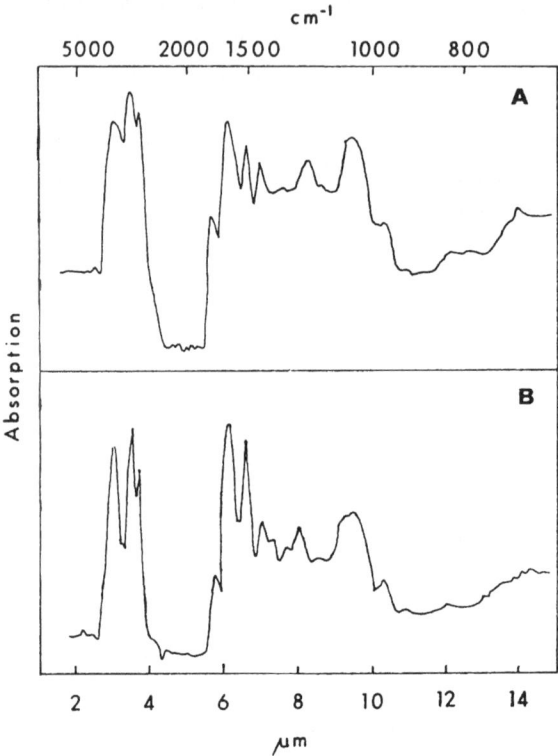

Fig. 2.9.   Infrared spectra of myelin (**A**) and of erythrocyte ghosts (**B**). (Adapted from Chapman and Wallach, 1968.)

2.3.1.3.   *Ultraviolet spectroscopy*

Electron transitions between the ground state and the electronically excited states underlie the absorptions observed in the spectral region from 150 to 600 nm, as well as the optical activity of various groupings.

Both classical absorption spectroscopy in the ultraviolet region and the more recent optical rotatory dispersion (ORD) and circular dichroism (CD) have been exploited in the study of membrane composition and structure, but restricted mostly to the investigation of proteins.

2.3.1.3.1. Absorption spectra

The UV absorption spectra are usually expressed in terms of absorbance A, which is defined as $-\log (I/I_0)$, I_0 being the intensity of incident light, I that of transmitted light. To provide a more standard basis for comparison between samples, the molar absorptivity ε (previously called the molar extinction coefficient) has been defined as A/bc where b is the path in cm and c the solute concentration in mol dm^{-3}, the overall dimension of ε thus being $dm^3 \, mol^{-1} \, cm^{-1}$. The use of the term optical density (O.D.) for absorbance should be restricted to cases where not only absorption but also light scattering contribute to the attenuation of the beam as it passes through the sample (for turbidimetry of cell suspensions, for instance).

In the case of membrane proteins, two contributions to the UV absorption may be recognized. The first is the $\pi^\circ \to \pi^-$ (nonbonding to antibonding; in the transition of the peptide N—C—O atomic group) bond (absorption at about 190 nm, $\log \varepsilon$ being $3.7-4.0$) accompanied by the $n \to \pi^-$ (nonbonding to antibonding, in the O atom) transition (absorption at 220 nm, $\log \varepsilon < 2.5$).

The second contribution arises from the presence of amino acid side chains, the highest absorptivities being displayed at the following wavelengths: $\log \varepsilon$ 4.78 (188 nm, phenylalanine, $\pi \to \pi^*$), 4.68 (193 nm, tyrosine, $\pi \to \pi^*$). 4.67 (219 nm, tryptophan, $\pi \to \pi^*$), 4.30 (197 nm, tryptophan, $\pi \to \pi^*$), 4.15 (185 nm, arginine, $n \to \sigma^*$), 3.75 (280 nm, tryptophan, $\pi \to \pi^*$), 3.15 (274 nm, tyrosine, $\pi \to \pi^*$).

In spite of their high absorptivities, these absorptions contribute relatively little to the overall protein spectrum because of the preponderance of the —NH—CO— group in the sample.

Among lipids, it is particularly the conjugated diene ($\log \varepsilon$ $3.47-4.0$, λ_{max} $230-260$ nm), the ketone ($\log \varepsilon$ $3.30-3.40$, λ_{max} $190-220$ nm) and the conjugated enone ($\log \varepsilon$ $3.95-4.18$, λ_{max} $220-240$ nm) that contribute to the diagnostic nature of UV absorption spectrum.

In addition to the UV-absorbing properties of natural membrane components, use has been made of external probes which attach to membranes and whose optical characteristics can be examined directly in the membrane material.

2.3.1.3.2. *Optical rotatory dispersion*

As a monochromatic, linearly polarized beam of light passes through an optically active sample (using UV wavelengths and hence quartz cuvettes), the plane of vibration of its electrical vector will be rotated. The magnitude of rotation can be evaluated by analyzing prisms. The molar unit of rotation Φ is given by $\alpha M'/100cl$, where α is the angle of rotation in degrees*, M' the molecular weight of the solute, expressed numerically as g/cmol (for proteins the average weight of one amino acid residue is taken), c the solute concentration in $g\,cm^{-3}$ and l the path in dm, the overall dimension of Φ being $rad\,cm^2\,dmol^{-1}$. Since the solvent itself may be polarizable, a Lorentz correction term is used to account for this. Then

$$\Phi' = \frac{3}{n^2 + 2}\,\Phi \qquad (2.17)$$

where n is the index of refraction of the solvent. The ORD spectrum has usually the form of a plot of Φ against λ but, obviously, if a preparation of undefined molecular weight is examined, one can plot α against λ. The wavelengths applied to ORD studies are usually in the region from 180 to 240 nm where $\pi^\circ \rightarrow \pi^-$ and $n_1 \rightarrow \pi^-$ electron transitions occur.

In membrane studies it is mainly the protein which yields useful ORD spectra and in this connection, they have been used particularly for distinguishing between the α-helical and other structures of peptide chains (Fig. 2.10). The diagnostic wavelength is 233 nm, the empirically (as well as theoretically; *cf.* Yang, 1967) calculated $[\Phi']_{233}$ being $-15\,000$ for a perfect α-helix, $-2\,000$ for a disordered chain, and $-3\,000$ to $-4\,000$ for the pleated-sheet conformation, taking a mean residue molecular weight of 115. Thus, for $[\Phi']_{233}$ more negative than, say, $-6\,000$, a considerable amount of α-helix is indicated.

The ORD spectra have been generally interpreted to indicate a prevalence of the α-helix in membranes but there are factors that may influence the quality of the spectra and, at least partly, affect the deductions made on their basis. The four important ones are the pos-

---

* This angle in radians is defined by

$$\alpha = \frac{\pi l(n_L - n_R)}{\lambda} \qquad (2.17')$$

where l is the optical path length through the sample, λ the wavelength used and n_L and n_R the indices of refraction for the left and right circularly polarized light.

sible contributions by side chains or prosthetic groups; the lack of equivalence between synthetic polypeptides used as standard for random-coil conformation and nonrepeating secondary structure found in proteins; the differences in the effect of longer helical structures as compared with a sum of shorter ones; possible existence of distorted helical regions in native proteins.

2.3.1.3.3. Circular dichroism

The value of circular dichroism, usually designated as ellipticity θ is actually the difference in absorbance by a sample of left and right circularly polarized light. The ellipticity has the same units as optical rotation, viz. rad cm^2 dmol^{-1}. It is related to the difference in absorptivity $\Delta\varepsilon$ of the left and right circularly polarized components as follows:

$$[\theta] = (2.303 \times 18\,000/4\pi)\,\Delta\varepsilon = 3\,298\,\Delta\varepsilon \qquad (2.18)*$$

$\Delta\varepsilon$ is obtained from

$$\Delta\varepsilon = \varepsilon_L - \varepsilon_R = \frac{A_L - A_R}{cl} \qquad (2.19)$$

where A_L and A_R are the actual measured absorbances, c is the molarity of the solution and l is the optical path length (in cm). Like with the optical rotatory dispersion, the molar ellipticity $[\theta]$ can be corrected for polarizability of the solvent by the Lorentz factor $3/(n^2 + 2)$.

Thus, while ORD reflects the difference in the indices of refraction (eq. 2.17'), the CD is related to the difference in absorbances. It follows from a fundamental consideration of polarizability that they are mutually related (see Moffitt and Moscowitz, 1959), a suitable expression for this fact being in the Kronig–Kramers transform

$$[\theta'(\lambda)] = -\frac{2}{\pi\lambda} \int_0^\infty [\Phi'(\lambda')] \frac{\lambda'^2}{\lambda^2 - \lambda'^2}\,d\lambda' \qquad (2.20a)$$

$$[\Phi'(\lambda)] = \frac{2}{\pi} \int_0^\infty [\theta'(\lambda')] \frac{\lambda'}{\lambda^2 - \lambda'^2}\,d\lambda' \qquad (2.20b)$$

* 2.303 is the conversion factor from natural to base-ten logarithms, the factor $4\,500\pi$ derives from the definition of absorbance $A = (4\pi\varepsilon/\lambda) \log e$ and from conversion of units of radians/cm to angular degrees/dm (cf. Djerassi, 1960, and Beychok, 1967).

where λ is the wavelength at which Φ or θ is determined and λ' is the current integration coordinate. Although the transformation is valid strictly only for the limits of 0 and ∞, even over finite regions of the spectrum useful information can be obtained. From this consideration it follows that both ORD and CD spectra yield

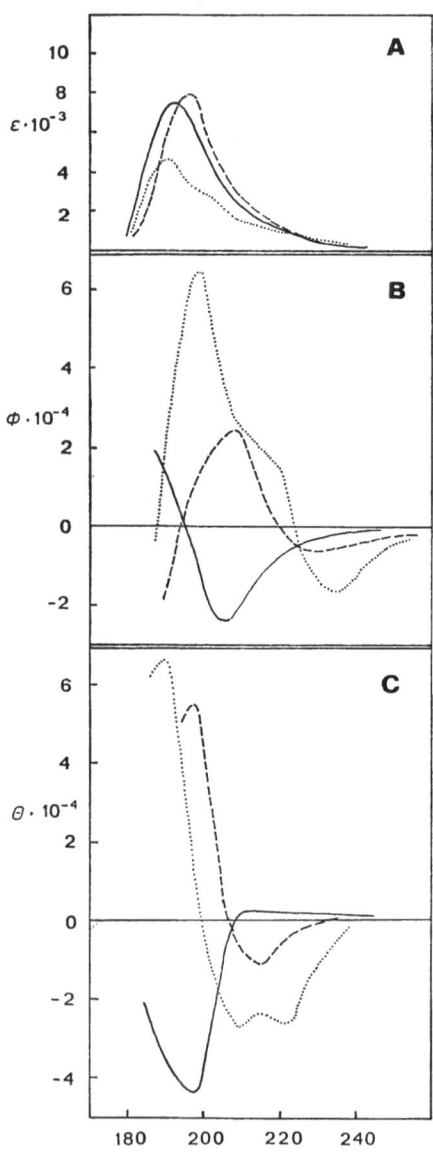

Fig. 2.10. Comparison of ultraviolet absorption (**A**), optical rotatory dispersion (**B**) and circular dichroism (**C**) spectra of poly-L-lysine with different types of secondary structure. (Adapted from Rosenheck and Doty, 1961; Holzwarth, 1972; Greenfield *et al.*, 1967.)

much the same information. However, ORD can be measured even at wavelengths where no absorption band is present (unlike CD). On the other hand, the number of Cotton effects* is much more readily obtained from CD measurements.

Like the ORD spectra discussed above, the CD curves have been used mostly for supporting the prevalence of one protein secondary structure over another. The value of ellipticity at 225 nm, $[\theta]_{225}$, is highly negative for an α-helix ($-40\,000$) but only less so for a pleated sheet (-7000) and is positive for a completely random coil ($+1500$).

Fig. 2.10 compares the absorption, ORD and CD spectra of synthetic poly-L-lysine of three different secondary structures.

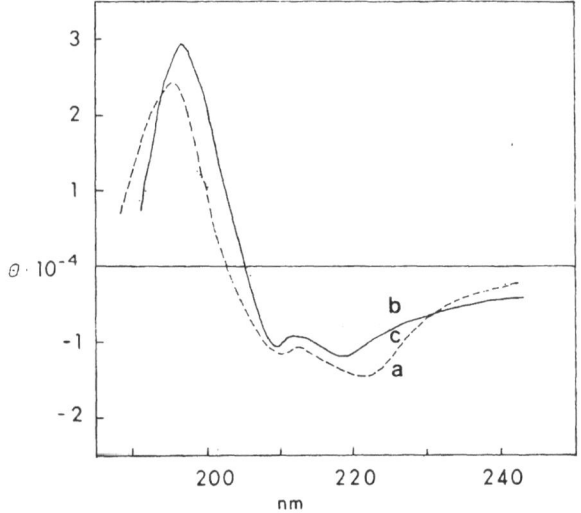

Fig. 2.11. Circular dichroism spectra of various membrane preparations. a Human red cell ghosts, b Ehrlich ascites cell plasma membranes, c supernatant at 20 000 g of sonicated mitochondrial fragments. (Adapted from Wallach, 1969; Ji and Urry, 1969.)

In working with membranes, several difficulties are encountered that complicate interpretation of CD and ORD spectra, mainly the rather low amplitudes and a red shift of the CD bands. The principal reason for this is the particle inhomogeneity of membrane suspensions.

* The Cotton effect is the term used to describe the deviations from monotony of the ORD and corresponding CD curves.

While the flattening of the curves is accounted for by the so-called absorption statistics (also Duysens' effect*, sieve effect, etc.), the red shift is probably due to light scattering on the particles (for a scholarly treatment of the subject consult Holzwarth, 1972).

Fig. 2.11 compares the CD spectra of several membrane preparations, showing the predominance of the α-helical structure, where the mentioned effects of particle size can be recognized.

2.3.1.4. *Nuclear magnetic resonance*

The technique of nuclear magnetic resonance is now in common use for a variety of analytical purposes. Their application to the investigation of lipids, both isolated and *in situ* in membranes, has yielded important information particularly on the fluidity of membrane lipid sheets.

The principle of the technique is the measurement of energy required to change the alignment of atomic nuclei in an external magnetic field. A nucleus with a nonzero spin (a paramagnetic nucleus) will absorb the energy of electromagnetic radiation applied at right angles to the external magnetic field, at a frequency defined by

$$v = \mu H_o / hI = \gamma H_o / 2\pi \qquad (2.21)$$

where μ is the nuclear magnetic moment, H_o is the external magnetic field strength, h is the Planck constant, I the nuclear spin and γ the so-called gyromagnetic ratio, specific for a given nucleus. Some of the characteristic values for biologically important isotopes are given in Table 2.13. The resonant frequency is usually found by varying the magnitude of the applied magnetic field at a constant frequency of electromagnetic radiation.

The value generally referred to in NMR spectrometry is the chemical shift δ defined by

$$\delta = \frac{v_S - v_R}{v_R} \cdot 10^6 \qquad (2.22a)$$

where v_S is the resonant frequency actually measured in a sample, v_R the resonant frequency of a reference compound (often tetramethyl-

---

* The absorption-flattening coefficient (Q_A), relevant to the Duysens effect is defined as the ratio of suspension absorbance to the true solution absorbance.

TABLE 2.13. **Characteristics of some nuclei used in NMR studies (in a magnetic field of 1 T)**

Nucleus	Spin I	NMR resonant frequency ν (MHz)	Magnetic moment μ (in nuclear magnetons[a])	Relative sensitivity at constant field	frequency
^{1}H	1/2	42.57	2.793	1.00	1.00
^{2}H	1	6.54	0.857	0.009 64	0.409
^{7}Li	3/2	16.55	3.257	0.294	1.94
^{13}C	1/2	10.71	0.702	0.015 9	0.251
^{14}N	1	3.077	0.404	0.001 01	0.193
^{15}N	1/2	4.316	—0.283	0.001 04	0.101
^{17}O	5/2	5.772	—1.893	0.029 1	1.58
^{19}F	1/2	40.07	2.627	0.834	0.941
^{23}Na	3/2	11.262	2.217	0.092 7	1.32
^{31}P	1/2	17.24	1.131	0.066 4	0.405
^{33}S	3/2	3.267	0.643	0.002 26	0.384
^{35}Cl	3/2	4.173	0.821	0.004 71	0.490
^{43}Ca	7/2	2.864	—1.315	0.063 9	1.41
^{127}I	5/2	8.519	2.794	0.093 5	2.33

[a] A nuclear magneton is analogous to the Bohr magneton (referring to an electron) and is defined as $eh/4\pi mc$ (e = elementary electrical charge, h = Planck's constant, m = mass of the proton, c = speed of light). When expressed in the absolute electromagnetic cgs system, it has the value of $5.05 \cdot 10^{-24}$ erg/gauss; in the SI system it is $5.05 \cdot 10^{-27}$ J T^{-1}.

silane or sodium 2,2-dimethyl-2-silapentane-5-sulfonate). δ may also be defined as

$$\frac{H_S - H_R}{H_R} \cdot 10^6 \qquad (2.22b)$$

the H's referring to the field strength at which resonance is observed. The chemical shift is brought about by the fact that the local magnetic field around the nucleus is different from that applied externally due to various shielding effects, dipole interactions and through the bond coupling with different neighboring nuclei. Because of undesirable effects of dissociable solvent hydrogen, measurements are usually performed in D_2O solutions (or in $CDCl_3$ in organic chemistry).

Instead of the δ-scale, a τ-scale is sometimes used, the relationship between the two being $\delta = 10 - \tau$. The chemical shifts due to various groupings are shown schematically in Fig. 2.12.

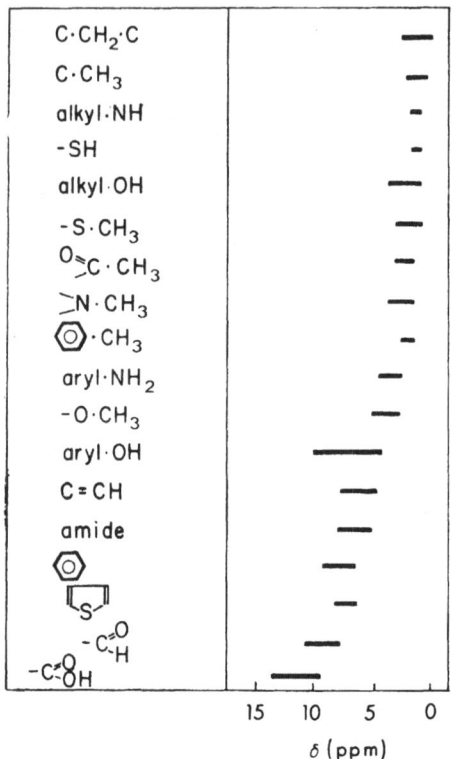

Fig. 2.12. Chemical shift values of various chemical groups in nuclear magnetic resonance.

The character of the NMR spectrum is determined by the life-time of the spin states, by the dipole interactions with neighboring nuclei, by magnetic field inhomogenities and by various minor factors.

In describing the lifetime of the spin states, let us consider a proton which can exist in two spin states: parallel and antiparallel to the magnetic field applied. The originally equal distribution in the two states will change if a radiation of proper frequency is operative (resonant frequency of the nucleus). The time required for attaining 63.2 % of the new equilibrium state ($= (e - 1)/e$) is called the spin-lattice relaxation time (for analogies with chemical relaxations see, e.g., Gutfreund, 1972). The frequency of the field surrounding the given nucleus is determined by fluctuating magnetic dipoles in the vicinity, the strength of interaction being inversely proportional to r^6 (r being the distance between interacting dipoles) so that intra-molecular interactions are much greater than intermolecular ones. The effect on the lattice-spin relaxation is the greater the larger the magnetic dipole of the neighboring nucleus (paramagnetic molecules, for instance).

The line width in the NMR spectrum is also affected by the so-called low-frequency dipole interactions between nuclei with a nonzero magnetic moment, the underlying effect being a change in the field surrounding a given nucleus. Like the spin-lattice relaxation, it is more powerful in the presence of interaction of large magnetic dipole moments.

Another important factor in changing the line width is the magnetic field inhomogeneity. The total line width is proportional to the reciprocal of the transverse relaxation time T_2, thus:

$$\pi \Delta v = 1/T_2 = 1/2T_1 + 1/T_{2m} + 1/T_{2d} \qquad (2.23)$$

where T_1 is the relaxation time for the spin-lattice relaxation process, T_{2d} that for the dipole interaction between nuclei and T_{2m} that for the inhomogeneity of the magnetic field (these last two are computed using the uncertainty principle; cf. Abragam, 1961). Δv for a Gaussian-type line (bell-shaped) is the peak width at maximum slope, for a Lorentzian-type line (fine-tipped, with long segments of practically invariant slope) is the peak width at half-maximum height.

NMR spectra are usually obtained at suitable applied frequencies (40, 60, 100, 120, 220 MHz) and show the absorption of energy (as a peak) versus magnetic field strength H, in either of two ways: If the sample is exposed to the frequency for a longer period of time we speak of the continuous-wave method; if intense pulses of the field are applied we have to do with the pulse method. This second method has the advantage that it requires much less time but, on the other hand, requires computer assistance to achieve results comparable to the continuous-wave method. The sensitivity of the method (generally 1 ml sample of a 1 mM solution) can be increased by sweeping over the entire field strength many times and computing the averaged transients (the CAT method).

Depending on the homogeneity of the applied magnetic field we speak of wide-line NMR and of high-resolution NMR. The first of these requires a homogeneity of about $1 : 10^6$ and is suited for studies where motion is relatively slow and anisotropic. The second technique, with magnetic field homogeneity of at least $1 : 10^8$ is technically more demanding. An advantage in resolution may be obtained by spinning the sample at the "magic" angle, defined as $\sec^{-1} \sqrt{3}$ ($= 54.74°$), to the magnetic field.

An example of NMR spectrum is shown in Fig. 2.13.

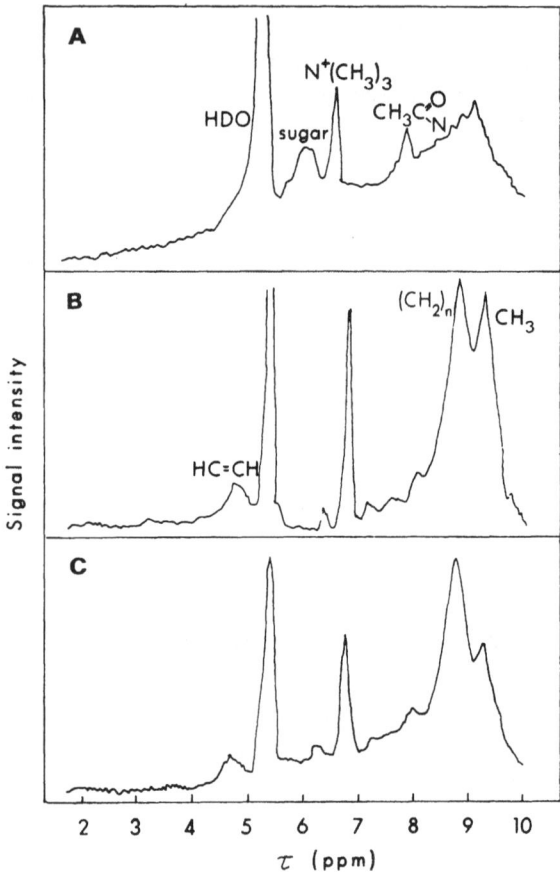

Fig. 2.13. Nuclear magnetic resonance spectra of 5 % sonicated dispersion in heavy water. **A** Erythrocyte membrane fragments, **B** total membrane lipid (phospholipids plus cholesterol), **C** total membrane phospholipid. (Redrawn from Chapman *et al.*, 1968.)

2.3.1.5. *Electron spin resonance*

Whereas all the above-described methods studied or could study the membrane components either *in situ* or after suitable extraction without any alteration of their composition*, electron spin resonance is a technique which requires the introduction into the membrane of a probe, generally a molecule with an unpaired electron, a paramagnetic recorder. Practical application of the technique rests on the possibility to synthesize compounds containing a nitroxy group where, at the nitrogen atom, there is an unpaired electron in the $2p\pi$ orbital confined practically to the nitrogen atom. The basic structure of a nitroxide is

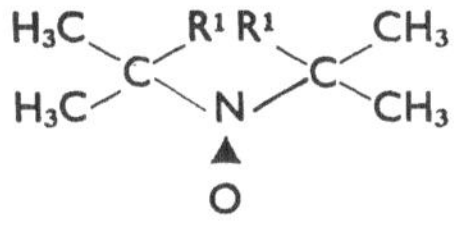

$$
\begin{array}{c}
H_3C \\
\quad \searrow C \\
H_3C \nearrow \quad \searrow N \nearrow \quad \searrow CH_3 \\
\blacktriangle \\
O
\end{array}
$$

and the following typical derivatives have been prepared:

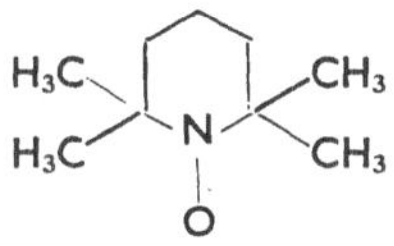

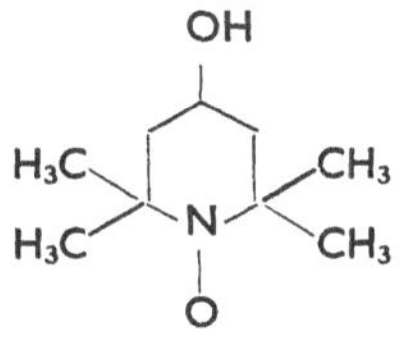

Tempol

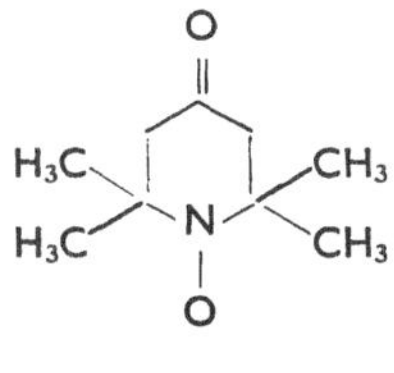

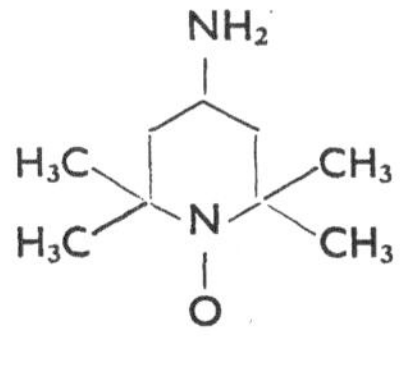

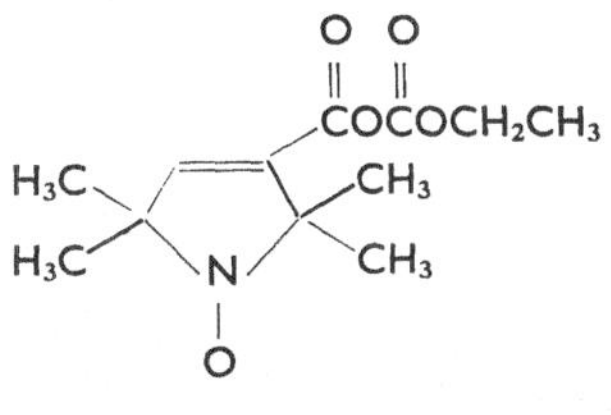

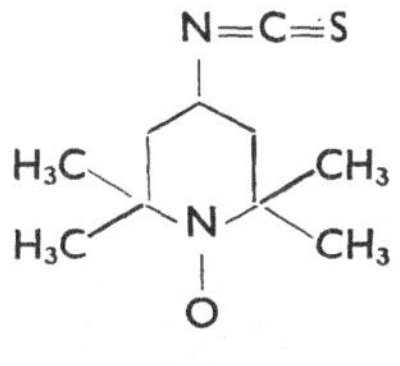

* Although UV and fluorescence external probes are also in use.

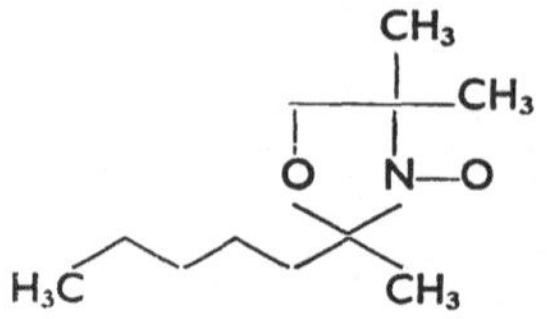

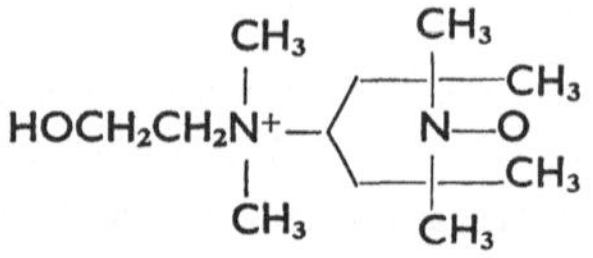

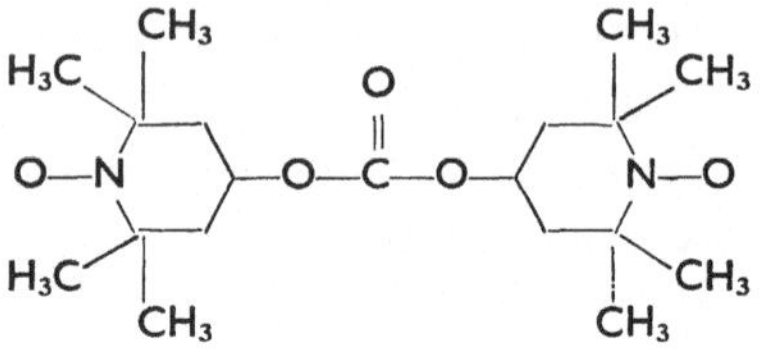

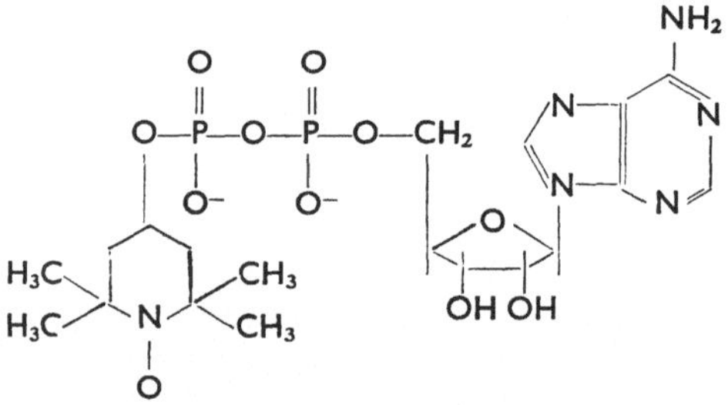

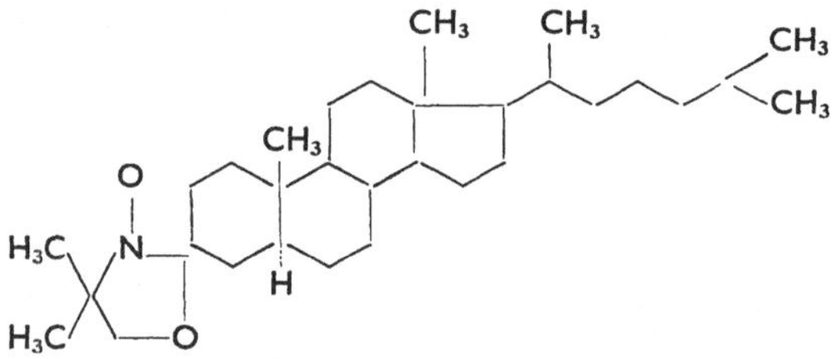

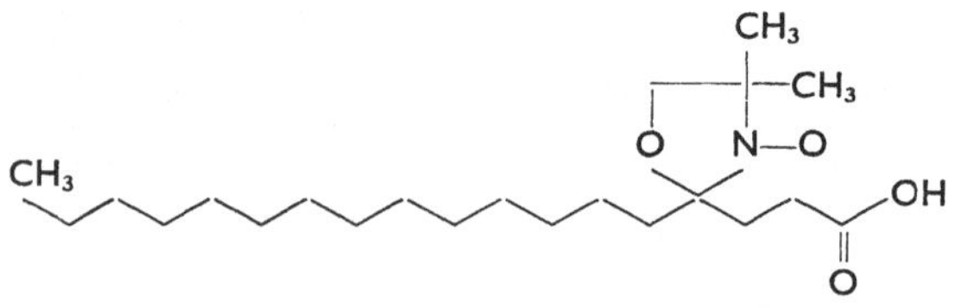

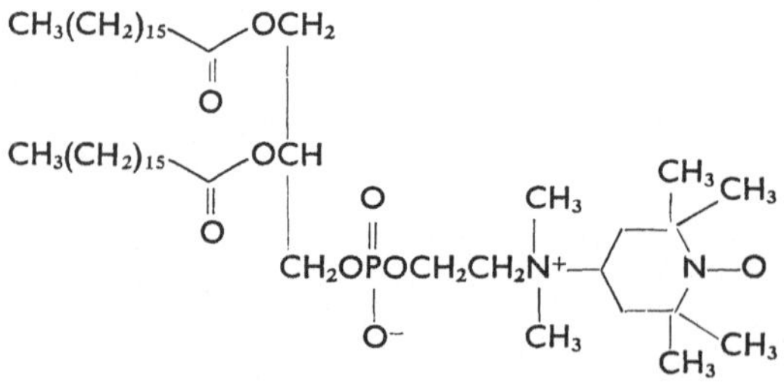

$$CH_3$$
$$|$$—$$CH_3$$
$$O \quad N—O$$

CH_3 ……………………… OCH_2

$$||$$
$$O$$

$$OC \quad O$$

CH_3 ………………… $||$
$$O \quad CH_2OPOCH_2CH_2N^+CH_3$$
$$|$$
$$O^-$$

The usefulness of the ESR method is in providing information about the viscosity and polarity of local areas of membranes, the fluidity of these areas, the degree of molecular ordering and rotational and translational freedom.

The underlying theory proceeds from the fact that a magnetic moment μ precesses in a magnetic field of intensity H with a characteristic frequency (the Larmor precession frequency) defined by

$$\omega = 2\pi v = 2\pi\mu H/h. \qquad (2.24)*$$

For paramagnetic atoms, this equation may be written as

$$\omega = 2\pi g\beta H/h \qquad (2.25)$$

where β is the Bohr magneton (cf. Table 2.13), h is the Planck constant and g is a characteristic value of free radicals (2.0023 for pure electron spin). The value of g has a tensorial order, being 2.0088 in the x direction, 2.0062 in the y direction and 2.0027 in the z direction (the example is taken from di-tert-butyl nitroxide).

If the above frequency is applied externally (it lies in the microwave region) the magnetic moments are made to precess coherently so that a magnetic field is induced in the sample at right angles to the constant external field. At this frequency an absorption of energy is recorded.

Practically speaking, in an ESR spectrum where the magnetic field H (in gauss or fractions of tesla, such as mT) is varied at a fixed frequency (usually 9.5 or 35 GHz), a peak is observed when the condition of eq. (2.25) is reached. An example of ESR spectrum is shown in Fig. 2.14. This spectrum is often electronically converted to

---

* This principle is the same as that underlying the use of NMR; cf. eq. (2.21).

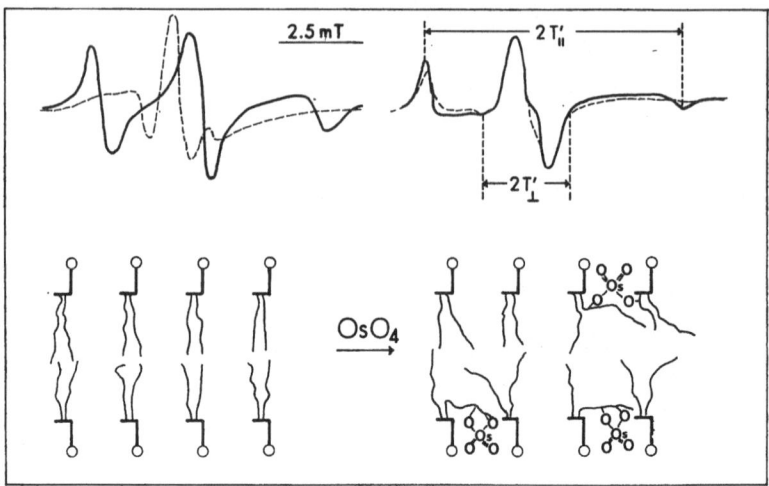

Fig. 2.14. Electron spin resonance spectrum of 5-[4′,4′-dimethyloxazolidine-*N*-oxyl]stearic acid in a lecithin multilayer. At left, normal membrane; at right, after treatment with osmium tetroxide. The heavy lines in the spectra correspond to the supporting slide being perpendicular to the preparation, the dashed lines to the slide being parallel with the preparation. The values of $T'_{\parallel}$ and $T'_{\perp}$ refer to equation (2.27). The lower part of the figure shows schematically the arrangement of phospholipid molecules before and after treatment with OsO_4. (Adapted from Jost *et al.*, 1971.)

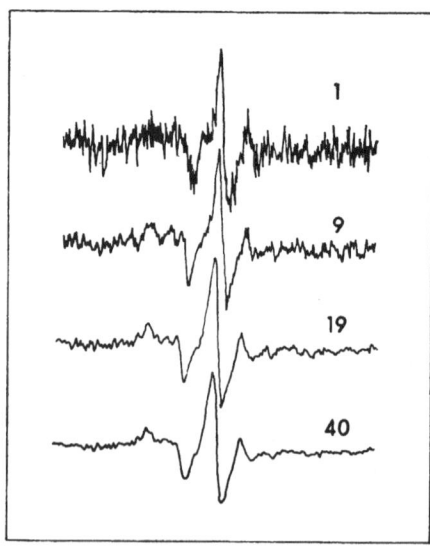

Fig. 2.15. Increasing the signal-to-noise ratio by repeated scanning (the number of sweeps is shown at each curve). The sample used was $3.75 \cdot 10^{-5}\ M$ sodium 5-[4′,4′-dimethyloxazolidine-*N*-oxyl]stearate in an aqueous solution of 5 % egg lecithin. (Redrawn from Jost *et al.*, 1971.)

its first or second derivative to provide further information. The interpretation (and simulation) of ESR spectra has been closely linked to the use of computers. There exist methods making it possible to predict the shape of an ESR spectrum on the basis of various assumptions about rotation of the spin-labelled molecule (e.g. Itzkowitz, 1967). Another computer application lies in the processing of repeated scans of the same spectrum to differentiate between background noise and true signal, the improvement obtained being proportional to $n^{0.5}$ where n is the number of sweeps (*cf.* Fig. 2.15).

The fine structure of the spectral lines contains information on the effects of rotation, interaction between molecules, transfer of spin labels across membranes and changes due to temperature (*cf.*, e.g., Jost *et al.*, 1971).

2.3.1.6. *Fluorescence*

Application of fluorescence techniques to membrane investigation has been scanty and probably more may be expected in the future. Still, some important information on subtle perturbations of conformation has been obtained by using both intrinsic protein fluorescence and the fluorescence of extrinsically added labels.

The principle of fluorescence is easy to grasp. A molecule in the ground state of energy may interact with light of an appropriate wavelength (its energy E is equal to hv, where h is the Planck constant and v ($= c/\lambda$) is the frequency) and be thus raised to a higher energy level, represented either by an excited singlet state (where all electrons in the molecule have paired spins) or by an excited triplet state (where one electron pair has unpaired spins). The triplet state energy level is somewhat lower than that of the corresponding singlet state, no matter whether the transition is of the type $\pi \rightarrow \pi^*$ or $n_1 \rightarrow \pi^*$. The energy of the excited state can then be released in various ways. (1) A very frequent one is the internal conversion which is not accompanied by any except thermal radiation; it may proceed with rate constants k_c of $10^2 - 10^{11}$ s^{-1}. (2) An important one is fluorescence characterized by a rate constant k_f of 10^8 s^{-1} where light of a lower frequency than that used for excitation is given off. (3) A rare one is phosphorescence, occurring from a triplet state where the rate constant k_p is of the order of 10^2 s^{-1} and which persists for appreciable periods after the excitatory radiation is turned off (*cf.* Fig. 2.16). The characteristics of some fluorescent molecules are shown in Table 2.14.

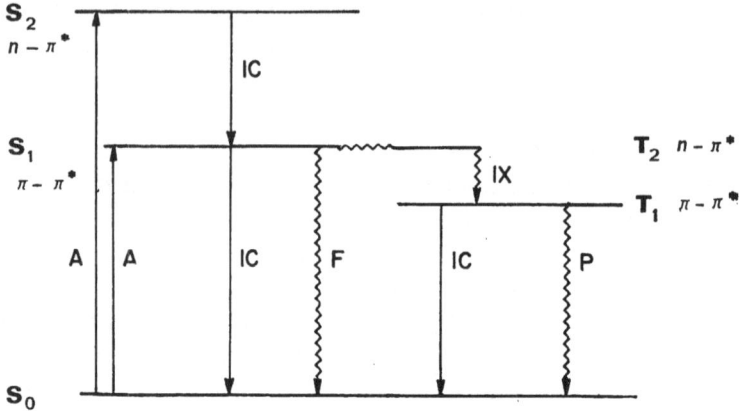

Fig. 2.16. Scheme of energy transitions during luminescence processes. The ground level is shown by S_0, the excited singlet levels by S_1 and S_2, the excited triplet levels by T_1 and T_2. A Absorption, IC internal conversion, IX intersystem crossing (with reversal of electron spin), F fluorescence, P phosphorescence.

TABLE 2.14. Wavelengths used in fluorescence assay

Wavelength (nm)	Energy (kJ mol^{-1})	Color	Region of typical excitation maxima	Region of typical emission maxima
200	598.7			
250	478.9		Phenylalanine	
300	399.0		Tryptophan, tyrosine	
350	342.1		NADH, ANS[a], vitamin A	Tryptophan, tyrosine
400	299.4	Visible limit		
450	265.9	Violet-blue	Flavins	NADH
500	239.5	Green		Flavins, vitamin A, protein-bound ANS
550	217.7	Yellow		Free ANS
600	199.7	Orange		
650	184.2		Chlorophyll	Chlorophyll
700	170.8	Red		

[a] 1-Anilinonaphthalene-8-sulfonate.

An important characteristic of the efficiency of fluorescence is the so-called quantum yield, defined as

$$Q_f = \frac{\text{number of quanta emitted}}{\text{number of quanta absorbed}} = \frac{k_f}{k_f + k_p + k_c} \quad (2.26)$$

Of the naturally occurring membrane components, only proteins show an intrinsic fluorescence which is due to the presence of phenyl-

TABLE 2.15. Fluorescence of amino acids

Amino acid	Absorption		Emission	
	λ_{max} (nm)	$\log \varepsilon$ $(M^{-1} cm^{-1})$	λ_{max} (nm)	Q_f
Phenylalanine	257	2.30		
	206	3.95	282	0.035
	187	4.76		
Tyrosine	275	3.08		
	222	3.90	303	0.21
	192	4.67		
Tryptophan	280	3.74		
	220	4.51	350	0.20
	196	4.32		

alanine, tyrosine and tryptophan. However, fluorescence quantum yield of these amino acids (as shown in Table 2.15) is markedly depressed in proteins, apparently due to quenching through collisions of the high-energy excited rings with protons from dissociable protein groups ($Q_f = 0.008 - 0.07$). Moreover, the fluorescence of proteins (except those without tryptophan) is due to the presence of tryptophan alone, the position as well as the height of the maximum depending both on the degree of hydrophobicity of the tryptophan milieu and on the quality of the solvent.

In membranes as such, the fluorescence behavior resembles that of proteins but there the various subtle changes in the emission spectrum may be interpreted in terms of interactions with other components (hydrophobic vs. polar) and of conformational changes. An example of this last-named phenomenon, described for membrane proteins in *vitro* is provided by the glutamine-binding protein from *Escherichia coli* (Weiner and Heppel, 1971).

A rather greater impact on our knowledge of membrane protein-lipid interactions has come from the use of so-called extrinsic probes which can be bound to the protein either covalently or, more frequently, adsorbed through noncovalent bonds.

Covalent probes include 4-acetamido-4'-isothiocyanostilbene-2,2'-disulfonic acid, diazosulfanilate, formylmethionylsulfone methyl phosphate and 1-dimethylaminonaphthalene-5-sulfonyl chloride.

The indicator of choice for the noncovalent type of binding has been sodium 1-anilinonaphthalene-8-sulfonate (ANS). In solution, it has a Q_f of 0.004 but when bound to a protein or a lipoprotein membrane the Q_f may rise to 0.8 and the emission λ_{max} is shifted (*cf.* Table 2.14). Moreover, it is highly sensitive to minute changes in mutual subunit positions, such as occur in allosteric proteins and possibly in biological membranes.

Other noncovalent fluorescent probes used are 1-toluidinonaphthalene-5-sulfonate (TNS), N-phenyl-1-naphthylamine (NPN), arsanilinochloromethoxyacridine and pyrene-3-sulfonate. Other fluorescent molecules come into use particularly as membrane markers and will be mentioned in the chapter on fractionation of membranes (2.6).

Fluorescent probes have also been frequently used for the observation of transport across cell or mitochondrial membranes (retinol, ostruthin, umbelliferone), a sensitive fluorescence microscope or microspectrofluorimeter being of essence for attaining useful data (e.g., Rotman and Papermaster, 1966).

2.3.1.7. *Calorimetry and related techniques*

There are two principal methods based on measuring the thermal properties of membranes or their components: the differential thermal analysis (DTA) and the differential scanning calorimetry (DSC). In the first of these, a sample is heated and then left to cool spontaneously while its temperature is being recorded. A generally smooth decrease is obtained in regions where no phase changes occur but a break or a kink is observed at the point of a thermal transition (solidification, for example). If another, inert reference sample is processed in parallel, it is possible to record the differences in the temperature of the two and these are then plotted against temperature (*cf.* Fig. 2.17).

In differential scanning calorimetry, which is superior to the

preceding technique, both the experimental and the reference sample
are kept at the same temperature (or a constant temperature difference)
while being heated by a regulated heat flow. The differences in current

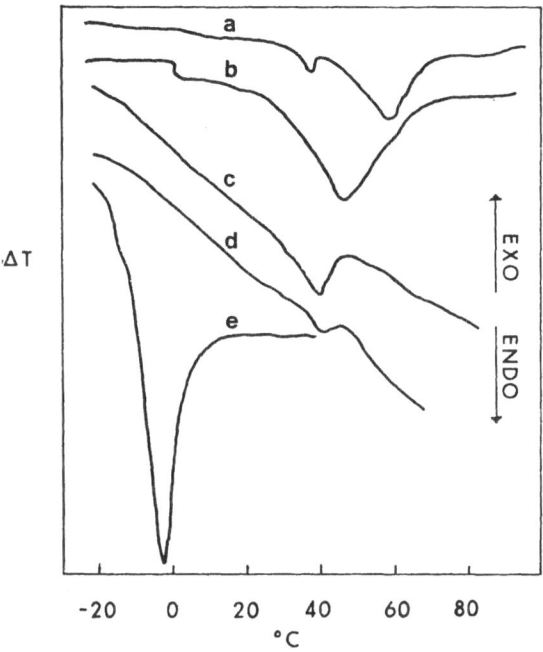

Fig. 2.17. Differential thermal analysis of myelin at different relative humidities.
The percent water content was as follows: a 3 %, b 5 %, c 10 %, d 15 %, e 30 %.
(Redrawn from Ladbrooke *et al.*, 1968.)

required for maintaining the samples at the same temperature are then
plotted against temperature as shown in Fig. 2.18. Whereas in DTA an
endothermic transition is plotted downward, in DSC it is shown as
an upward deflection.

The techniques have been used particularly to study the melting
of membrane lipids, the effects of lipid-protein interactions, composition, presence of water, lipid-metal interactions, etc. Dependence of
various secondary marker effects (fluorescence probes in particular)
on gradually increasing temperature has been widely used as a diagnostic for membrane structure but this is not actually a thermal
technique and it will be discussed in connection with the structure of
membrane lipids.

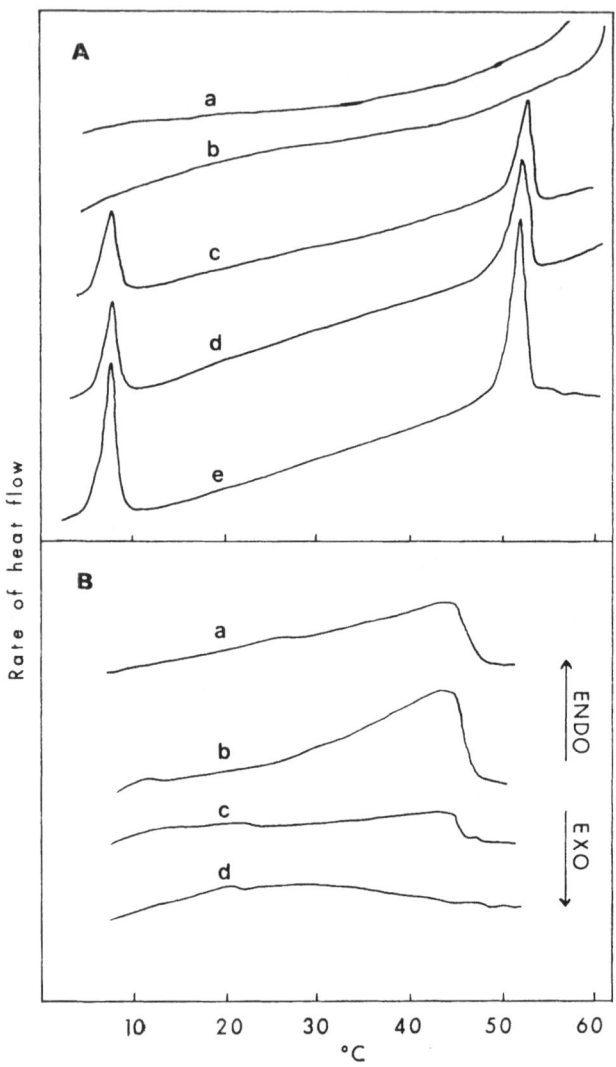

Fig. 2.18.   Differential scanning calorimetry. **A** Mixtures of water with distearoyl-lecithin. Water content (% w/w): a 10, b 20, c 25, d 30, e 40. (Adapted from Chapman and Dodd, 1971.) **B** 50 % dispersions of lipids from *Acholeplasma* (*Mycoplasma*) *laidlawii* B membranes. a Total lipids, b glycolipids, c phospholipids, d neutral lipids. (Adapted from Chapman and Urbina, 1971.)

2.3.1.8. Other techniques

Dilatometry has been little used so far but it is a suitable auxiliary technique to fluorescence and calorimetric methods. In analogy with changes in the specific heat content of a sample, its volume will change (this is true particularly of lipids; see Träuble and Haynes, 1971). Using a dilatometer, a volumetric device that can be designed in

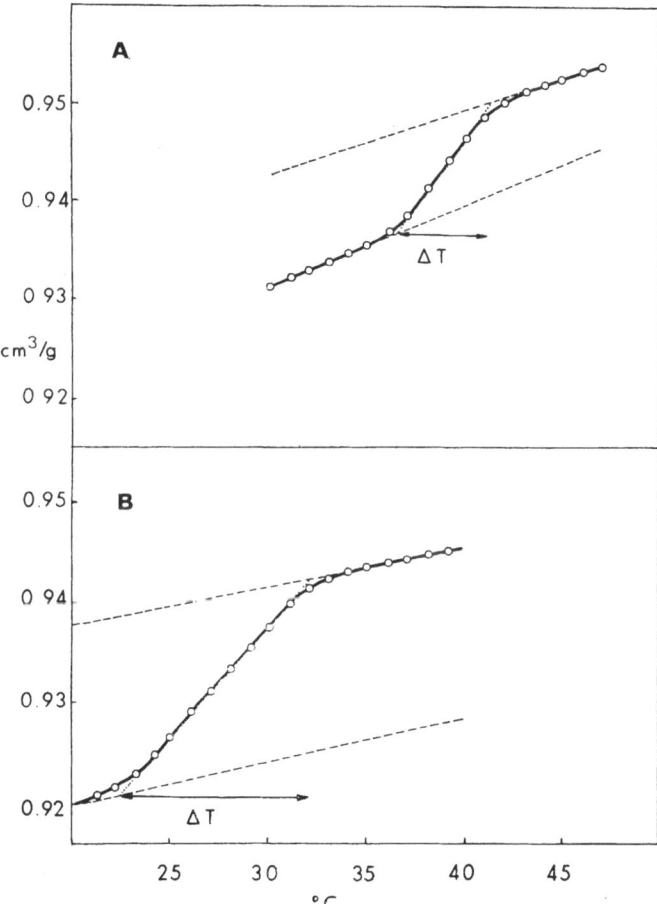

Fig. 2.19. Dilatometry of 18.5 % (v/v) mixtures of trans-18 : 1-phospholipids **A** and of trans-16 : 1-phospholipids **B** in water. The ordinate shows the specific volume of the lipid at increasing temperature after subtraction of water expansion. The transition temperature T_t lies half-way between the two extrapolated straight lines. (Redrawn from Overath and Träuble, 1973.)

different ways to measure small volume changes (of 0.1 % or so) one can compare the volume expansion of a mixture of lipid with water, with that of distilled water alone. It will be seen from Fig. 2.19 that during the gel – liquid transition of lipids a relatively greater increase in the specific volume of the dispersed lipid is found.

A basically identical information can be obtained from the measurement of light scattering of a lipid – water dispersion. Since the amount of light scattered from particles increases with their size, a plot of scattered light intensity *vs.* temperature will show steeper negative slopes in the region where volume suddenly increases.

2.3.2. Organization of lipids

2.3.2.1. Biological membranes

The amount of data pertaining to the structure and organization of membrane lipids is enormous and rather heterogeneous but the last few years have witnessed a certain convergence of opinion among lipid chemists and physical chemists dealing with membranes so that a fairly clear outline of the situation can be offered.

The majority of membrane lipids are probably in the bilayer arrangement, such as formed spontaneously in mixtures of phospholipids with water (p. 32). Thermodynamic considerations as well as physicochemical evidence show rather convincingly that the polar heads of the lipids face outward, the hydrophobic chains pointing into the membrane. Molecular models of lipids have been employed to bring analytical data into agreement with the steric requirements of such a bimolecular leaflet (Finean, 1958; Vandenheuvel, 1963; Jost *et al.*, 1971). One of the possibilities, based partly on X-ray diffraction data, is shown in Fig. 2.20.

As indicated in the lowest panel of the figure, there is a more or less stiff region of the hydrocarbon chains near the polar heads and a more flexible region in the center of the bilayer. The thickness of the double layer depends, among other things, on the presence of negatively charged phospholipids in the juxtaposed layers, mutual repulsion resulting in greater hydration. Addition of electrolytes with strongly binding positive counterions (Na^+, K^+) shrinks the membrane; addition of Ca^{2+} may result in bridges being formed between the neighboring phosphatides with the result of closer packing and decreased permeability.

The biological bilayers are far from being symmetrical in composition. Of the various examples available, the erythrocyte membrane is best understood (Bretscher, 1972; Verkleij *et al.*, 1973). Phosphatidyl choline and sphingomyelin appear predominantly in the outer layer, phosphatidyl serine and phosphatidyl ethanolamine mainly in the inner one.

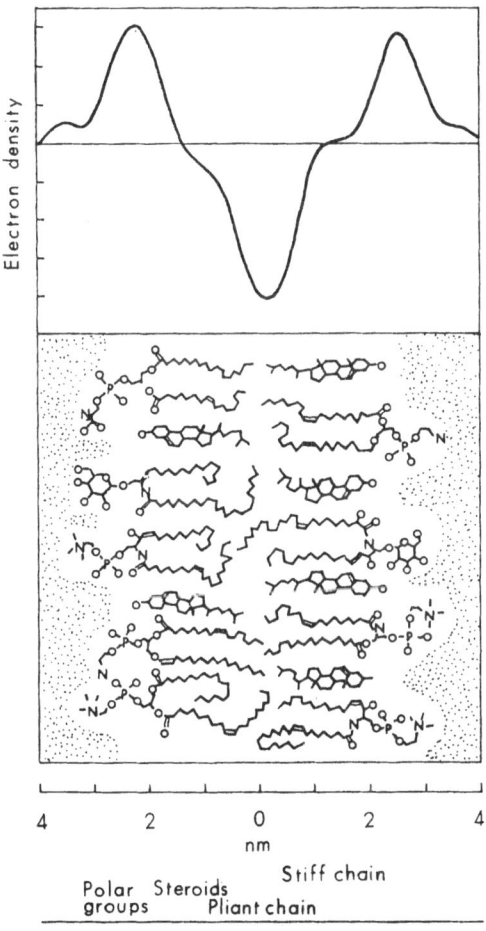

Fig. 2.20. The electron density profile of myelin from the rabbit sciatic nerve (upper panel) and a schematic representation of the myelin structure, showing the rigid and flexible parts of the phospholipid + cholesterol bilayer. (Adapted from Caspar and Kirschner, 1971.)

The closeness of packing of the phospholipids is greatly influenced by the quality of the fatty acyl residues and by the presence of non-polar lipids, in particular of sterols. As shown in Table 2.16, several

Table 2.16. Packing areas of fatty acids and phospholipids (adapted from O'Brien, 1967) in nm^2/molecule

Compound	Native	After adding cholesterol 1 : 1
Myristic acid (14 : 0)	0.37	—
Palmitic acid (16 : 0)	0.24	—
Stearic acid (18 : 0)	0.23	—
Oleic acid (18 : 1)	0.48	—
Erucic acid (22 : 1)	0.40	—
16-Methylheptadecanoic acid	0.32	—
1,2-Dimyristoyllecithin	0.72	0.61
1,2-Distearoyllecithin	0.45	0.45
1,2-Dioleyllecithin	0.83	0.77
1,2-Dilinoleoyllecithin	0.99	0.99

effects may be recognized. Straight-chain fatty acids alone, as well as their corresponding lecithins, take up less area the longer the hydrophobic chain (due to stronger hydrophobic interactions); branching and particularly *cis*-unsaturated bonds cause a substantial increase in packing; cholesterol condenses the packing particularly of mono-unsaturated phospholipids. There is steric justification for this since phospholipids (especially PC and PS) have surface head groups greater than their hydrophobic ends while cholesterol is frayed so that they fit nicely (Israelachvili and Mitchell, 1975). In comparable membranes, such as those of various mammalian erythrocytes, the closeness of packing is directly proportional to the relative impermeability of the membrane. Thus, in the series of erythrocytes of rat, rabbit, man, pig, ox, the content of arachidonic (20 : 4) acid decreases (*cf.* Table 2.9) in the same order as the susceptibility to hemolysis (measured by the rate of bursting in glycol, glycerol or thiourea; Jacobs *et al.* 1950). The reducing effect of cholesterol on permeability observed both in *Mycoplasma* and in erythrocytes is in keeping with these arguments (McElhaney *et al.*, 1970; Bruckdorfer *et al.*, 1969).

A high rigidity of the packing, however, may not be the desirable property of lipid membranes. On the contrary, as shown by growing

Acholeplasma laidlawii in the presence of stearic acid plus small amounts of various loosely packing fatty acids, cell lysis ensuing in the presence of stearic acid alone can be prevented the more effectively the greater the spatial demands of the accompanying fatty acid (*cf.* Tourtellotte, 1972).

At "biological" temperatures, the membrane lipids (at least in the center of the bilayer) appear to be in a liquid-crystalline or rather defective ordered state which, on lowering the temperature, passes into a crystalline state. The transition point, often designated as the melting temperature T_m or transition temperature T_t, can be defined by various techniques (*cf.* Fig. 2.17, 2.18, 2.19) including extrinsic fluorescence, its value depending greatly on the composition of the lipid bilayer as well as on various external factors such as pH and, obviously, on the interactions with other membrane components.

As to the composition effects, the more unsaturated or branched or cyclized the fatty acyl residues, the lower the T_m. This may be of critical importance for psychrophilic organisms which, to be able to thrive at low temperatures, contain high amounts (up to one-third of the total) of *cis*-unsaturated acids.

An interesting example of a regulatory effect of lipid composition is provided by the isomerization of pinifolic acid present in relatively large amounts in conifer chloroplasts. In the frost-resistant state in winter it occurs as

while in summer the three-ring form is present, rendering the chloroplasts sensitive to a lowering of temperature (Bervaes and Kuiper, 1975)

(Position of the carboxyl group is not definitely established.)

Artificially prepared lecithins show the following transition temperatures in dependence on the type of acyl present:

Acid	T_t, °C
14 : 0	23
16 : 0	41
18 : 0	58
22 : 0	75
18 : 1 (*cis*)	−22

In qualitative agreement, *Escherichia coli* membranes enriched by growth on different fatty acids, show the following T_t:

Enriched with	T_t
16 : 1 (*trans*)	27
18 : 1 (*trans*)	35
18 : 1 (*cis*)	15

The effect of pH on the transition temperature is apparently associated with the degree of protonation of phospholipids and hence their electrostatic interaction with the surrounding milieu and with

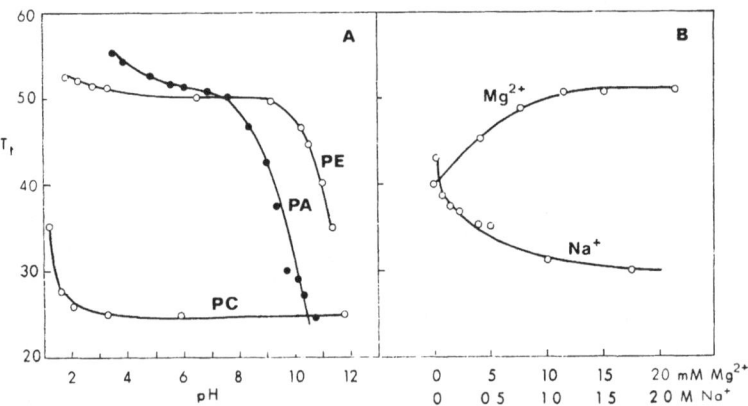

Fig. 2.21. Effect of pH and of cations on the transition temperature of different phospholipids, using fluorescence measurement of $2 . 10^{-6}$ M N-phenylnaphthyl-amine. The transition temperatures shown were recorded at increasing temperature. **A** Effect of pH on the transition temperature of phosphatidyl ethanolamine (**PE**), phosphatidyl choline (**PC**) and phosphatidic acid (**PA**), all $2.5 . 10^{-4}$ M, with myristic acid as the constituent acyl residues. **B** Effect of Na^+ and Mg^{2+} on the transition temperature of $2.5 . 10^{-4}$ M C_{14}-phosphatidic acid at pH 9.7 (for Na^+) and 9.2 (for Mg^{2+}). (Adapted from Träuble and Eibl, 1974.)

neighboring molecules. Similarly, divalent cations Mg^{2+} and Ca^{2+} increase the transition temperature by binding to negative charges of phospholipids. Somewhat surprisingly, the effect of univalent cations Li^+, Na^+ and K^+ is to decrease the transition temperature. These ions do not interact with negatively charged phospholipids and their effect is probably due to an increase of ionic strength which reduces the surface potential and thus brings about further ionization of phospholipids (Fig. 2.21).

Events taking place at the transition temperature lend further support to the bilayer model. Above T_t the apparent cross-section area per lipid increases in the case of dipalmitoyllecithin from 0.48 to 0.58 nm^2 while the thickness of the bilayer decreases from 4.6 to 4.0 nm, in agreement with the view of kinks being formed in the array of lecithin molecules (Träuble and Haynes, 1971).

The importance of the "fluid" state of membrane lipids for various membrane processes can be documented by comparing the temperature dependence of various transports with the lipid composition of the membrane. The Arrhenius plot of a transport rate shows usually

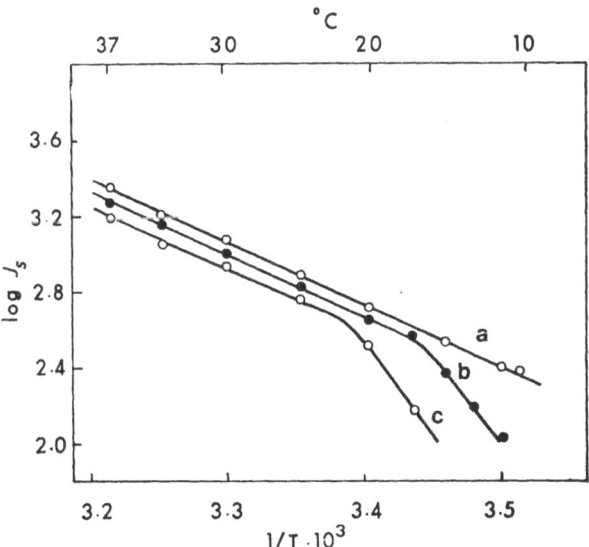

Fig. 2.22. Arrhenius plots of the rate of uptake of 2-deoxyglucose by *Acholeplasma laidlawii* grown in a medium containing extra oleic acid (a), without any supplement (b) and without supplement but cells containing exceptionally many saturated lipids (c). (Adapted from Tourtellotte, 1972).

a break at the transition temperature—this holds for thiomethyl-β-galactoside efflux, o-nitrophenyl-β-galactoside uptake in *Escherichia coli* (Overath et al., 1971a), for Na,K-adenosinetriphosphatase of sheep kidney medulla (Grisham and Barnett, 1973), for the Mg^{2+}-adenosinetriphosphatase of *Acholeplasma laidlawii* (de Kruyff et al., 1973) and for 2-deoxyglucose uptake by the same microorganism (Tourtellotte, 1972) as shown in Fig. 2.22.

The more general importance of the fluid state of the membrane for transport processes is best demonstrated by the fact that the transition temperatures for both β-galactoside and β-glucoside uptake by *Escherichia coli* are identically related to the membrane lipid composition (influenced by supplementing the medium with one or another fatty acid).

Supplement to medium	Break in Arrhenius plot °C
Elaidic acid (*trans*-Δ^9-18 : 1)	30
Oleic acid (*cis*-Δ^9-18 : 1)	13
Dihydrosterculic acid $(CH_3-(CH_2)_7-CH-CH-(CH_2)_7COOH$ $\diagdown CH_2 \diagup$	11
cis-Vaccenic acid (*cis*-Δ^{11}-18 : 1)	10
Linoleic acid (*cis*, *cis*-$\Delta^{9,12}$-18 : 2)	7

The temperature characteristic of ion transport processes (studied mainly in artificial membranes) is an important tool for defining the action of various ionophore antibiotics (cf. section 5.3.6.). Thus, the abrupt cessation of the conductivity effect of nonactin, valinomycin, and the like, as the lipids "crystallize" at a certain temperature, indicates the mobility of the ionophores to be instrumental for their function. On the other hand, the channel-forming ionophores, such as gramicidin, remain operative even well below the transition temperature.

In a number of instances (cf. Linden et al., 1973) two breaks were found in the temperature dependence both of transport rate and of spectral parameters of the Tempol spin label. The higher-temperature transition denotes the moment where solid patches of membrane lipids are first detected and when an increase in lateral compressibility is to be expected, the lower-temperature one denotes the end of lateral phase separation when all of the lipid is in a solid phase. The data suggest that the transport agency lies perpendicularly in the membrane and is sensitive to lateral compression (an example is shown in Fig. 2.23).

A major component of cell membranes that bears on the fluidity of cell lipids is cholesterol (and possibly ergosterol and other sterols) but not epicholesterol and other 3α-OH isomers. By interacting

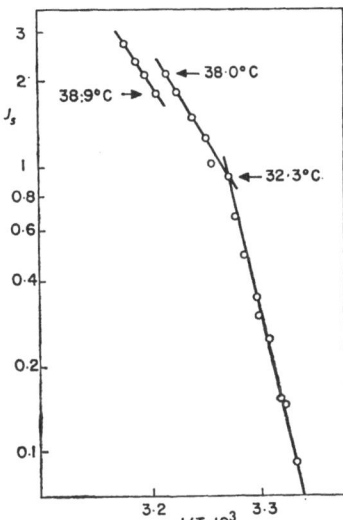

Fig. 2.23. Arrhenius plot of β-galactoside transport in *Escherichia coli*, using cells grown in elaidate-supplemented medium. (Redrawn from Linden *et al.* 1973.)

particularly with the liquid-crystalline lecithin, cholesterol affects the ordered arrangement of hydrocarbon chains and hence its increased concentration results in a lowering of T_t both in artificial mixtures (Chapman, 1969) and in biological membranes (de Kruyff *et al.*, 1973, studying the membrane Mg^{2+}-adenosinetriphosphatase).

Using data of electron spin resonance, one can characterize the fluidity of a membrane by the so-called order parameter S (Seelig, 1970; Hubbell and McConnell, 1971) which may be defined as

$$S = (T'_{\parallel} - T'_{\perp})/(T_z - T_x) \qquad (2.27)$$

$T_z - T_x$ being 2.5 mT, $T'_{\parallel}$ is half the distance between two outer side bands in the ESR spectrum, $T'_{\perp}$ is half the distance between two inner lines of the spectrum (*cf.* Fig. 2.14). S defines the average orientation of the long axis of the $2p\pi$ orbital of the spin-labelled molecule with respect to the direction normal to the membrane. It can be used for computing the flexibility of a lipid bilayer (*cf.* Seelig *et al.*, 1973).

The fluidity of the lipid phase does not signify a completely random movement of lipid molecules in the membrane but still the value of lateral diffusion is remarkably high, being of the same order of magnitude in an artificial dipalmitoyllecithin membrane $(D = 3 . 10^{-8} \text{ cm}^2 \text{ s}^{-1})$ as in an *Escherichia coli* membrane $(D = 3.25 . . 10^{-8} \text{ cm}^2 \text{ s}^{-1})$. This corresponds to an average distance traveled per second of 3 μm, on the basis of Einstein's equation (eq. 4.26). This corresponds to about 1/10 the distance around an average cell. The lateral diffusion coefficient was arrived at by an elegant method of electron spin resonance using N-oxyl-4′,4′-dimethyloxazolidine derivatives of stearic acid (Sackmann *et al.*, 1973). The exchange interaction between spin-labelled molecules is related to the rate of lateral (two-dimensional) diffusion and brings about a broadening of the ESR spectrum if the label-to-lipid ratio is greater than 0.025. From the broadening of the spectrum (ΔH_{ex}) which is a component of the total width of the central line

$$\Delta H = \Delta H_0 + \Delta H_{\text{dipole}} + \Delta H_{\text{ex}} \qquad (2.28)$$

(ΔH_0 is the width at an extremely low label concentration, ΔH_{dipole} is the broadening due to dipole-dipole interactions and can be neglected for ratios less than 0.1), the ΔH_{ex} can be obtained. This value is related to the exchange frequency W_{ex} ($W_{\text{ex}} = 1.4 . 10^6 H_{\text{ex}}$; in Hz) and this, in its turn, to the lateral diffusion coefficient

$$D = 3/4 \frac{F\lambda W_{\text{ex}}}{d_c} \cdot \frac{1 + c}{c}. \qquad (2.29)$$

where F is the area per lipid molecule ($0.5 - 0.6 \text{ nm}^2$), d_c the critical distance for the occurrence of spin exchange (about 2 nm) and λ the length of one diffusional jump in the lipid lattice (about 0.8 nm); c designates the label-to-lipid ratio. Now, in this case

$$D = 1.82 . 10^{-15} W_{\text{ex}} \frac{1 + c}{c} \qquad (2.29')$$

The magnitude of lateral diffusion is of paramount importance in considering processes of biogenesis of membranes and immunological response.

Besides lateral diffusion, there is the possibility of a flip-flop movement of lipids in the membrane. However, in contrast with lateral diffusion, the flipping is extremely improbable, with half-times of the order of hours.

The significance of the phase transition of lipid membranes for the regulation of cellular processes remains to be explored. Although it is obvious that the operational state of membrane lipids is the fluid state there may be instances when a sudden decrease of a membrane function (generally a transport process) is required – a case in point is the transmission of the nerve impulse when a change in permeability for Na^+ vs. K^+ is held responsible for its propagation (cf. p. 284). One could perhaps speculate that local changes in cation concentrations could bring about a phase transition either of the whole membrane or, more likely, of an area surrounding the transport agency or governing the permeability to a given solute. This mechanism would be of general applicability and might account for some of the rapid permeability responses particularly of excitable membranes.

However, the fluidity of lipids in membranes is not a sufficient condition for transport processes to occur. It should be made clear that the presence of proteins in general and of specific proteins in particular is essential, as will be shown in the following sections of the book.

2.3.2.2. *Artificial membranes*

As was stated several times before, molecules of amphipathic lipids dispersed in a high enough concentration in an aqueous medium tend to aggregate to a bilayer structure with the hydrocarbon chains pointing inward and the polar heads interacting with the medium. Not only are the membranous structures thus formed stable but they can be fairly easily formed as a unit bilayer membrane that is amenable to experimental manipulation. There are two main forms of such membranes: (1) The classical (Mueller *et al.*, 1962; Mueller and Rudin, 1968) *flat membrane* (often called the black lipid membrane because of negative interference effects causing it to appear black under the microscope). Its formation is shown in Fig. 2.24 and some of its properties appear in Table 2.17.

(2) The *spherical membrane,* either with the same or with different media outside and inside. These membranes range in size from lipid vesicles (e.g., Bangham, 1968) as little as several tens of nm across, to larger (hydrated and dehydrated) vesicles (e.g., Mueller and Rudin, 1969) about 1 μm across, to large "bubble" membranes as large as a centimeter in diameter (Mueller *et al.*, 1964). The small-size vesicles can be formed in a multilayered version and this is the type usually

called liposomes. The formation of these types of membranes is shown in Fig. 2.25 and 2.26. A recent comprehensive treatise of liposomes may be found in Bangham *et al.* (1974).

Table 2.17. Physical characteristics of biological and lipid bilayer membranes (adapted from Wallach, 1972)

Property	Biological membrane	Lipid bilayer
Thickness (nm)	4—13	4.6—9.0
Resistance (Ω cm^2)	10^2—10^5	10^3—10^9
Capacitance (μF cm^{-2})	0.5—1.3	0.3—1.3
Resting potential difference (mV)	10—200	0—140
Index of refraction	1.55	1.37
Interfacial tension (mN m^{-1})	0.03—3.0	0.2—6.0
Permeability to water (10^{-4} cm s^{-1})	25—33	5—10
Dielectric breakdown (mV)	100	150—200

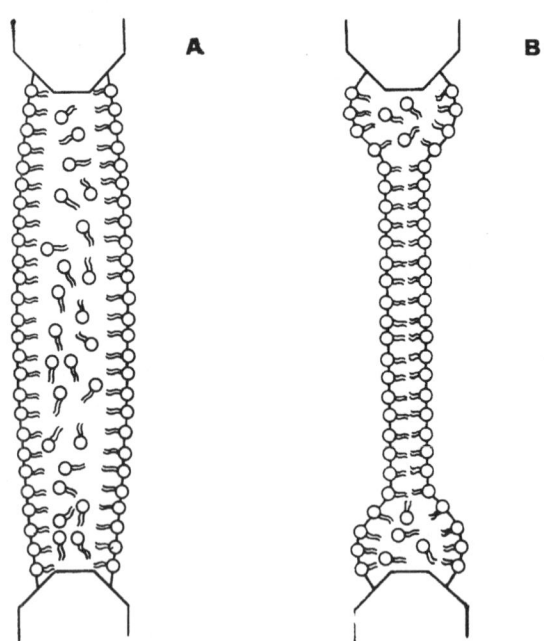

Fig. 2.24. Schematic representation of black lipid membrane formation. In **A**, the lipid layer is several molecules thick and shows various interference colors. In **B**, it has thinned spontaneously to a bilayer which appears black when viewed in a microscope.

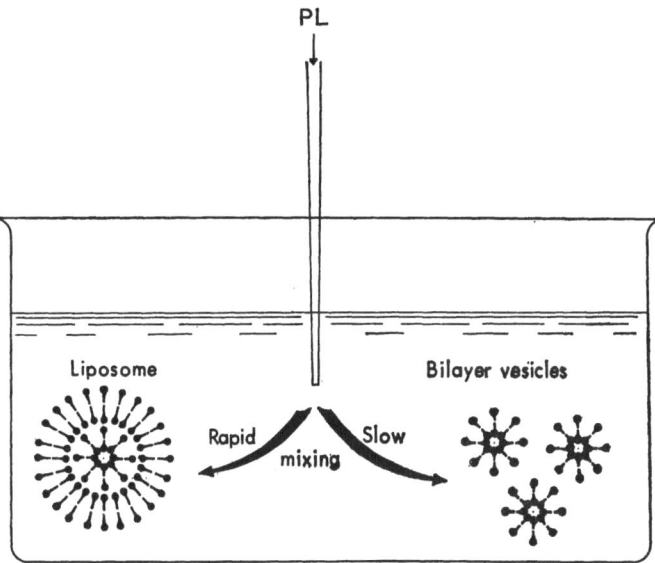

Fig. 2.25. Schematic representation of formation of multi-layer liposomes and bilayer vesicles, depending on the rate of stirring the phospholipid (PL) emulsion. Liposomes can also be formed from bilayers by sonication.

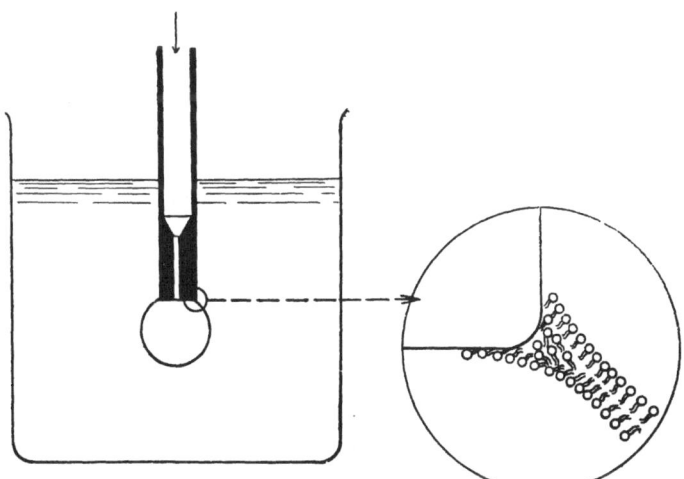

Fig. 2.26. Schematic representation of bubble membrane formation. The pipette at left carries a drop of phospholipid at its tip and a solution within. By gently blowing at arrow the solution is expelled from the pipette and a bubble containing it is formed, ideally composed of a lipid bilayer.

There is an analogy between these round vesicles and soap bubbles but bubbles in air have their hydrophobic parts exposed on the surface, with polar parts pointing into the bilayer film.

Spherical vesicles behave like small osmometers and transport of water across their membranes brings about volume changes which can be readily monitored, e.g., by light-scattering measurements.

Materials used for the preparation of artificial membranes include various phosphatides as well as neutral glycerides and combinations with cholesterol and various minor additives, such as α-tocopherol. Many lipid mixtures of biological origin have been used, such as extracts of erythrocytes, brain fractions, synaptic vesicles, mitochondria, chloroplasts, etc. The stability of the membranes formed (up to several hours) is difficult to predict and, besides the quality of the major components, it is subject to subtle effects due to oxidation (work in oxygen-free atmosphere is of fundamental importance), presence of contaminants in the solvents used (hydrocarbons completely free of other homologues are relatively costly and difficult to obtain) and some other poorly defined factors.

If cholesterol alone is used, it should be partly oxidized in air, whereupon black membranes are formed that are stable and show permeability properties like phospholipid membranes do (Tien *et al.*, 1966).

In most cases, the symmetrical phospholipid bilayers were studied. However, asymmetrical membranes can also be constructed with, say, phosphatidyl choline in one film and phosphatidyl ethanolamine in the other (Montal and Mueller, 1972). This is done by using a Teflon partition partly immersed into water so as to separate two different lipid films on the surface. The partition is provided with an opening and, as it is immersed deeper, it carries the lipid films with it so that at the opening they are apposed and form a true black membrane.

In principle, it is also possible to prepare membranes with a single film of lipid and an adherent single film of protein (Frommhertz, 1970). The technique, consisting in transferring a lipid monolayer onto free medium, can be applied to the formation of multilayer structures of defined layer sequence and composition.

The usefulness of existing artificial lipid membranes is now beyond any doubt but their potential significance depends on the success of incorporating native proteins, particularly of transport function, into their structure, something that has been possible only in a few rather special cases, e.g., Na,K-adenosinetriphosphatase (Jain *et al.*, 1969; Redwood *et al.*, 1969) or sucrase-isomaltase involved in intestinal glucose transport (Storelli *et al.*, 1972).

Perhaps the greatest value of artificial membrane models derives from the application of agents affecting their permeability to both ions and nonelectrolytes. The results will be treated in some detail in the respective chapters (p. 296 and p. 204).

2.3.3. Organization of proteins

Although proteins represent usually more than a half of total membrane weight no general pattern of their organization in membranes appears to exist. Various optical methods (in particular ORD and CD) have shown the proteins to be $30-50\%$ in the α-helical configuration, with a predominance of random-coil arrangement in the rest. Perhaps only the mitochondrial protein contains less, if any, α-helix. There is no known membrane protein with predominant β-configuration (the antiparallel pleated sheet), an exception being the proteins of the outer membrane of *Escherichia coli* (Nakamura *et al.*, 1974).

The overall structure of a membrane is probably governed by the lipids but associations among proteins may play a role in maintaining the membrane shape. Thus, practically delipidated mitochondrial membranes, myelin and *Mycoplasma* membranes retain their appearance after OsO_4 staining in the electron microscope, this supporting the possibility of a structural framework of proteins.

In mitochondrial cristae, chloroplast lamellae and bacterial cytoplasmic membranes one would perforce expect some sort of defined mutual arrangement of proteins that ensures the highest efficiency of their function, namely the immediate transfer of the product of one reaction to act as substrate for the next enzyme in the functional sequence. (The dependence of these enzyme activities on the presence of lipids is another matter and will be dealt with on p.108).

Indirect support for a wide protein structuration comes from experiments indicating the lateral cooperativity of some membrane phenomena, such as the effect of a few molecules of hormones on the entire cell membrane (Sonnenberg, 1971) or a similar effect of colicin El on the character of the *Escherichia coli* envelope (Phillip and Cramer, 1973). Although treated theoretically for an array of lipoprotein subunits (for a review see Wallach, 1971), a cooperativity among lipids (in the form of a pressure wave) might account for these findings as well.

Membrane proteins are now generally assumed to be of two types, *viz.* the peripheral and the integral proteins. Peripheral proteins are those that are easily removed from the membrane by mild agents or even distilled water and are assumed to blanket the surfaces of the membrane, albeit asymmetrically. When solubilized, they generally form true solutions and are free of lipid. They include, for instance, cytochrome *c*, spectrin, α-lactalbumin, the heat-stable protein HPr (p. 234), aldolase, nectin, the oligomycin-sensitivity conferring protein (OSCP) and the periplasmic binding proteins.

Integral proteins, on the other hand, are embedded to a greater or lesser degree in the lipid continuum and may span the entire thickness of the membrane. They require detergents or chaotropic agents for solubilization and are usually associated with lipids, not forming true solutions. They include cytochrome b_5, various phycoproteins, hormone receptors, lectin receptors, some eukaryotic binding proteins, and others.

The two types of proteins must differ in the distribution of their hydrophobic amino acid residues. Whereas the peripheral proteins are to a large extent hydrophilic on their surface (they are mostly water-soluble) with their hydrophobic amino acids buried inside, the integral proteins have the hydrophobic residues exposed on their surface to allow for maximum interaction with the nonpolar milieu of the membrane. Still, parts of the integral proteins should be polar on the surface to permit interactions with the polar groups of lipids and with the peripheral proteins.

The best-known membrane proteins are globular: Rhodopsin from the rod outer segment in the retina has a molecular weight of 28 000 and is spherical with a diameter of over 4 nm; mitochondrial cytochrome *c* oxidase has a molecular weight of 20 000 − 25 000 and is cylindrical with dimensions of $9.8 \times 5.0 \times 5.0$ nm^3 (Vanderkooi and Sundaralingam, 1970).

The lateral diffusion of proteins in the membrane is slower than that of lipids, an estimate for the lateral diffusion coefficient of a protein of 100 000 mol. wt. being 3.10^{-10} cm^2 s^{-1} (Sackmann *et al.*, 1973) with the corresponding average distance traveled per second being 0.3 μm.

If several hundred protein molecules are clustered together the diffusion in the plane becomes rather slow so that the cluster may not move much throughout the cell's life.

Beside the lateral translational movement some proteins are known to rotate in the plane of the membrane. Rhodopsin from the retinal rod outer segment rotates with a relaxation time of 0.3 ns, bacteriorhodopsin from *Halobacterium halobium* with $\tau = 10$ ms.

Proteins can apparently flip from one membrane face to the other (particularly those involved in transport) but so far no direct measurement of this movement exists. Judging from the maximum transport rates and from the number of binding protein molecule per cell one can estimate a frequency of movement of about $10^4\,s^{-1}$.

2.3.4. Lipid-protein interactions and membrane structure

It will have become apparent from the two preceding sections that neither lipids nor proteins alone can account for the various membrane phenomena observed and that they must be considered in mutual context to arrive at a plausible membrane model.

Interactions between lipids and proteins must be considered in the first place. There can be covalent binding, electrostatic attraction, polar interaction, and hydrophobic interaction. While covalent binding is little probable, the other three factors may play their roles, the final outcome being the attainment of a thermodynamically most favorable structure, i.e., minimum free energy content of the system. This is achieved most plausibly by allowing for both polar and hydrophobic interactions. It should be noted that hydrogen bonds, although certainly playing a role in maintaining the structure of proteins, are probably of no special consequence for lipid-protein organization since a hydrogen bond is endowed with the same relative free energy in an aqueous and in a lipid environment. Some of the reasons behind this minimalization of free energy content have been described already on p. 32.

On the molecular level, this means that hydrophilic (polar) groups should be in contact with each other and with water while hydrophobic (nonpolar) groups should be in contact with each other but not with the polar groups or the external aqueous medium.

This requirement is most easily met by a simple lipid bilayer as described above (Gorter and Grendel, 1925), with proteins spread on its surfaces, such as suggested by Davson and Danielli (1943). This model, strongly supported and documented by electron microscopic evidence by Robertson (e.g., 1959), in its principles has survived for a long time and has been amended several times to greater sophistication (e.g., Hechter, 1965).

However, there are several properties of biological membranes that are not accounted for by a simple lipid bilayer (*cf.* Table 2.17), in particular the electrical resistance and water permeability. To explain the greater "leakiness" of natural membranes, hydrophilic pores were introduced by Stein and Danielli (1956), polar polypeptides permitting passage through the bilayer. A variant of the lipid bilayer, trying to account for the fact that there is not enough lipid in the membrane to form a continuous double sheet organized as in the classical Danielli models, is that by Hybl and Dorset (1970) or Fettiplace *et al.* (1971).

Electron-microscopic observations of lipid-water mixtures (*cf.* Fig. 2.6), as well as of ultrathin tangential sections of membranes (Fig. 2.27) and of freeze-etched plasma membrane preparations

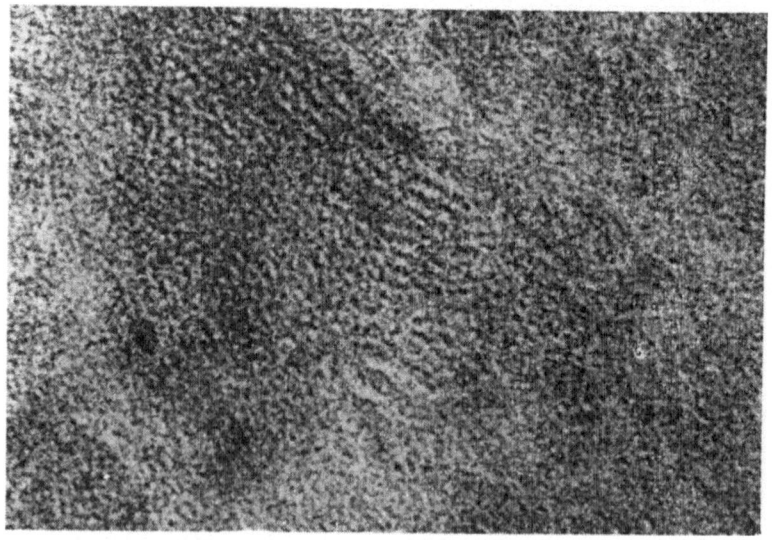

———————— 50 nm

Fig. 2.27. Globular structures seen in an ultrathin tangential section through the Golgi membranes of pancreatic acinar cells. (Reproduced with permission from Sjöstrand F. S. (1964). *Symp. Soc. Chem. Biol. Suppl.* **14**, 103.)

(*cf.* Fig. 2.37)* have brought in a new possibility of organization, *viz.* one of globules or micelles or lipoprotein units that would form the membrane sheet. To retain the optimum requirements for a membrane containing amphipathic components, the units could be arranged either as lipid globules in a protein matrix, the polar heads forming the periphery of the globules (the model of Lucy and Glauert, 1964) or, alternatively, as protein globules arranged in a lipid continuum (Vanderkooi and Green, 1970), this model retaining the lipid bilayer concept and being in fine agreement with electron-density distribution across membranes. The globular micelle models for proteins floating in a lipid bilayer were refined to accommodate some particular physicochemical data, e.g., that by Wallach and Gordon (1968) or by Zahler and Weibel (Zahler, 1969).

Another model, doing away with the bilayer concept altogether, is that due to Green and Perdue (1966) and to Benson (1968) where the membrane is composed entirely of lipoprotein units organized in a curved plane. An even more elaborate model along the same lines was proposed by Sjöstrand (1968).

All these models, although some of them rather complicated, failed to take account of the fact that the functions of membranes as well as their morphology are heterogeneous and hence a model including regularly repeating units in an array cannot meet the requirements placed on a universally acceptable model. Moreover, the now well recognized lateral mobility of membrane components and apparent phase transitions of lipids in membranes led to two types of membrane structure models.

The one by Singer and Nicolson (1972) assumes the existence of a lipid matrix broken up at places by individual protein molecules or their clusters (integral proteins), the whole sheet then covered with a layer of peripheral protein patches. The one by Green (for the latest version see Green *et al.*, 1972) assumes two continuous domains in the membrane, one of lipids (in which some proteins, particularly glycoproteins, may be immersed), one of proteins (the "intrinsic" proteins), the whole sheet again covered by an inhomogeneous sheet of extrinsic protein.

Each of these two models has its advantages and both must be viewed with these important aspects in mind: (1) the structures

---

* However, the size of the particles seen here is substantially greater than would be possible to accommodate within a native membrane.

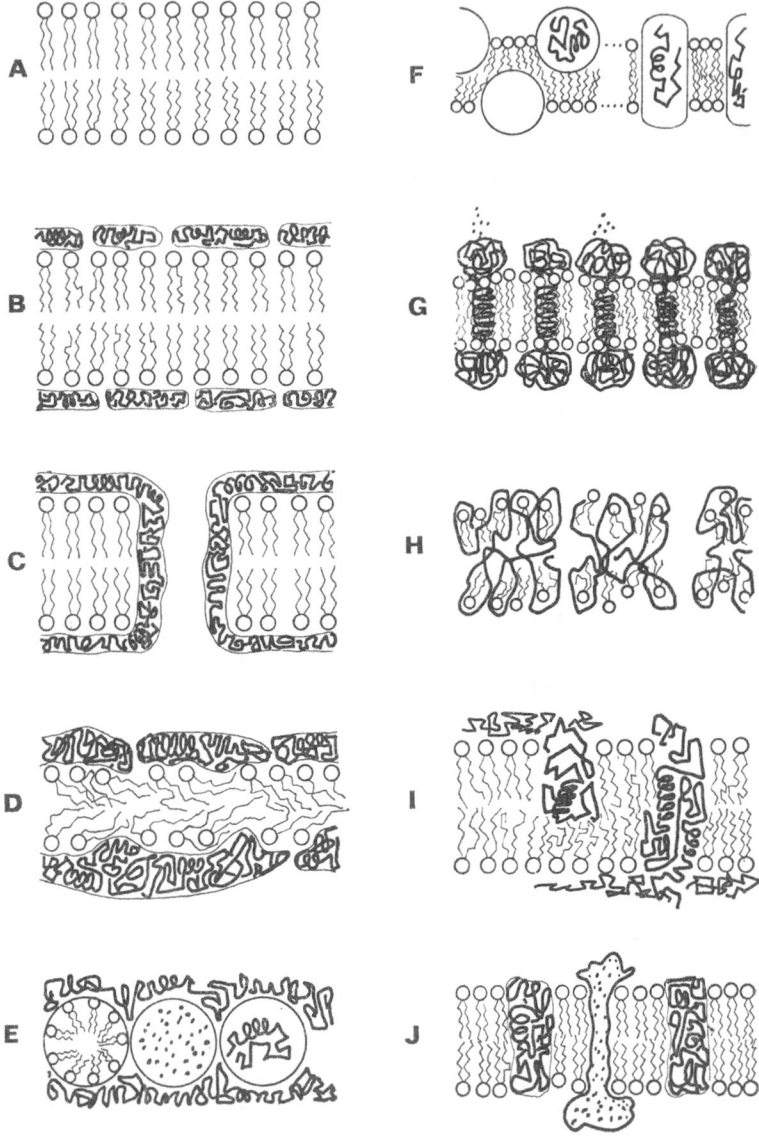

Fig. 2.28. A synopsis of various historical and modern models of cell membrane structure. An attempt was made to use the same graphical means for all the models, a circle with two wavy lines representing a phospholipid, heavy twisted line a polypeptide chain, dotted areas a glycoprotein. **A** Gorter and Grendel, 1925; **B** Davson and Danielli, 1943; **C** Stein and Danielli, 1956; **D** Hybl and Dorset, 1970; **E** Lucy and Glauert, 1964; **F** Vanderkooi and Green, 1970; **G** Zahler, 1969; **H** Benson, 1968; **I** Singer and Nicolson, 1972; **J** Green *et al.*, 1972.

depicted are not rigid but are in a state of constant reorganization; (2) no two membranes are identical in respect of the relative participation of proteins and lipids and the extent of local domains. Extreme cases, such as the inner mitochondrial membrane or chloroplast thylakoids, may be highly ordered with a definite repeat distance between subunits (perhaps with alternating protein and lipid) allowing for efficient transmission of substrate material. At the other end of the scale, lymphocyte plasma membranes may be in a completely random state of (dis)organization, allowing for swift transmission of accidental signals. By the same token, the various structures, both seen and surmised to protrude from the membranes of mitochondria as well as plasma membranes (*cf.* Fig. 2.48 and 2.35**D**) must be viewed in the context of membrane heterogeneity, one of its aspects being asymmetry about the central plane, the other the ability to accomplish different functions at different areas (such as in polar cells of various epithelia, etc.). The models discussed are summarized in the composite Fig. 2.28 and an overall view, incorporating most of the known data, is presented in Fig. 2.29.

Whatever the actual local arrangement in the membrane, there exist well-founded estimates of the number of integral (intrinsic) proteins in relation to the number of lipids. In a brilliant sequence of

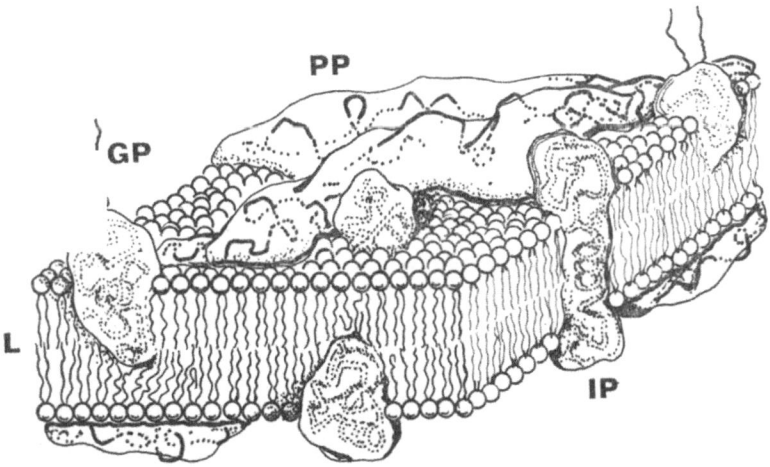

Fig. 2.29. A comprehensive view of a typical plasma membrane, incorporating most of the recently available information. **L** Lipid continuum, **IP** integral proteins, **PP** peripheral proteins, **GP** glycoproteins. The number of polar heads of phospholipids should be roughly 10 times greater with respect to the number of integral proteins.

deductions based on the use of fluorescent markers with *Escherichia coli* membranes, Träuble and Overath (1973) arrived at a value of one integral protein per 600 lipid molecules, 130 of these being closely associated with the protein molecule and not taking part in the phase transition described on p. 89. In terms of membrane area, the integral proteins take up roughly 10% of the area (the value is subject to variation with the cross-sectional area of the protein). Peripheral proteins, on the other hand, take up in this case 47% of the surface (an average for both membrane faces).

The association of some lipid molecules with particular proteins has its corollaries in the finding of (*a*) interaction between particular proteins and (*b*) with lipids at an air—water interface (London *et al.*, 1974). While the so-called Folch-Lees protein interacts with various phospholipids but most powerfully with cholesterol, the other major protein, the *A* basic protein, interacts practically only with cholesterol sulfate.

The second argument has been demonstrated for a number of enzymes, the role of the lipid being of several types: (*1*) It can activate the substrate for the enzyme (in the case of lipopolysaccharide synthesis; p. 139); (*2*) it can activate the enzyme by stabilizing a certain conformation (probably quite frequent); (*3*) it can organize a multi-enzyme system (possibly in mitochondria); (*4*) it can act as a covalently linked cofactor in a reaction (the carrier lipids). Table 2.18 lists some of the lipidactivated enzymes. The effects of phospholipids may be surprisingly subtle. Thus, adenylate cyclase sensitivity to glucagon is restored by phosphatidyl serine, its sensitivity to noradrenaline by phosphatidyl inositol.

The direct effect of membrane fluidity on membrane-bound enzymes is nicely seen in the recent report by Bloj and co-workers (1973) who found that the Hill coefficient (an indicator of allosteric cooperativity) of the inhibition by fluoride of rat erythrocyte Na,K-adenosinetriphosphatase is raised from 2.07 in rats fed corn oil (containing a high proportion of saturated fatty acids) to 3.60 in animals on a cholesterol supplement (cholesterol decreases the T_t; see p. 93). Even if the transition temperature is not affected by diet or cultivation medium (oleic *vs.* linolenic acid in *Escherichia coli*), there may be effects on the Hill coefficient (1.60 *vs.* 2.12 for Ca^{2+}-ATPase (Siñeriz *et al.*, 1973).

2.4. ASSEMBLY OF MEMBRANES

Whereas lipid components in a suitable aqueous medium tend to form bilayer structures spontaneously, the assembly of biological membranes is a rather complicated process that is only poorly understood. It may be viewed from two different aspects which we shall discuss briefly.

The first concerns the reconstitution of solubilized membrane components to a membrane that would be morphologically and functionally equivalent to the original, a feat that is relevant to the structure and strength of interactions of the components.

There are a number of proteins that can be bound again to membranes from which they have been extracted (*cf.* Razin, 1972). They include first of all various adenosinetriphosphatases that have been solubilized by washing with dilute Tris buffer or sonication of *Streptococcus faecalis*, *Micrococcus lysodeikticus*, *Bacillus megaterium* and the inner mitochondrial membrane. They can be reattached in the presence of Mg^{2+} or Ca^{2+} and usually several coupling factors. Other proteins thus reattached include mitochondrial flavin-containing amine oxidase, mitochondrial cytochrome *c*, mitochondrial hexokinase, hydrogen hydrogenase from *Vibrio succinogenes* (Razin, 1972), the phosphate-binding protein from *Escherichia coli* (Medweczky and Rosenberg, 1969, 1970), the glutamate-binding protein from *Escherichia coli* onto depleted cells (Willis and Furlong, 1974), the glutamate-binding protein from *Bacillus subtilis* to vesicles prepared from a transport-negative mutant (Diesterhaft and Freese, 1974).

Like proteins, extracted membrane lipids may be added to depleted membranes to restore enzyme activity, a number of examples being found in Table 2.18.

There are ways of reconstituting solubilized membranes simply by adding an extracted protein or lipid to the remaining membrane. Generally, this is achieved by removing the detergent that has been used for the solubilization. If sonication is used for membrane disruption (it usually does not lead to real solutions) the reconstitution is achieved by manipulating the ion content of the mixture. Likewise, if restoration is to be accomplished from separated proteins and lipids (which offers the possibility of forming "hybrid" membranes) the presence of divalent cations appears to be of essence. It is quite characteristic that many reconstituted systems form vesicles.

Among the most important achievements in membrane re-

Table 2.18. Membrane enzymes requiring lipids for activity (adapted from Coleman, 1973; Finean, 1973; Machtiger and Fox, 1973)

Enzyme	EC number	Source	Reactivating lipid
3-Hydroxybutyrate dehydrogenase	1.1.1.30	Beef heart mitochondria	PC
Malate oxidase	1.1.3.3	*Mycobacterium avium*	DPG
Pyruvate oxidase	1.2.3.3	*Escherichia coli*	LPE, PC, FA, TG
Succinate dehydrogenase	1.3.99.1	Beef heart mitochondria	DPG PE PC
Acyl-CoA dehydrogenase	1.3.99.3	Various	Mixed TG, FA
Cytochrome *c* oxidase	1.9.3.1	Beef heart mitochondria	PE DPG PC
UDP-glucuronosyltransferase	2.4.1.17	Guinea-pig liver microsomes	Mixed lipids
UDP-galactose-lipopolysaccharide galactosyltransferase	2.4.1.44	*Salmonella typhimurium*	Dioleyl PE PG PS[a]
UDP-glucose-lipopolysaccharide glucosyltransferase I	2.4.1.58	*Salmonella typhimurium*	PE PA, PG[a]
Isoprenoid-alcohol kinase	2.7.1.66	*Staphylococcus aureus*	PG, DPG LPG, LDPG PC, PE

Phospho-enol-pyruvate-HPr-phosphotransferase (EII)	2.7.3.9	Escherichia coli	PG PS, DPG
Cholinephosphate cytidylyltransferase	2.7.7.15		LPC PC
Phospho-N-acetylmuramoyl-pentapeptidetransferase	2.7.8.13		PG, PS, PE, PI
Phosphatidate phosphatase	3.1.3.4		Mixed lipids
Glucose-6-phosphatase	3.1.3.9	Rat liver microsomes	LPC PC
Alkenyl-glycerophosphinicocholine hydrolase	3.3.2.2	Rat liver microsomes	PC, SP PE
Ca^{2+}-adenosinetriphosphatase	3.6.1.3	Rat muscle microsomes	PI, mixed lipids
Na$^+$,K$^+$-adenosinetriphosphatase	3.6.1.3	Various sources	Cholesterol; or PS; or PC LPC PA
Adenylate cyclase	4.6.1.1	Rat liver plasma membrane	PS PC
Acyl-CoA synthetase (GDP-forming)	6.2.1.10	Rat liver mitochondria	

[a] For reconstruction, the protein must be combined first with substrate, then with lipid. Abbreviations used: PC, phosphatidyl choline; PE, phosphatidyl ethanolamine; PS, phosphatidyl serine; PI, phosphatidyl inositol; PG, phosphatidyl glycerol; DPG, cardiolipin; PA, phosphatidic acid; SP, sphingomyelin; LPC, lysophosphatidyl choline; LPE, lysophosphatidyl ethanolamine; LPG, lysophosphatidyl glycerol; LDPG, lysodiphosphatidyl glycerol; FA, fatty acids; TG, triglycerides.

constitution is the functional electron-transport chain put together from soluble succinate dehydrogenase, cytochromes b, c and c_1, cytochrome c oxidase, phospholipids and coenzyme Q_{10} (Yamashita and Racker, 1969) – the recovery of activity took several hours in this case. Another is the almost complete reconstitution of oxidative phosphorylation (e.g., Racker et al., 1969; Kagawa and Racker, 1971) and another still the partial restoration of photosynthetic activity of chloroplast membranes (e.g., Takacs and Holt, 1971). A long way toward reconstituting all the functions of sarcoplasmic reticulum membrane has been made (e.g., Martonosi, 1968). A truly remarkable feat is the identification of all the components of the phosphotransferase transport system (see p. 233) and their joining together in a definite sequence (enzyme IIB + Ca^{2+} + phosphatidyl glycerol + + enzyme IIA) to form a functional system (Kundig and Roseman, 1971).

Functional systems can be constructed even from solubilized enzymes and artificial bilayer membranes or liposomes. Many proteins decrease the electrical resistance of these membranes without being enzymically active or endowing them with a specific transport function. The interactions may be merely electrostatic, as suggested by the necessity of low pH and/or a basic protein (e.g., lysozyme, cytochrome c, ribonuclease). Transport systems successfully attached so far include adenosinetriphosphatase (Redwood et al., 1969; Jain et al., 1969), sucrase-isomaltase (Storelli et al., 1972) and the receptor protein for acetylcholine (De Robertis, 1971). Romeo et al. (1970a, b) reconstituted a functional galactosyltransferase in a phospholipid monolayer containing lipopolysaccharide.

The second aspect from which membrane assembly can be considered is the biogenesis of various cell membranes. Here the field abounds with hypotheses but already a number of rather definite points can be made.

The whole process of biogenesis can be split into three stages: (1) Biosynthesis of the component molecules; (2) assembly of the component molecules; (3) modification and migration of the components.

1. The first of these points is a matter of the enzyme equipment of the cell. Phosphatidic acid is synthesized in bacteria from glycerol by phosphorylation (ATP + glycerol kinase), monoacylation (acylcarrier protein + glycerolphosphate acyltransferase) and diacylation (acyl-carrier protein + monoacylglycerolphosphate acyltransferase)

whereas in eukaryotic cells the usual pathway is from dihydroxy-acetonephosphate which is reduced to glycerol phosphate and this then reacts with two molecules of acyl-CoA to the phosphatidic acid.

From here on, the pathways differ again in bacteria and in eukaryotes as indicated in the following scheme:

Pathway to triglycerides, cationic and zwitterionic lipids of animals and fungi

Pathway to all bacterial phospholipids and all anionic phospholipids

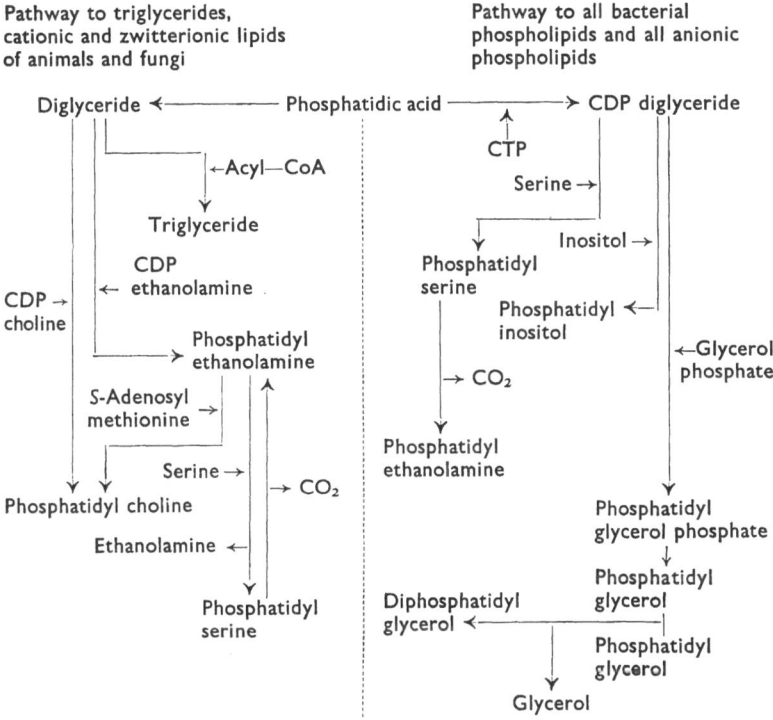

For the synthesis of sterols and minor lipids the reader is referred to specialized biochemical texts.

The site of lipid synthesis is presumably the endoplasmic reticulum (in bacteria, the plasma membrane carries all the enzymes required) although, obviously, not all cells are properly equipped so that in many cases the lipids formed on one membrane may have to travel through the cytoplasm to the locus of their insertion into another membrane.

The insertion is assisted by two mechanisms. (*1*) Action of lyso-phosphatides which solubilize parts of the phospholipid membrane by forming small micelles and thus permit the entry of another lipid into the vacated place. (*2*) Operation of the so-called lipid exchange proteins which can be isolated from various animal tissues (e.g., Wirtz *et al.*, 1972; Helmkamp *et al.*, 1974) and which are specific carriers for certain types of phospholipids, such as lecithin or phosphatidyl inositol, effectively assisting in an exchange of a membrane phospholipid for this particular lipid species.

Proteins are synthesized, as is well known, by the highly complex process taking place on the ribosomes. It is not clear at present what part of protein is synthesized on the endoplasmic reticulum ribosomes and what part of the total arises on membrane-associated ribosomes or ribosomes included in other organelles (mitochondria and chloroplasts). At any rate, some of the specific membrane proteins must similarly travel through the cytoplasm to the point of their destination.

2. The assembly step is the crucial one in the sequence. One may envisage several possibilities at the molecular level (*cf.* Overath *et al.*, 1971a): Lipids are incorporated (*a*) in patches near the enzymes of their synthesis, or (*b*) in a random fashion. Proteins are incorporated (*a'*) into new lipid areas, (*b'*) into both new and old ones, (*c'*) in a random fashion. Evidence listed below supports the possibilities under (*b*) and (*c'*). The main arguments in favor of a random insertion of lipids as well as proteins into the membrane framework derive from estimates of the turnover rate of the components.

The half-lives of phospholipids (using ^{32}P as a label) were found to differ from type to type. Thus, phosphatidyl inositol of rat brain had a half-life of 12.5 days, sphingomyelin of 40 days (Freysz *et al.*, 1969). In bacteria, the same label shows the half-life of phosphatidyl glycerol to be 1 h, that of diphosphatidyl glycerol perhaps twice that while phosphatidyl ethanolamine shows practically no turnover at all (Kanemasa *et al.*, 1967).

The turnover rate of some phospholipids can be markedly increased by specific stimulation of cells. Thus, Lapetina and Michell (1973) report enhanced phosphatidyl inositol turnover in cerebral cortex sympathetic ganglia, the vagus nerve, pancreas, submaxillary glands and other exocrine tissues on exposure to acetylcholine, propranolol or electrical shock.

The relatively slow turnover of some membrane lipids (it may come to a complete stop when bacteria undergo extensive growth; McElhaney and Tourtellotte, 1969) contrasts sharply with the capabilities of many cells to support fantastically rapid new membrane formation. The paradigm of this phenomenon is the behavior of amoebae which, when pierced or even severed in two pieces, surround the injured areas within tens of seconds with a membranous sheath appearing trilaminar under the microscope (e.g., Szubinska, 1971).

Similarly striking differences in the rates of turnover were found among membrane proteins. Although not enough comparative data are available on particular plasma membrane proteins, different enzymes of rat endoplasmic reticulum have the following half-lives: hydroxymethylglutaryl-CoA reductase $2-3$ h, NADH dehydrogenase $3-4$ days, cytochrome b_5 reductase 5 days, and NAD^+ nucleosidase 16 days.

The rate of turnover of membrane proteins is in a striking relationshop to their size. The larger the molecule, the faster it turns over — this being perhaps the single rather universal law valid not only for membranes but for soluble proteins as well (cf. Dehlinger and Schimke, 1971). The experimental protocol supporting this statement deserves to be described. Animals (rats) were fed ^{14}C-labeled amino acid; after 4 days, they were given the same, but ^{3}H-labeled, amino acid. Several hours after this administration they were killed, the plasma membranes of their liver were separated, extracted with 1% Triton X-100 ("soluble" proteins) and then with 0.5% sodium dodecyl sulfate ("insoluble" proteins). The extracts were chromatographed on Sephadex and the radioactivity due to ^{14}C and ^{3}H was assayed in the individual fractions (Fig. 2.30). In this experimental schedule, the ^{3}H/^{14}C ratio is seen to be twice as high for very large molecular-weight proteins as for the small molecules (such as cytochrome c).

Among other things, this relationship suggests that membrane proteins are degraded probably only after dissociation from the membrane matrix — what the underlying control mechanism is remains to be explored.

Another set of data supporting the rather random incorporation of membrane components derives from experiments with bacterial auxotrophs requiring fatty acids for full growth (cf. Cronan et al., 1969) or using mutants defective either in the reduction of dihydroxyacetone phosphate (Hsu and Fox, 1970) or in the acylation of either glycerol-3-phosphate (Cronan et al., 1970) or of monoacylglycerol-3-phosphate (Hechemy and Goldfine, 1971). In the fatty acid auxotrophs one can alter the fatty acid composition of the membrane lipids (and hence the transition temperature) at will, in the latter mutants one can stop and start lipid synthesis by removing or adding glycerol or glycerol phosphate to the medium. By growing cells first

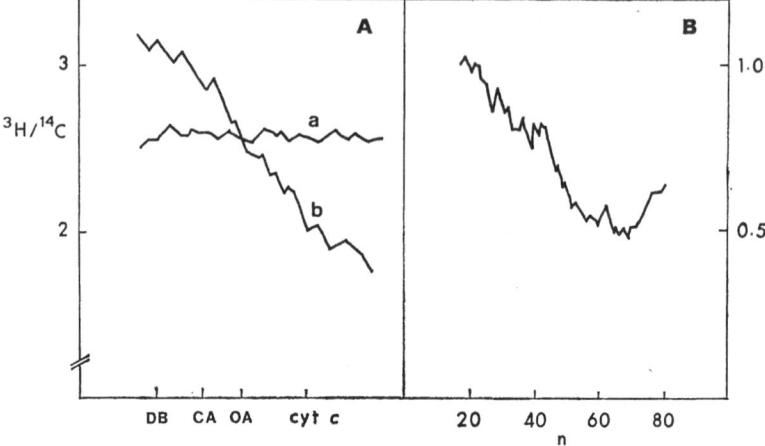

Fig. 2.30. **A** Relative rate of degradation of "soluble proteins" of rat liver as a function of molecular weight. ^{14}C-Labeled leucine was administered to rats four days before ^{3}H-labeled leucine. The animals were killed 4 h later and the proteins chromatographed on Sephadex (curve b). Control animals (curve a) received both isotopes at the same time. The chromatography was calibrated by dextran blue (DB), conalbumin (CA, 78 000), ovalbumin (OA, 45 000) and cytochrome c (13 000). **B** A similar pattern is obtained with double-labeled plasma membrane, the number of fractions (n) being plotted on the abscissa. (Adapted from Dehlinger and Schimke, 1971.)

in the presence of oleic acid (transition temperature for β-galactoside transport 14.9 °C) and shifting to palmitelaidic acid (*trans*-16 : 1), or first in palmitelaidic acid (transition temperature for the same process 33.4 °C) and then shifting to oleate, and by exposing the cell to a transport inducer at different intervals after the shift, one obtains a smooth gradual increase or decrease, respectively, of the transition temperature. Moreover, a completely analogous situation is observed when the inducer is added for a short period before the shift from one to the other fatty acid (Overath *et al.*, 1971*b*; Tsukagoshi and Fox, 1973*a*, *b*). Thus, under all conditions it is the average composition of membrane lipids which determines the total membrane transport characteristics, this militating against any localized, patchy assembly of either lipids or proteins.

Finally, evidence based on membrane growth supports the absence of localized and permanent domains of synthesis and incorporation. Using a density label for lipid synthesis (bromostearic acid) makes it possible by sucrose gradient centrifugation to distinguish between cells or subcellular bodies containing lipids synthesized in the presence and in the absence of bromostearic acid (Fox *et al.*, 1970). If a rod-shaped bacterium synthesized its membrane only at one pole, within a single generation time (after one division) there should exist a group of heavy and a group of light cells. If the growth zone were at both poles or equatorially located, two divisions should produce an analogous result. However, a complete uniformity of weight was always found in *Escherichia coli* (Tsukagoshi *et al.*, 1971). Likewise, in the *Escherichia coli* strain P678-54, forming tiny spherical cells apically during division, a localized growth zone should be reflected in unequal density of the daughter and mother cells but no such differences were found (Wilson and Fox, 1971*b*).

Thus the conclusion may be drawn that there are no localized sites of membrane assembly. However, one should be wary of making hasty judgments for several reasons.

Lateral diffusion of lipids is so rapid that even if there are special lipid-synthesizing domains we would be unable to detect them. The rapid spreading of membrane components over the entire surface is also documented by the distribution of antigens (mouse and human) as shown by cell fusion techniques (Frye and Edidin, 1970). However, the fact remains that replacement of an "old" phospholipid with a new one is specific and errors probably rarely occur.

On the other hand, in the case of proteins at least, there is much indirect evidence that, in spite of the large diffusion coefficients, there exist areas in the membrane that contain different proteins for various special purposes. In the case of polarized cells (e.g., in epithelial tissues) this is most obvious but there are indications that even in micro-organisms, the entire membrane area is not statistically homogeneous. To cite one example, morphological cell division starts in the membrane at a particular site which is then probably equipped with different proteins (and possibly lipids) as compared with the rest of the membrane.

Moreover, highly specialized membranes, such as the mitochondrial inner membrane or chloroplast lamellae, are certainly not randomized since they could not fulfil their functions as efficiently as they do.

There is still room for speculation on how the heterogeneity of membrane proteins is maintained. A likely explanation is in an intimate contact between the plasma membrane and the intracellular organelles, in particular with the endoplasmic reticulum, whereby information and possibly even building material is transmitted to the plasma membrane. An interaction through microtubules would explain a great deal in this respect. Microtubules are filamentous cylindrical structures, 18−25 nm in diameter, now known to exist in practically all cell types, composed of globular subunits of the protein tubulin. The structures may carry side chains or bridges to other microtubules. The protein may exist in the soluble state and then undergo self-assembly whenever need arises. This process is prevented by colchicine, vinblastine, colcemid and other antimitotic agents. Microtubules are well known from the locomotion organs of cells (cilia, flagella) where they function under the energy input from ATP by sliding along each other like muscle fibrils do. Beside this function, they may be instrumental in maintaining cell structure, directing cytoplasmic transport, separating chromosomes during mitosis, governing secretion and many other vectorial processes. Their implication in the immobilizing of membrane proteins was documented by the action of colchicine which randomizes the distribution of, for instance, lectin and receptor proteins or transport proteins in lymphocytes which otherwise are held in place so as to prevent clustering during concanavalin A binding, on the one hand, and to protect transport proteins from digestion during phagocytosis, on the other.

3. Much has been said about the translocation of both lipids and proteins in the foregoing section. Let it be added that there is evidence for transmission of material from one membrane to another. Lipopolysaccharide synthesized at the inner cytoplasmic membrane of Gram-negative bacteria is rapidly transferred to the outer membrane. Lipid components are exchanged between mitochondria and microsomes (Wirtz and Zilversmit, 1968). Fusion of lysosomes with plasma membranes and analogous processes necessarily involve transmission of membrane material.

To conclude this section, it should be stressed that although the magnitude of lateral diffusion coefficients offers enough time for lipids and proteins to spread themselves homogeneously over the entire membrane surface so as to mask possible localized assembly lines there is much evidence on the nonhomogeneity of cell membranes, both with regard to their asymmetry about the inner hydrocarbon core and with regard to special features localized at special sites of the membrane plane.

2.5. ELECTRON MICROSCOPY

Our present knowledge of the morphology of subcellular organelles, most of which are of membranous character, is due to the extensive use of electron microscopy of various cell preparations. The now classical trilaminar appearance of membranes in ultrathin sections derives mainly from the use of electronic "stains", such as potassium permanganate, osmium tetroxide and uranyl acetate. The problem is not satisfactorily solved as to the chemical basis for the attachment of these compounds: Proteins are known to bind some of them but so are the polar heads of phospholipids and so are the double bonds of unsaturated fatty acids (*cf.* Korn, 1968).

One of the possibilities, fairly well documented for osmium tetroxide, is that during fixation the hydrocarbon chains are contorted and swung near the plane of the polar head groups where they are held by interaction with OsO_4 (*cf.* Fig. 2.14).

The thickness of the membranes apparent in ultrathin sections is not the same for different membranes (Table 2.19). The three component lines (two dense enclosing an electron-transparent one) themselves are of different thickness; proceeding outward the values in nm (after OsO_4 fixation) are $3.5 - 3.1 - 2.4$ (cat pancreas), $3.7 - 3.3 - 2.7$ (intestinal epithelium striated border) and $3.3 - 2.2 - 2.5$ (intestinal epithelium lateral cell surface).

TABLE 2.19. **Thickness (in nm) of various membranes in ultrathin sections of mouse kidney proximal tubule cells (Sjöstrand, 1963)**

Fixation	Plasma membrane	Smooth cytoplasmic membranes (Golgi apparatus)	Mitochondrial membranes
$KMnO_4$	9.6	7.7	6.2
OsO_4	9.3	6.2	5.1

Much effort has been expended to show the ultrastructure of the railroad-track pattern and in many cases the results have been rather convincing. Thus, mitochondrial membranes (*cf.* Fig. 2.48), chloroplast lamellae (*cf.* Fig. 2.51), retinal rod outer segments (Fig. 2.31) show a clear substructure. If radially symmetrical, the substructural elements may be visualized more clearly by using the technique of Markham *et al.* (1963). An enlarged microphotograph is carefully centered on a rotating disc which is then turned either manually by a certain angle (such that the total circle is divided into an integral number of angles *n*) and photographed in every position and the pictures then superimposed; or it is arranged like a fast-rotating stroboscope and exposed by an electronic device at suitable intervals corresponding to the above angles.

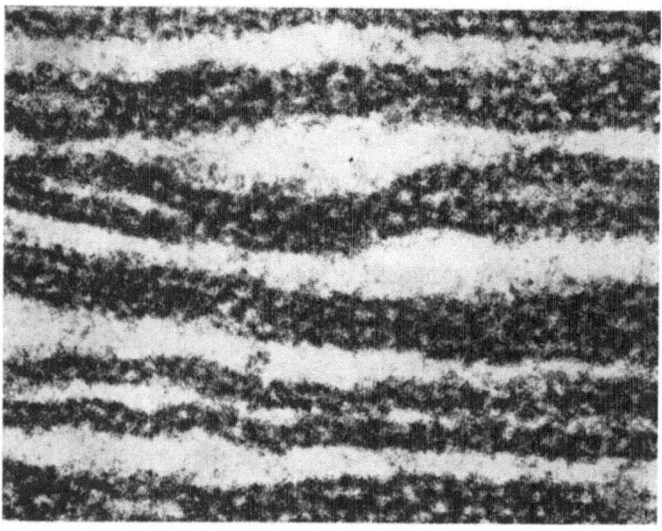

 2 nm

Fig. 2.31. Globular structure in the membranes of the retinal rod outer segment of tadpole. (Taken with kind permission from Sjöstrand, F. S. (1968). In: *The Membranes* (ed. by Dalton, A. J. and Haguenau, F.), p. 151. Academic Press, New York—London.)

On the other hand, much of the granular structure purported to be present in electron micrographs of membranes is apparently due to (*a*) coarse grain of the film used, (*b*) peripheral, surface elements of the membrane rather than the intrinsic core of it which appears to be mostly planar.

The structure of the membrane surface is best visualized through the application of a more "native" technique, doing away with fixation and staining, *viz.* freeze-etching and -fracturing. Cells or their subfractions are rapidly frozen to below −150 °C in liquid freon (maintained at that temperature by liquid nitrogen), placed in a vacuum at −100 °C and fractured with a deeply cooled knife (Fit. 2.32**A**). Subsequently, some of the ice may be sublimed (etched away) for several minutes to expose the surrounding unfractured layers (Fig. 2.32**B**). Immediately after that, the object is shadowed with carbon and platinum to obtain a replica of the original sample. Examples of freeze-etched membranes may be seen in Fig. 2.37.

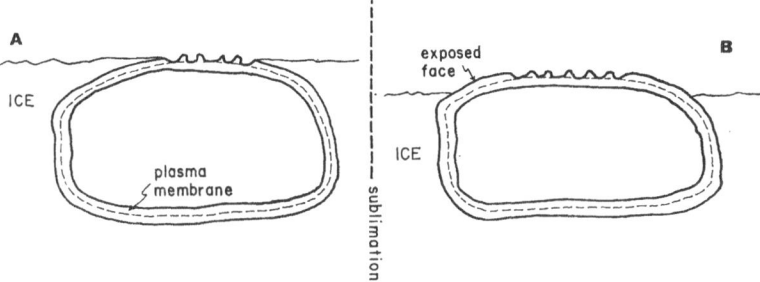

Fig. 2.32. Schematic drawing of the technique of freeze-fracturing (**A**) and freeze-etching (**B**), the latter technique including sublimation of the superficial ice layer and exposition of parts of unfractured material.

Various arrangements of globules are often found in the inner core of membrane split open by freeze-etching and often interpreted as evidence for integral proteins spanning the membrane (e.g., Tourtellotte, 1972).

2.6. ISOLATION OF MEMBRANES

The separation from one another of the various cell membranes is no simple task and, rigorously speaking, it has never been accomplished completely when starting from the whole cell. The situation is much simpler, say, with mammalian erythrocytes where osmotic bursting releases the entire cell content (with no intracellular membranes present) so that all one has to do is to centrifuge and wash the stromas (ghosts), which are in fact plasma membranes.

With other cells the procedure is usually to break up the cells by one of the following techniques.

1. Application of a hypotonic solution which effects an osmotic lysis of tissue cells with consequent emptying of the cell content into the medium.

2. Disruption of cells in a mechanical device. The one most commonly used is a glass or Teflon pestle fitting smoothly into a glass tube with the cell suspension in which it is rotated while the tube is moved up and down to ensure thorough mixing and shearing of the content (this is the Potter—Elvehjem type) (Fig. 2.33**A**).

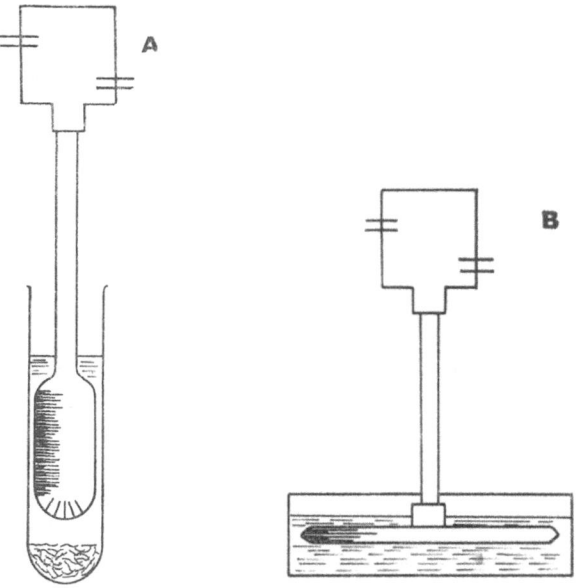

Fig. 2.33. Schematic drawing of a Potter—Elvehjem homogenizer (**A**) and a horizontal glass disc homogenizer (**B**). The former is drawn approximately half its size, the latter is reduced about five times.

Another rotating device used with excellent results for thick-walled microorganisms derives from a stirrer used in the paint and enamel industry. A smooth thick glass disc is fixed at the lower end of a vertical driving shaft and is immersed slightly into a cell suspension containing lead-free glass beads (Balottini) of the appropriate size (this is determined mostly empirically; a mixture of grade 12 and 14 will do well for yeast and fungal cells). With external cooling and without access of oxygen, cells are broken up in a few minutes at most (Fig. 2.33**B**).

3. A sudden change of pressure can be applied in different modifications for bursting cells. In a French or Hughes press a frozen cell pellet is pushed through a tiny opening, in the process of which it melts and the cells are disrupted.

In a procedure due to Wallach (1967) cells are compressed with a nonoxidant gas at 5—7 MPa and rapidly decompressed.

4. Freezing and thawing will reduce the cell to fragments but these are difficult to separate into any defined membrane types.

5. Sonication will break up some cells but here, again, the danger of progressive damage to individual membranes is high.

The technique generally employed for separation of the material obtained by cell disruption is centrifugation at different centrifugal forces. The rationale for the procedure is that particles of different size and different density sediment in the centrifuge at different rates, reflected in the average time t (in seconds) required for the particle to reach the bottom of the tube:

$$t = 843\eta K^{-1}(\text{r.p.m.})^{-2}D^{-2}(d_p - d_m)^{-1} \log (r_{max}/r_{min}) \quad (2.30)$$

where η is the viscosity of the liquid in poises ($\text{g cm}^{-1}\,\text{s}^{-1}$), which for water at 0 °C is 0.017 93, at 20 °C 0.010 09, at 40 °C 0.006 57; D is the particle diameter in cm, d_p the density of the particle, d_m that of the liquid in g cm^{-3}, r_{max} and r_{min} are the radii (cm) of rotation of the bottom and of the meniscus in the centrifuge tube, respectively; r.p.m. is the number of revolutions per min, K is the particle shape factor (0.222 for a sphere, 0.19 for a disc, 0.06 for a cylinder). Even if one can thus calculate the time required for a given particle to sediment, it is readily seen that a clear separation of related particles is impossible because the sedimentation time of a particle in a lower layer of the suspension is obviously shorter than that of a particle near the meniscus.

The above formula is related to the expression (for a spherical particle)

$$s = D^2(d_p - d_m)/18\eta \quad (2.31)$$

where s (in seconds) is the sedimentation coefficient per unit centrifugal field. This is usually expressed in Svedberg units $S(= 10^{-13}\,\text{s})$.

For calculating the centrifugal field of a given centrifuge run, the well-known formula is used:

$$G = 1.12 \cdot 10^{-5} (\text{r.p.m.})^2\, r \quad (2.32)$$

where G is the multiple of the value of acceleration due to gravity (g), r is the centrifuge radius in cm. There are two ways of indicating the centrifugal acceleration applied: (a) in multiples of g and the duration of run, thus 5000 g for 10 min or 10 000 g for 90 min; (b) in time-integrated units, thus $5 . 10^4 g$ min or $9 . 10^6 g$ min. Although both expressions are intrinsically equivalent, it should be appreciated that the time-integrated presentation must be understood within reasonable limits – it would be absurd to spin a sample either at $10^7 g$ for 1 min or at $10^3 g$ for 10^4 min.

The sedimentation properties of some important cell fractions of rat liver are shown in Table 2.20. The purity of a fraction obtained

TABLE 2.20. Sedimentation of rat liver homogenate (adapted from Steck and Fox, 1972)

Fraction	Diameter (μm)	Pelleting force (g min)	Sedimentation coefficient (S)	Equilibrium density ($g\ cm^{-3}$)
Whole cells	15—20	10^3	10^7	$c . 1.20$
Nuclei	5—11	$5—10 . 10^3$	$10^6—10^7$	$c . 1.32$
Golgi apparatus	1—3	$4 . 10^4$	10^5	1.06—1.14
Mitochondria	0.7—1.1	10^5	10^4	1.18—1.21
Lysosomes	0.4—0.6	$4 . 10^5$	$5 . 10^3$	1.20—1.22
Peroxisomes	0.4—0.6	$4 . 10^5$	$5 . 10^3$	1.22—1.24
Rough microsomes	0.05—0.2	$3—10 . 10^6$	$0.3—2 . 10^3$	1.13—1.25
Smooth microsomes	0.05—0.3	$3—10 . 10^6$	$0.3—2 . 10^3$	1.10—1.20
Plasma membranes				
Bile fronts	2—10	10^4	10^6	1.16—1.18
Vesicles	0.1—0.7	$10^5—10^7$	$0.3—5 . 10^3$	1.12—1.15
Soluble proteins	< 0.01	$> 10^8$	2—20	$c . 1.30$

in differential centrifugation is never very high and, to achieve better results, the desired fraction should be recentrifuged several times (cf. Cotman et al., 1970) and a suitable density gradient employed for final separation.

The density gradient to be used depends greatly on the objective of separation and on the type of membrane involved. Various inorganic and organic compounds are in use (Table 2.21).

There are two basic approaches to density gradient centrifugation. One uses a shallow gradient whose function is mainly to prevent

TABLE 2.21. Density of gradient solutions at 25 °C (adapted from Wallach, 1972)

Solute	Concentration (weight percent)					
	10	20	30	40	50	60
LiCl	1.054	1.113	1.178	1.250		
LiBr	1.073	1.160	1.261	1.281	1.529	1.716
KBr	1.072	1.158	1.257	1.371		
NaBr	1.078	1.172	1.281	1.410		
RbBr	1.079	1.174	1.285	1.419	1.582	
CsCl	1.079	1.174	1.286	1.420	1.582	1.785
CsBr	1.081	1.180	1.297	1.440	1.616	
Potasium acetate	1.048	1.100	1.155	1.213	1.242	1.333
Potassium citrate	1.066	1.140	1.221			
Potassium tartrate (20 °C)	1.066	1.139	1.218	1.305	1.400	
Glycerol	1.021	1.045	1.071	1.097	1.124	1.151
Sucrose	1.038	1.081	1.127	1.176	1.230	1.289
Polysucrose	1.034	1.068	1.102	1.136		
Polyglucose	1.038	1.076	1.114	1.152		
Urografin	1.043	1.087	1.130	1.173	1.217	1.260

convection during the centrifuge run. The homogenate is placed on the top and briefly spun – various particles will distribute themselves according to their sedimentation rates but they would all reach the bottom if centrifuged for a more extensive period of time. This is the technique known as rate zonal centrifugation.

The other approach used now more extensively is called equilibrium or isopycnic density gradient centrifugation. Here the gradient is steeper and its bottom layer is denser than any of the particles to be separated. After a sufficiently long run (generally 10^8 g min) the particles come to a stop at the gradient layer equal in density to their own. An example of such a separation in a discontinuous sucrose gradient is shown in Fig. 2.34.

With some cell organelles one can change their buoyant density artificially. If animals are fed with Triton, it accumulates in lysosomes and decreases their density quite substantially; likewise, polystyrene particles phagocytized by leukocytes are eventually enclosed in phagosomes which are in fact plasma membrane derivatives and can again be readily separated on the basis of altered density (*cf.* Wallach and Lin, 1973).

Although some membranes are readily recognized morphologically (e.g., mitochondria), with others (the "smooth microsomal fraction") the degree of purity of a given separation must be de-

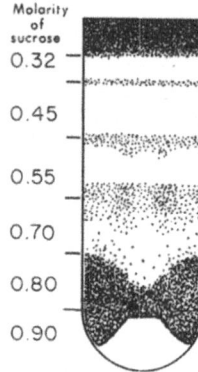

Molarity of sucrose
0.32
0.45
0.55
0.70
0.80
0.90

Fig. 2.34. Distribution of various subcellular fractions of a rat lung homogenate in a discontinuous sucrose gradient.

termined by a different method. The so-called enzyme markers are most often used, the rationale of the approach being in the fact that some cell enzymes are located almost exclusively in a given organelle or rather its membrane. More will be said about the enzyme equipment of the various organelles in the subsequent section; here we shall list only some of the most common enzyme markers used (Table 2.22).

Beside the intrinsic enzyme markers specific to the individual membrane types, plasma membrane lends itself to a different type of labeling. Intact cells are exposed to various fluorescent, paramagnetic or radioactive markers which attach themselves covalently to the cell surface and, after homogenization and fractionation, can be detected by suitable devices. There are two classes of such compounds, small and large.

The small ones (4-acetamido-4' isothiocyanostilbene-2,2'-sulfonic acid, diazonium salt of sulfanilate, formylmethioninesulfone methyl phosphate, 2,4,6-trinitrobenzene, 1-dimethylaminonaphthalene-5-sulfonyl chloride) are readily available but they penetrate after a longer exposure across the plasma membrane and can thus label even intracellular membranes.

The large compounds are superior because they do not penetrate across the plasma membrane. They include *N*-(3-mercuri-5-methoxy-

TABLE 2.22   Enzyme markers of various cell membranes

Membrane	Enzyme
Plasma membrane	Apyrase
	Na, K-adenosinetriphosphatase
	5′-Nucleotidase
Endoplasmic reticulum	Nucleoside phosphatase (acting on IDP)
	Glucose-6-phosphatase
	NADH dehydrogenase
Mitochondria	Choline dehydrogenase
	Glutamate dehydrogenase
	Amine oxidase (flavin-containing)
	Cytochrome c oxidase
	Succinate dehydrogenase
	Isocitrate dehydrogenase
Thylakoids	$NAD(P)^+$ transhydrogenase
	Ferredoxin-$NADP^+$ reductase
Lysosomes	Acid phosphatase
	Ribonuclease
	Deoxyribonuclease
	Arylsulfatase
Golgi apparatus	UDPgalactose-N-acetylglucosaminegalactosyl transferase
Peroxisomes	Catalase
	D-Amino-acid oxidase

propyl)poly-D,L-alanyl amide-[203]Hg (Yariv *et al.*, 1969), *p*-chloro-mercuribenzoate coupled with aminoethyl dextran of molecular weight of 250 000 (Ohta *et al.*, 1971) and substituted anilinoflavazole attached to an isomaltose chain (Himmelspach *et al.*, 1971) which can be diazotized to tyrosyl residues in the membrane

Furthermore, plasma membranes can be labeled in the presence of peroxidase, NaI and H_2O_2 whereby surface tyrosine residues are iodinated (Phillips and Morrison, 1971; Hubbard and Cohn, 1972).

Plasma membranes also can be tagged with antibodies (radioactive or fluorescent) to membrane antigens and with so-called lectins which are compounds of plant origin known to attach to surface glycoproteins and glycolipids (concanavalin A, phytagglutinins from beans, etc.).

2.7. MORPHOLOGY AND FUNCTION OF DIFFERENT BIOLOGICAL MEMBRANES

Although this is a book primarily on membrane transport, it may be useful to give an overview of the broad-ranging functions of various membranes in general. There may be more emphasis placed here on plasma membranes than on organelle membranes but this is due merely to the fact that, from the cell's point of view, plasma membrane is the locus of decisive transport functions which, for a limited period at least, can perform this role even in the absence of other cell organelles (in some cells, like anucleate erythrocytes, there is actually no other membrane present).

2.7.1. Plasma membrane

2.7.1.1. Morphology

The exterior surface of practically all animal cells and of some microorganisms that live in body fluids is formed by a relatively thick (about 10 nm at the least) membrane which may be quite flat (in its classical form in Fig. 2.35A), or corrugated (Fig. 2.35B), or deeply invaginated (Fig. 2.35C) or form finger-like projections (Fig. 2.35A). A special kind of plasma membrane is the myelin sheath of various

Fig. 2.35. Different modifications of plasma membrane. **A**. Human red blood cell membrane, fixed with $KMnO_4$. (Taken from Robertson, J. D. (1964). In: *Intracellular membraneous structures* (ed. by Seno, S. and Cowdry, E. V.), p.

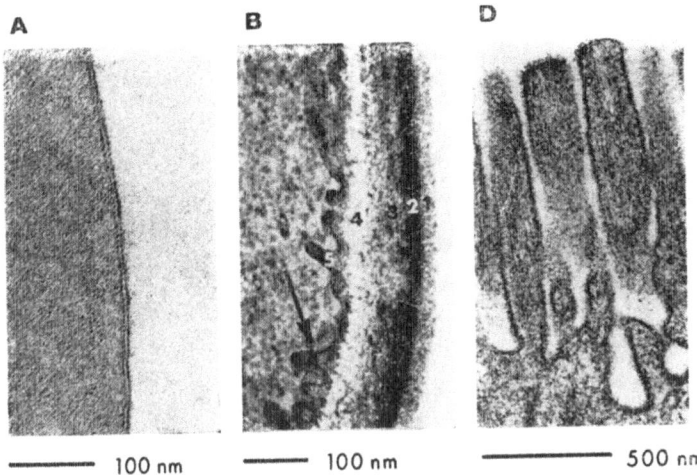

A

B

D

—————— 100 nm   ———— 100 nm   ———————— 500 nm

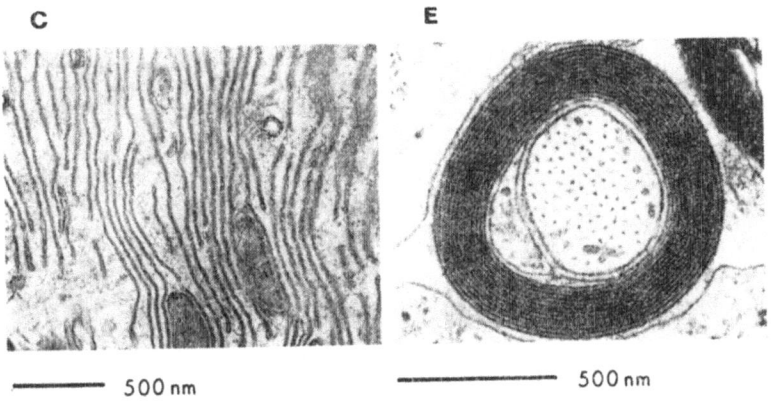

C

E

———— 500 nm   ————————— 500 nm

379. Japan Soc. Cell Biol., Okayama.) **B**. Cross section through the periphery of *Candida albicans* showing corrugated plasma membrane (arrow) and several layers of the cell wall (1—5). The cell was fixed with acrolein-tris-(1-aziridinyl) phosphine-OsO_4 and stained with lead and uranyl acetate. (Reproduced with kind permission from Djaczenko, W. and Cassone, A. (1972). *J. Cell Biol.* **52**, 186.) **C**. Ultrathin section of a cell from mosquito larva anal papilla. The membranes running vertically through the photograph are projections of basal plasma membrane. (Reproduced with permission from Copeland, F. (1964). *J. Cell Biol.* **23**, 253.) **D**. Intestinal brush border modification of the plasma membrane showing also the fuzzy glycocalyx. The preparation is from mouse jejunum (Reproduced with permission from Sjöstrand, 1963.) **E**. Membrane sheaths of a nerve central axon. (Taken with permission from Dean, G. (1970). *Scient. Amer.* **223**, 40.)

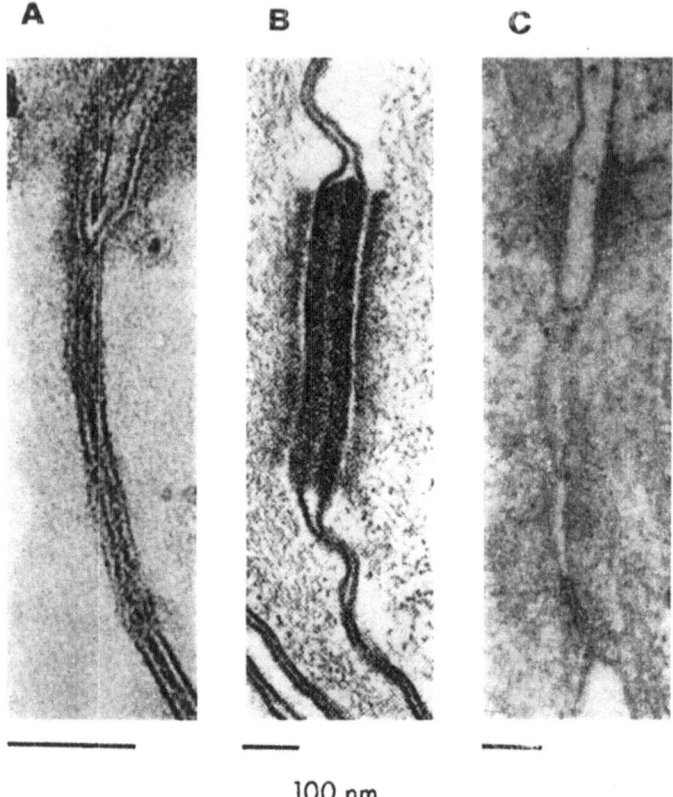

A   B   C

100 nm

Fig. 2.36. **A**. A tight junction (arrow) in isolated rat liver plasma membranes. (Taken with permission from Benedetti, E. L. and Emmelot, P. (1968). In: *The Membranes* (ed. by Dalton, A. J. and Haguenau, F.), p. 33. Academic Press, New York—London.) **B**. A desmosome in muscle tissue. (Taken from Curtis, A. S. G. (1967). *The Cell Surface: Its Molecular Role in Morphogenesis*. Logos Press and Academic Press, Woking and London.) **C**. A desmosome-like structure at the adjacent cell membranes of mouse jejunum. (Courtesy of Dr. J. Ludvík, Institute of Microbiology, ČSAV, Prague.)

Fig. 2.37. Surface views of various plasma membranes. **A**. Fractured outer membrane of *Mycoplasma* grown on oleate-supplemented medium (the particles are assumed to lie in the hydrophobic plane of the plasma membrane). (Taken with permission from Tourtellotte, M. E., Branton, D. and Keith, A. (1970). *Biochemistry* **66**, 909.) **B**. The plasma membrane of *Saccharomyces cerevisiae* with

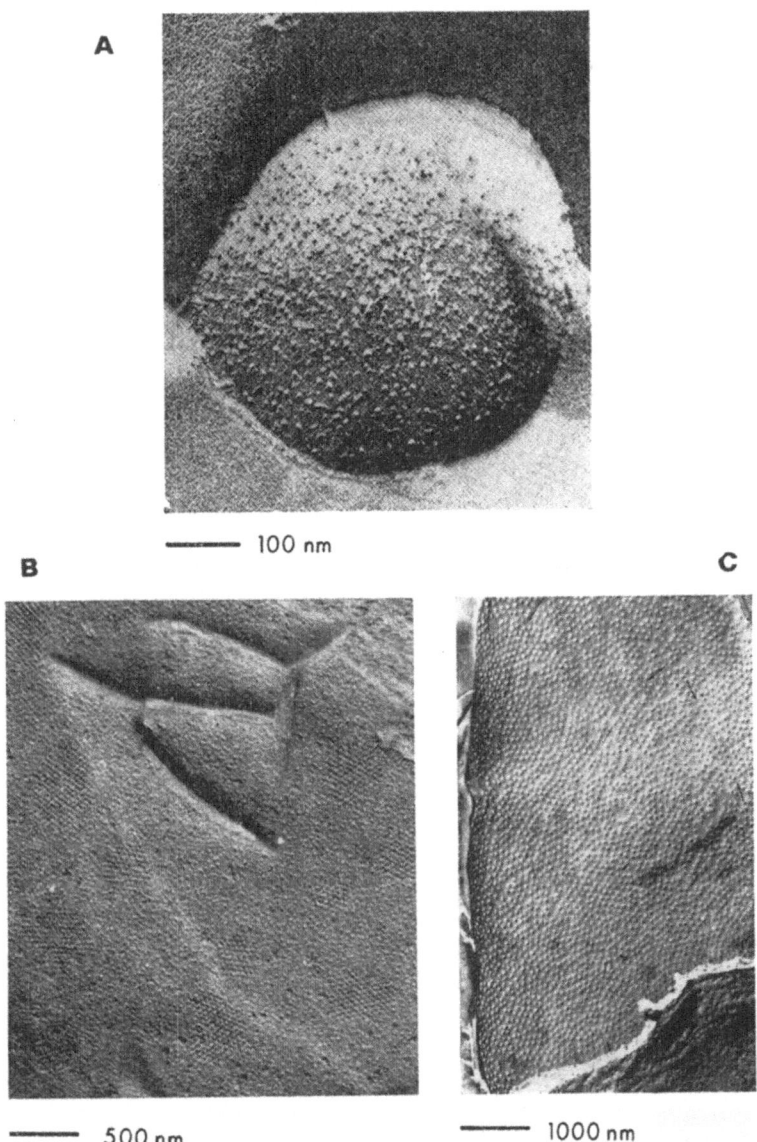

A

─────── 100 nm

B

C

─────── 500 nm

─────── 1000 nm

typical grooves and hexagonally arranged globules. (Courtesy of Prof. O. Nečas, J. E. Purkyně University, Brno, Czechoslovakia.) **C**. Plasmalemma of a myocardium capillary with numerous openings often of crater-like appearance. Note the much greater size of these surface features than of the globules or pits in **A** and **B**. (Taken with permission from Simionescu, M., Simionescu, N. and Palade, G. E. (1974). *J. Cell Biol.* **60**, 128.)

nerve fibers, arising from the enveloping Schwann cell which winds itself several times about the axon (Fig. 2.35**E**).

Where two adjacent organ cells come into close contact, the juxtaposed plasma membranes can form junctions of three types: (*1*) the tight junction, (*2*) the intermediate junction, and (*3*) the desmosome. The tight junction appears to represent a continuous belt surrounding a part of the cell surface and separating the lumen of an organ from intercellular spaces between the cells lining it (Fig. 2.36**A**).

The intermediate junction and the desmosome are less tightly joined, appearing rather as two trilaminar membranes cemented together. In the case of the desmosome, there is much cytoplasmic material adhering from both sides even in isolated membrane preparations. It is surmised that at the desmosomes special communication channels between cells exist (Fig. 2.36**B**).

Viewed in a freeze-fractured preparation, the plasma membrane appears to be spattered with protein particles in *Mycoplasma* (Fig. 2.37**A**), deeply grooved at places in *Saccharomyces cerevisiae* (Fig. 2.37**B**) and broken up by fenestrae in a capillary membrane (Fig. 2.37**C**).

Plasma membranes are more difficult to isolate in a native state than most other membranes, an exception being the erythrocyte stroma. They tend to break up and form small vesicles spontaneously, the vesicles appearing in the microsomal fraction during differential centrifugation. They can be separated by their buoyant density (Steck and Wallach, 1970) which for the liver bile fronts in sucrose is 1.17 and for the yeast plasma membranes in Urografin about 1.15.

2.7.1.2. *Functional properties*

The functions of plasma membranes are manifold indeed (Table 2.23). The questions of transport of various compounds will be dealt with in more detail in the following chapters of the book; here we shall restrict ourselves to a survey of the functions that are distinct from transport.

2.7.1.2.1. *Antigenicity*

On its outer surface, the plasma membrane of many cells carries its antigenic determinants in the form of glycolipids and glycoproteins. These components also play a major role in the phenomena of cell

TABLE 2.23. **Functions of the plasma membrans in various cells**

Function	Comment
Selective diffusion of small molecules and ions	p. 192 and p. 256
Passive and active carrier transport	p. 207 and p. 222
Pinocytosis and phagocytosis	p. 318
Cell adhesion and fusion	in wall-less cells
Mechanical protection of cell interior	especially amoebas, mycoplasmas
Electrical insulation	myelin sheath
Site of surface antigens	in wall-less cells
Generation of nerve impulses	nerve plasma membrane
Site of various metabolic enzymes converting oxidative to phosphatebond energy	especially in bacteria
Site of enzymes involved in cell-wall synthesis	bacteria, fungi, plants (?)
Site of hormone receptors	p. 146
Site of absorption of light quanta	photosynthetic bacteria
Locale of piezoelectric effects	motile bacteria and protozoans
Biological clock	p. 147

recognition, contact inhibition and transfer of surface material. It has been known for some time that normal tissue cells coming into contact are suddenly immobilized by a "freezing" of cytoplasmic streaming and thus can develop in alignment with one another (Abercrombie and Heaysman, 1953). This property is absent in cancer cells. The alignment of normal cells can come about either through structural fit of the surface glycoproteins and glycolipids (Fig. 2.38**A**) or through an enzyme-substrate link where a surface galactosyl transferase is involved (Fig. 2.38**B**).

The sugar residues exposed on the surface of the plasma membranes are also responsible for a number of agglutination reactions, themselves caused by glycoproteins of plant origin, e.g., wheat germ agglutinin, soybean agglutinin and the jack bean agglutinin with two binding sites on the molecule, termed concanavalin A (the binding site appears to contain two tyrosines, one aspartic acid and one arginine; Hardman and Ainsworth, 1973). They combine specifically with N-acetyl-glucosamine, N-acetylgalactosamine and α-methyl-D-glucoside, respectively, the reaction being generally cryptic in non-transformed cells but very powerful in cells altered by viruses.

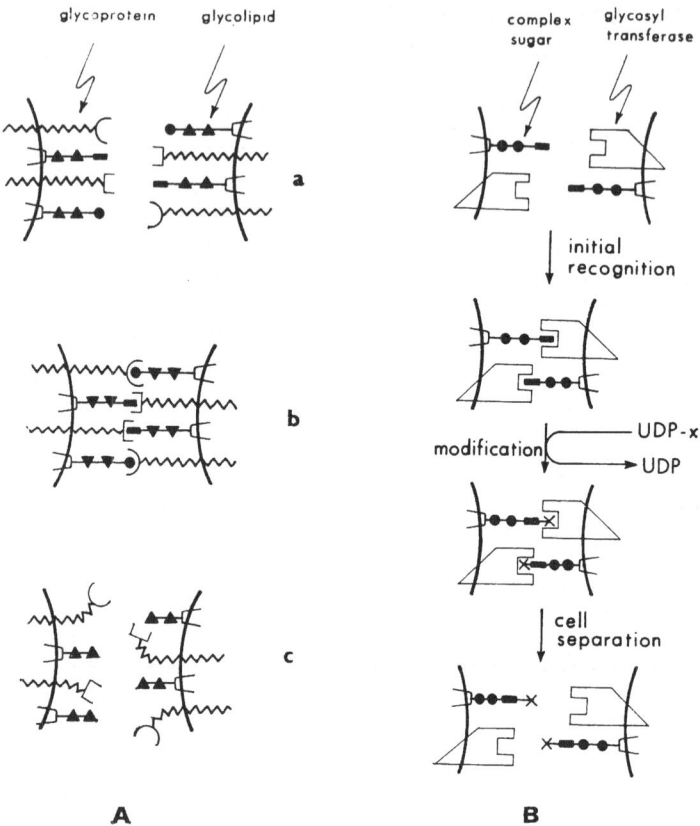

Fig. 2.38. **A**. Intercellular interaction through surface glycolipids and glyco-proteins. **a** Separated cells, **b** confluent cells displaying complementary linkage through noncovalent bonds or possibly Ca^{2+} bridges, **c** malignant cells with incomplete carbohydrates and hence no complementarity. **B**. The role of cell surface glycosyl transferase in cell-cell recognition and cell contact modification.

2.7.1.2.2. Enzyme content

The enzyme equipment of plasma membranes can be very extensive, there being substantial differences between various plasma membranes. A surprisingly varied assortment is found in liver plasma membranes, erythrocyte membranes being much poorer. Few enzymic activities were detected in yeast membranes while the bacterial membrane contains practically all the mitochondrial enzymes of eukaryotic cells plus a number of others (Table 2.24). It is noteworthy that even in the enzyme equipment of plasma membranes a definite

TABLE 2.24.   Major enzyme activities found in various plasma membranes

Rat liver	Erythrocyte	Baker's yeast	*Mitrococcus lysodeikticus*
Mg-ATPase	Mg-ATPase	Mg-ATPase	Mg-ATPase
Na, K-ATPase	Na, K-ATPase	Acid phosphatase	Cytochrome *c* oxidase
5'-Nucleotidase	Cathepsin I	Phospholipase	Catalase
Acetate kinase	Cathepsin II	Phosphatidyl-inositol kinase	NADH dehydrogenase
NAD^+ pyrophosphatase	Ribosephosphate isomerase	(β-Fructofuranosidase)	Lactate dehydrogenase
Phosphodiesterase	Acetylcholinesterase		Succinate dehydrogenase
Nucleosidediphosphatase (acting on IDP) Leucine aminopeptidase			Oxoglutarate dehydrogenase

asymmetry is apparent. Erythrocyte membranes contain acetylcholinesterase and NAD^+ nucleosidase on the outer surface but lipoamide dehydrogenase, protein kinase and adenosinetriphosphatase on the inner surface (Steck, 1974).

2.7.1.2.3.   Cell walls

An important function of membranes of unicellular organisms of plants is the secretion of cell-wall material and, possibly, the function as a template or primer for some of the wall components.

All the rigid cell walls of plants and microorganisms are mainly polysaccharide plus varying amounts of peptide. Starting at the lower end of the phylogenetic tree, we shall deal first with bacteria.

Bacteria

There appear to be two fundamental types with regard to the structure of the cell envelope, the Gram-positive and the Gram-negative bacteria, the latter being distinguished from the former by the fact that they possess two membranous structures, an outer membrane and an inner, cytoplasmic, membrane (Fig. 2.39).

The peptidoglycan or mucopeptide or murein layer appears practically in all bacteria and is built on a skeleton of alternating *N*-acetylmuramic acid and *N*-acetylglucosamine

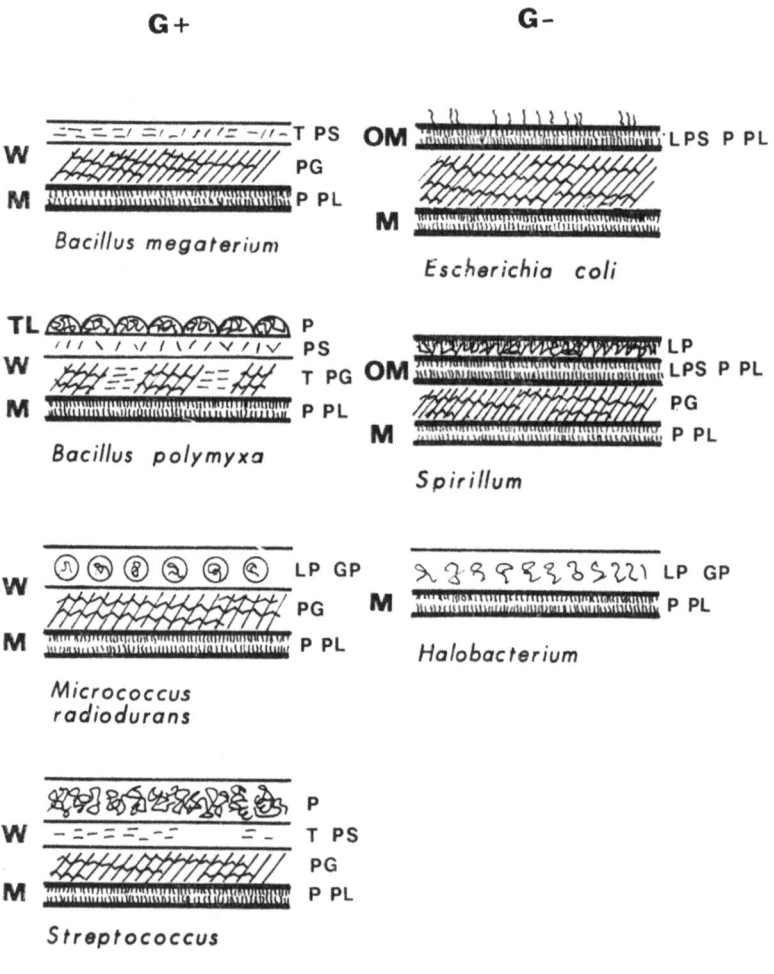

Fig. 2.39. Schematic models of bacterial cell envelopes both in Gram-positive and in Gram-negative species. M Plasma (cytoplasmic) membrane, OM outer membrane, W cell wall, TL T-layer; T teichoic acid, PS polysaccharide, PG peptidoglycan, P protein, PL phospholipid, LP lipoprotein, GP glycoprotein, LPS lipo-polysaccharide.

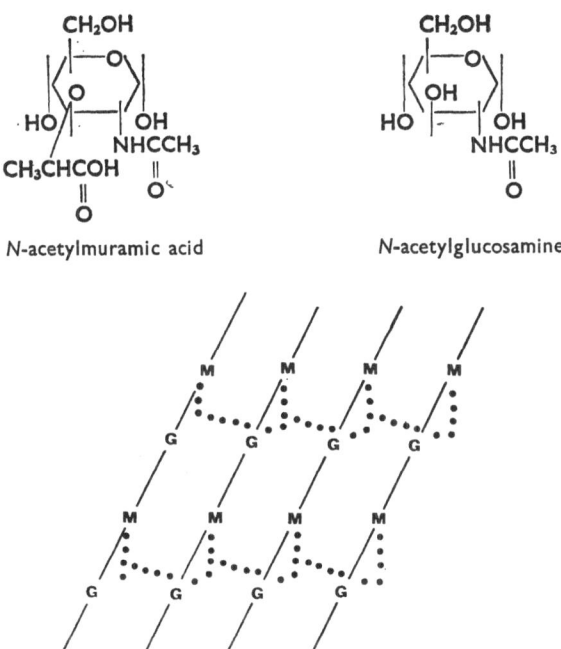

Fig. 2.40. Schematic representation of cell wall peptidoglycan, based on data from *Staphylococcus aureus*. The black dots represent amino acids of the peptide cross-links, **M** stands for *N*-acetylmuramic acid, **G** for *N*-acetylglucosamine. (Redrawn with permission from Ghuysen and Shockman, 1973.)

The *N*-acetylmuramic acid can be phosphorylated in position 6 or acetyl-ether linked or even cyclized between the carboxyl of the lactyl ether in position 3 and the nitrogen in position 2. Strands of *N*-acetylglucosamine and *N*-acetylmuramic acid are joined by oligopeptide links as shown in Fig. 2.40, the links being of several kinds:

L-ala→D-gln

γ →L-lys→D-ala→[Gly→Gly→Gly→Gly→Gly]→L-lys

(*Staphylococcus aureus*)

→[L-ala→L-ala→L-ala→L-thr]→

(*Micrococcus roseus*)

→[Gly$_3$, L-ser$_2$]→

(*Staphylococcus epidermidis*)

$$\cdots\cdots\cdots\rightarrow [\text{L-ser}\rightarrow\text{L-ala}]\rightarrow$$

(*Lactobacillus viridescens*)

$$\cdots\cdots\cdots\rightarrow [\text{L-ala}\rightarrow\text{L-ala}]\rightarrow$$

(*Streptococcus pyogenes*)

$$\cdots\cdots\cdots\rightarrow [\text{L-ala}]\rightarrow$$

(*Arthrobacter crystallopoietes*)

$$\cdots\cdots\cdots\rightarrow [\text{D-asn}]\rightarrow$$
$$\beta \mid \rightarrow$$

(*Streptococcus faecalis,*
Lactobacillus casei)

$$\downarrow$$
L-ala→D-glu   DAP
$$\gamma \mid \rightarrow\text{DAP*}\rightarrow\text{D-ala}\cdots\cdots\uparrow$$
$$\cdots\cdots\uparrow$$ (*Bacillus megaterium*)

$$\downarrow$$
L-ala→D-glu   ------L-lys------
$$\gamma \mid \rightarrow\text{L-lys}\rightarrow\text{D-ala}\underline{\quad\quad}\uparrow$$
$$\cdots\cdots\uparrow$$
(*Aerococcus, Gaffkya*)

$$\downarrow$$
L-ala→D-gln   DAP
$$\gamma \mid \rightarrow\text{DAP}\rightarrow\text{D-ala}\rightarrow\text{Gly}\underline{\quad}\uparrow$$
$$\cdots\cdots\uparrow$$
(*Clostridium perfringens,*
Streptomyces albus)

$$\downarrow$$
L-ala→D-glu→Gly   ------L-lys------
$$\gamma \mid \rightarrow\text{L-lys}\rightarrow\text{D-ala}\rightarrow[\text{L-ala}\rightarrow\text{D-glu}\rightarrow\text{Gly}$$
$$\cdots\cdots\uparrow \qquad\qquad\qquad \gamma \mid \rightarrow\text{L-lys}\rightarrow\text{D-ala}]_n\underline{\quad}$$
(*Micrococcus lysodeikticus*)

* DAP = diaminopimelic acid

$$\downarrow$$
Gly→D-glu ·······
       | γ |→Homoser→D-ala→D-orn
           ↑   ↑
         Gly→D-glu ···················
            γ |→

(Corynebacterium)

$$\downarrow$$
L-ser→D-glu ·······
       γ |→L-orn→D-ala →D-lys
              ↑

(Butyribacterium rettgeri)

    The biosynthesis of peptidoglycans proceeds from cytoplasmic uridine triphosphate and N-acetylglucosamine-1-phosphate to uridine-diphospho-N-acetylglucosamine, followed by reactions with phospho-enolpyruvate, NADPH and the various amino acids, each reaction being catalyzed by a special enzyme. The UDP-oligopeptide formed then reacts in the membrane with a 55-carbon isoprenoid alcohol phosphate (undecaprenyl-P) to undecaprenyldiphospho-oligopeptide. This alcohol phosphate is called the glycosyl carrier lipid and appears to be a rather wide-spread membrane entity, known to be active also in the transport of teichoic acid precursor in *Staphylococcus* (Baddiley *et al.*, 1968), of the lipopolysaccharide O-antigen of *Salmonella* (Osborn, 1969) and even in the glycosylation of glycoproteins (Caccam *et al.*, 1969).

    The undecaprenyldiphospho-oligopeptide reacts with UDP-N-acetylglucosamine to an undecaprenyldiphospho-disaccharide-oligopeptide and this is then translocated from the cytoplasmic membrane to the wall acceptor. The whole synthesis is shown in Fig. 2.41. together with the probable sites of action of several antibiotics. For a most up-to-date review see Ghuysen and Shockman (1973). It is rather likely that the peptidoglycan coat which maintains the shape of bacteria (rods, spheres, etc.) is not the determinant of gross cell morphology but rather a consequence of other factors in play (Henning and Schwarz, 1973).

    The other major component found in the walls of Gram-positive bacteria is teichoic acid, a mixture of polymers of ribitol 5-phosphate and glycerol 1-phosphate, the polyols being linked phosphodiesteric-

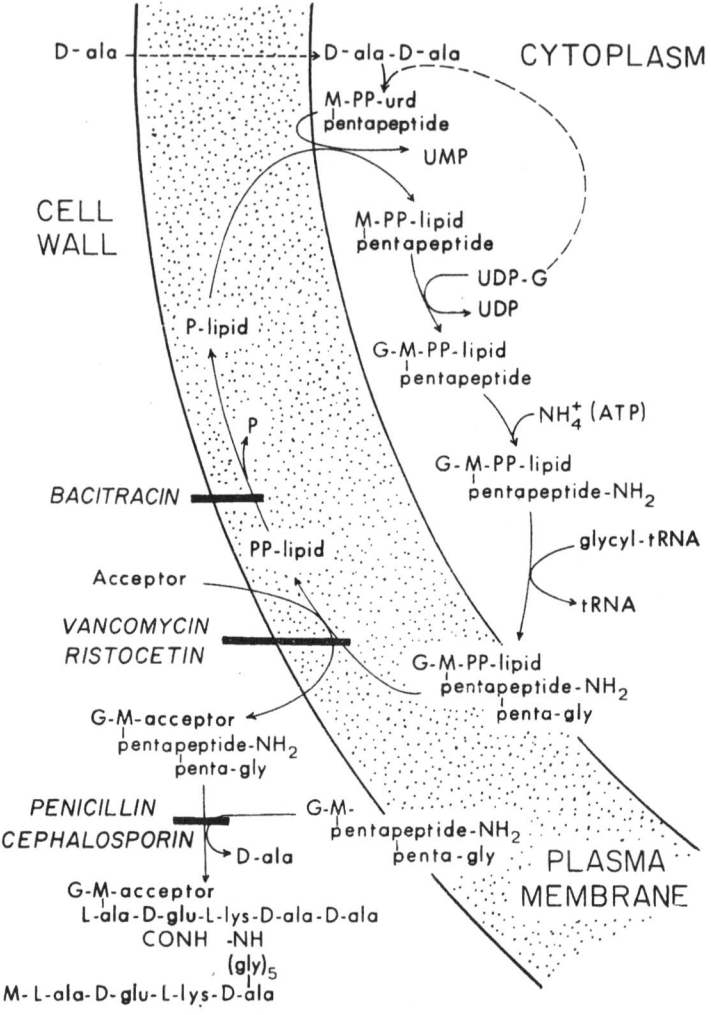

Fig. 2.41. An outline of the synthesis of peptidoglycan of type I (e.g. in *Aerococcus*) with the points of attack of various antibiotics. The dashed line at upper right represents the sequential building up of the UDP-M-pentapeptide from uridine-diphospho-*N*-acetylglucosamine. M *N*-acetylmuramic acid, G *N*-acetylglucosamine, lipid the undecaprenyl residue.

ally and frequently substituted with glucosyl, *N*-acetylglucosaminyl or ester-linked D-alanine residues.

$$\underset{\substack{\text{HOPOCH}_2 \\ \parallel \\ O}}{\overset{O^-}{|}} \cdots \overset{|}{\underset{RO}{|}} \overset{|}{\underset{O}{|}} \overset{|}{\underset{O}{|}} \text{--CH}_2\text{O} \cdots$$

The chains are synthesized by sequential transfer of polyol phosphate from cytidinediphosphoribitol or CDP-glycerol to the acceptor (apparently preexisting teichoic acids). *N*-Acetylglucosamine attachment requires UDP. In some cases, the glycosyl carrier lipid is also involved.

An important component of the outer envelope of Gram-negative bacteria is the lipopolysaccharide. An example of its structure is shown in Fig. 2.42. It will be seen that various unusual sugars are present in its molecule. In fact, the O-antigen from other species of *Salmonella* was found to contain also paratose (3,6-dideoxy-D-glucose) and tyvelose (3,6-dideoxy-D-mannose). The lipid A region contains C_{12}, C_{14}, C_{16} and C_{14}-hydroxy fatty acids in a 1 : 1 : 1 : 3 ratio.

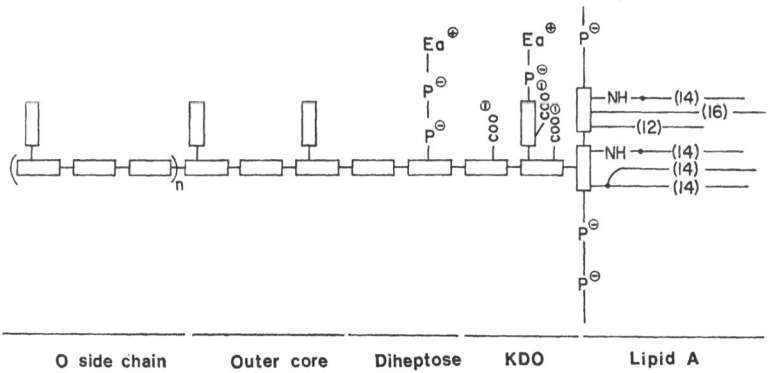

| O side chain | Outer core | Diheptose | KDO | Lipid A |

Fig. 2.42. A schematic outline of the lipopolysaccharide molecule. Ea Ethanolamine, P phosphate, KDO three molecules of 3-deoxy-D-manno-octulosonic acid. (Redrawn from Nikaido, 1973.)

Lipid *A* appears to be synthesized by the action of fatty acyl transferases on the diglucosamine skeleton. Once formed, it attaches the KDO residue, ATP being required for this step. Practically nothing is known about the attachment of the mannoheptoses but it is known that the glucose and galàctose residues are joined through the action of UDPglucosyltransferases under activation of Mg^{2+}, phospholipids (particularly phosphatidyl ˙ethanolamine) being required for the process (see Table 2.18).

The O-antigen part is synthesized separately under participation of the glycosyl carrier lipid and of diphospho derivatives of various nucleosides, *viz.* UDPgalactose, TDPrhamnose, GDPmannose and CDPabequose. The oligosaccharides are then transferred onto the

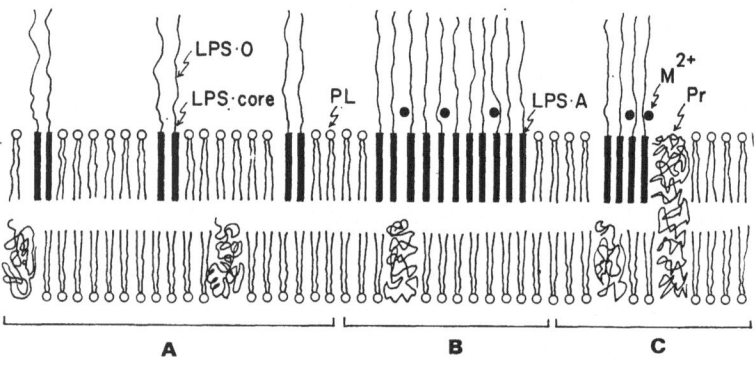

Fig. 2.43. Possible arrangement of molecules in the outer membrane layer of a Gram-negative bacterium. In **A**, lipopolysaccharide molecules are interspersed among phospholipids. In **B**, bivalent cations reduce the repulsion between lipopolysaccharide molecules so that these can lie next to each other. In C, proteins interact both with phospholipids and with lipopolysaccharide. LPS lipopolysacharide, PL phospholipid, O the O side chain, A lipid A, Pr protein. (Redrawn after Nikaido, 1973.)

core lipopolysaccharide again by mediation of the isoprenoid carrier lipid (all these facts, and more, are elegantly summarized by Nikaido, 1973).

The completed lipopolysaccharide must then be transported from the cytoplasmic membrane (*cf.* Fig. 2.39) to the outer membrane, apparently via the connections between the two membranes through the peptidoglycan layer. A model of the outer membrane is shown in Fig. 2.43.

The outer membrane is invested with several functions. Thus, the O-antigen chain has a survival value for the bacterium in a host animal, due to the possibility of enormous diversification of the sugar residues and hence relatively low levels of antibodies capable of coping with them.

The somatic O-antigen is not the only antigen carried by the surface of Gram-negative bacteria. Various mutants have been prepared with defects in the side-chain synthesis of the O-antigen or in their transfer to the lipopolysaccharide core — their antigens are usually termed R-antigens. There exist also transient forms designated as T-antigens (mainly in *Salmonella* and *Shigella*).

The wall contains also the common antigen CA, the C-antigen and the bacterial agglutinogen BA.

Moreover, in addition to the above somatic (or cellular) antigens, capsule-forming bacteria contain a variety of polysaccharide antigens in the outer capsule. These polysaccharides are pronouncedly negatively charged (due to phosphate groups, hexuronic acid and sialic acid) and often contain improbably high amounts of phosphorus. Thus, in the *Haemophilus* capsular antigen the sugar : P ratio is 1 : 1. The sugar composition of these polysaccharide antigens is not as varied as in the lipopolysaccharide, the predominant components being galactose, glucose, mannose, fucose, rhamnose, glucuronic acid and galacturonic acid.

The outer membrane as a whole apparently acts as a permeability barrier to various antibiotics and obviously other compounds although, generally, it is the inner, cytoplasmic membrane which carries specific transporting mechanisms.

Bactcrial walls also contain polysaccharides without any admixture of either peptides or lipids but they are relatively of minor occurrence and importance.

Yeasts

In comparison with bacteria, the exterior coat of yeast cells is rather less complex although it may reach a thickness of 0.2 μm, comparable to the diameter of a small bacterium.

In *Saccharomyces*, the cell-wall is over 80 % polysaccharide, one-half of it glucan, the other half mannan. It is generally assumed that mannan predominates in the outer part of the wall, together with some amorphous glucan, while fibrillar $\beta(1\rightarrow3)$glucan forms the innermost covering of the plasma membrane; the cross section through the wall is shown schematically in Fig. 2.44.

Other yeast genera, however, may contain no mannan at all but,

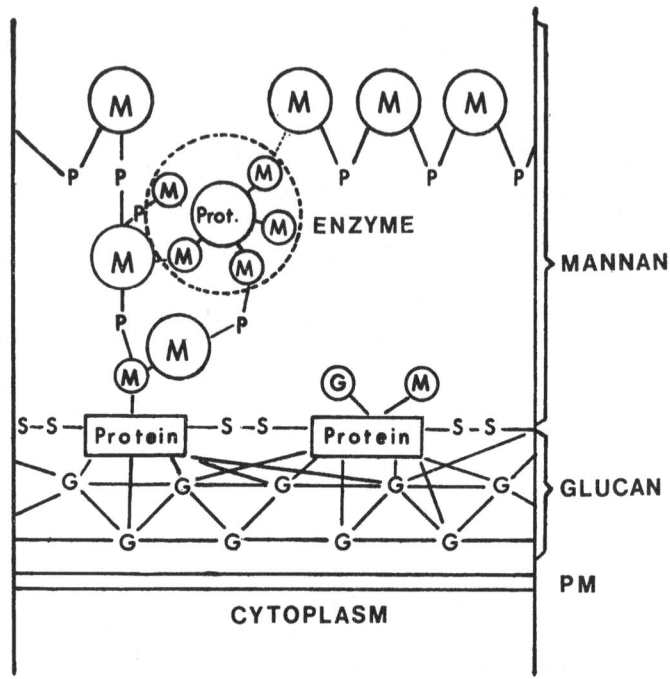

Fig. 2.44 A schematic representation of the structure of yeast cell wall. M Mannan, G glucan, P phosphate, ENZYME invertase, PM plasma membrane. (Adapted from Lampen, J. O. (1968). *Antonie van Leeuwenhoek, J. Microbiol. Serol.* **34,** 1.)

in compensation, large amounts of chitin (*Nadsonia, Rhodotorula, Endomyces*).

The *Saccharomyces* glucan contains mostly $\beta(1\rightarrow3)$ linkages in the fibrillar part but also $\beta(1\rightarrow6)$ linkages in the remaining part. The structure of the mannan is rather more varied, there being structural units of the following types present:

$$\rightarrow Man^1 \rightarrow {}^6Man^1 \rightarrow$$
$$\uparrow^2_1$$
$$Man$$
$$\uparrow^2_1$$
$$Man$$

$$\rightarrow Man^1 \rightarrow {}^6Man^1 \rightarrow {}^6Man \rightarrow$$
$$\uparrow^2_1$$
$$Man$$
$$\uparrow^2_1$$
$$Man$$
$$\uparrow^3_1$$
$$Man$$

The cell walls of yeast contain some protein (*c.* 10 %), possibly in a complex with glucan, and a small amount of lipid (1 − 10 %).

Both mannan and glucan are synthesized under participation of uridinediphosphosugar derivatives.

Like some yeast species, filamentous fungi contain major amounts of chitin which is a polymer of $\beta(1\rightarrow4)$-N-acetyl-D-glucosamine. The walls are organized usually as closely packed microfibrils, parallel to the underlying plasma membrane. Sugar residues found in various fungal polysaccharides are (besides glucosamine) mainly glucose, mannose and galactose.

Plants

Especially heavy cell envelopes are known to occur in algae and in higher plants. Little is known about the architecture of algal cell walls but they are known to contain, as major fibrous components, the following polysaccharides: $\beta(1\rightarrow4)$-mannan, $\beta(1\rightarrow3)$-xylan, two types of cellulose and glucomannan. Hemicelluloses which are also present have been found to contain also galactose, arabinose, rhamnose and glucose.

Higher plants are characterized by the overwhelming content of cellulose in their cell walls, cellulose being a polymer of $\beta(1\rightarrow4)$-linked molecules of D-glucose. However, only few cell walls are practically speaking pure cellulose, other polysaccharides being usually present. A common one is $\beta(1\rightarrow4)$-linked xylan, often substituted with L-arabinose linked usually $(1\rightarrow3)$. While it has been extremely difficult to arrive at a molecular weight of cellulose (owing ot the fact that the native state in a wall may represent a continuum of fibers, microfibers and ultramicrofibers) some reasonable estimates have been obtained for xylans (about $10-15\,000$, in general). Another polysaccharide of plant walls is a glucomannan with usually 2 or 3 mannose residues per glucose, linked $\beta(1\rightarrow4)$. Another polysaccharide is arabinogalactan, probably a branched galactan (linked $\beta(1\rightarrow3)$ in the backbone and $\beta(1\rightarrow6)$ in the side chains) with mostly terminal arabinose residues.

Finally, there are the pectins, i.e. $(1\rightarrow4)$-linked polygalacturonic acids.

A special group of plant cell wall polymers are the lignins, made up of networks of p-hydroxyphenylpropane, vanillin, syringaldehyde, derivatives of guaiacol and coniferol.

The architecture of the plant cell wall is rather intricate and no universal pattern has been arrived at but there is general agreement on the existence of a primary wall, formed during longitudinal growth of the cell, and of a secondary wall, formed by gradual layering of material from the cell interior moved through the plasmalemma onto the primary wall (Fig. 2.45).

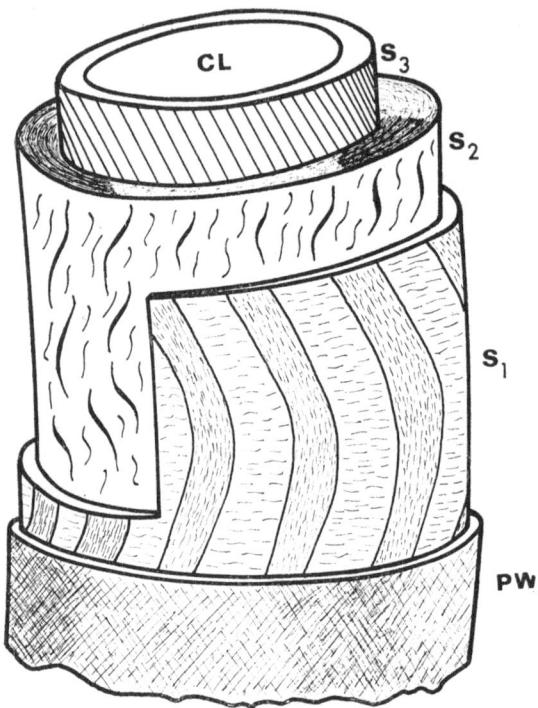

Fig. 2.45. A diagrammatic representation of the secondary wall of a mature cotton hair. **CL** Cell lumen, S_1, S_2, S_3 various layers of the secondary wall with different arrangement of the microfibrils, **PW** primary wall. (Redrawn from Roelofsen, P. A. (1959). *The Plant Cell Wall*. Gebr. Borntraeger, Berlin.)

There is little doubt that most (and perhaps all) wall polysaccharides are synthesized by attaching sugar moieties from uridinediphospho derivatives, there being specific enzymes to catalyze the attachment by a $(1\rightarrow3)$, $(1\rightarrow4)$ or $(1\rightarrow6)$ link. However, GDP derivatives are also known to operate under certain conditions. It is not clear by what means the UDP-sugar molecule is transported from the inside of the plasmalemma to the external polysaccharide primer.

Protozoans

An example of a nonplant extramembrane cell coat is the external envelope of amoebae, These protozoans have a thick (20 nm) unit membrane covered with filaments (100–200 nm long) extending into the surrounding medium (Fig. 2.46) and serving to adsorb nutrients which are then engulfed by pinocytosis. The composition of the whole coat is 45 % lipid, 34 % protein and 21 % carbohydrate. A rather surprising feature of this membrane is that it protects the cell against the strongly hypotonic external medium although it contains none of the generally necessary fibrillar meshwork.

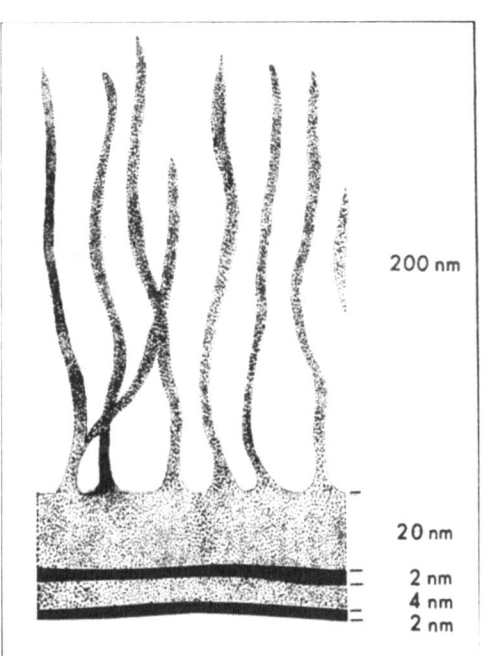

Fig. 2.46. A diagrammatic drawing of the surface coat of the amoeba *Chaos chaos*. The plasma membrane proper is represented by the three lowermost lines. (Adapted from Brandt, P. W. and Pappas, G. D. (1962). *J. Cell Biol.* **15**, 55.)

2.7.1.2.4. *Binding and receptor properties*

The surface of plasma membranes shows a distinct binding affinity for a number of ions, small molecules and polymers.

The cation-binding affinity of erythrocyte surface has been shown to decrease as follows: $Th^{4+} > UO_2^{2+} > La^{3+} > Cu^{2+}, Ni^{2+} > Ca^{2+},$

Sr^{2+}, Ba^{2+}, Mn^{2+}. On the other hand, agglutination of erythrocytes is caused with highest efficiency by Ca^{2+}, Cu^{2+}, Th^{4+}, Ti^{3+}, Al^{3+}, less by Ag^+, Cd^{2+} and much less by Mn^{2+}, Co^{2+} and Hg^{2+}.

There is practically no binding of anions on the plasma membrane surface.

The cations bound to the surface reflect the magnitude of the surface charge, usually measured electrophoretically. The charge density has been measured for a variety of cells and found to range from about $7 . 10^{-3}$ C m^{-2} in a hamster kidney cell to $2 . 10^{-2}$ Cm^{-2} in a sheep erythrocyte (Curtis, 1967). This relatively high negative charge results in the isoelectric points of most single cells lying between pH 2 and 4. In mammalian cells, the surface charge is composed of $10^7 - 10^8$ negative charges per cell and of $10^6 - 10^7$ positive charges per cell, depending on the surface area, the density being as indicated above.

One of the important functions residing in some plasma membranes is the susceptibility to hormone stimulation. It appears that the hormone recognizing or initiator systems contain three parts: the receptor which specifically recognizes the hormone, the transducer which couples the function of the receptor with the effector, and the effector which then brings about appropriate change in the cell function. Some of the hormone receptors have been partly purified. The insulin receptor from adipose plasma membrane has a molecular weight of 300 000 and has two types of sites with dissociation constants of 10^{-10} M and 10^{-9} M (there is competition with insulin binding by plant lectins, both of them inhibiting adenylate cyclase); there appears to be 1 site per μm^2. The glucagon receptor has a K_D of $4 . 10^{-9}$ M and a molecular weight near 50 000 (it stimulates adenylate cyclase). The catecholamines are bound to fat cells and to erythrocytes with K_D's of about 10^{-6} M, the erythrocytes containing about 5 sites per μm^2. Acetylcholine is bound with a K_D of 10^{-5} M and another one of 10^{-7} M in the *Electrophorus* eel electric organ, and with K_D's of 10^{-7} and 10^{-8} M in the *Torpedo* electroplax. The electric eel acetylcholine receptor has a molecular weight of about 360 000, and is composed of 50 000 subunits. The site density is extraordinarily high: about 33 000/μm^2 in the subsynaptic area, 100 000 per μm^2 in frog muscle, and 12 000/μm^2 in mouse diaphragm. Oxytocin is bound to frog-skin receptors in several ways, the most important one having a K_D of $5 . 10^{-3}$ M, the maximum capacity being $1-2$ pmol/g tissue.

The character of the transducer remains unknown but physical changes in the lipid organization may be involved. The effector system is most often adenylate cyclase (certainly for peptide hormones, such as insulin, glucagon, oxytocin, vasopressin, thyreocalcitonin, secretin, adrenocorticotropin; *cf.* Jost and Rickenberg, 1971). The specificity of action is generally ensured by the receptor but, at least in the case of insulin, appears to be rather complicated.

A plasma membrane function perhaps related to the hormone sensitivity is the ability of specialized cells (lymphocytes, in particular of the bone-marrow or B type, or plasma cells) to respond to antigen binding by starting the synthesis of protein antibodies that inactivate the antigen either by neutralizing it (*antitoxins*), by causing the antigen-producing organisms to clump together (*agglutinins*), by causing the invading cell to disintegrate (*lysins*) or by making it more susceptible to attack by phagocyting macrophages (*opsonins*).

A special susceptibility carried by some bacteria is toward bacteriocins, such as the various colicins produced by *Escherichia coli* strains. Colicins are single-chain proteins, ranging in molecular weight from 45 000 to 80 000 and causing the death of various bacteria. They are coded by extrachromosomal genetic elements (the so-called Col factor plasmids) and they bring about cessation of rather varied cell functions: Protein synthesis is blocked by E3, DNA structure is broken down by E2, the integrity of the plasma membrane is disrupted by K and E1. The mechanism of their action may resemble the three-step action of hormones.

A rather peculiar function has been recently proposed for the plasma membrane by Njus and co-workers (1974), *viz.* that of maintaining circadian rhythms of organisms, involving lateral diffusion of proteins. The function is temperature-insensitive due to compensation in fatty acid composition of the membrane.

2.7.2. Mitochondrion

These are ubiquitous organelles of eukaryotic cells, about 1 μm thick and $1-3$ μm long, of a typical architecture (Fig. 2.47), containing an outer and an inner unit membrane. From a disrupted cell, they can be separated after removal of nuclei by centrifuging at $5-20 \cdot 10^4 \, g$ min and then purified in isopycnic density gradients, forming bands at $1.18-1.21$ g cm^{-3}.

In many respects they resemble bacteria (they contain circular DNA, small ribosomes and protein synthesis sensitive to chloramphenicol; they divide autonomously, etc.) and a popular theory has it that they have invaded primitive eukaryotic cells early in evolution and provided them symbiotically with an efficient system of storing oxidation energy. (For an excellent treatment of the data, see Getz, 1972.)

Be it as it may, the two mitochondrial membranes are the sites of various enzymes and have characteristic properties (Table 2.25).

TABLE 2.25.   Some properties of mitochondrial membranes

Feature	Inner membrane	Outer membrane
Thickness (nm)	6.5	6.5
Fine structure	globular projections, 9 nm in diameter	flat, with 3nm pits
Buoyant density (g cm^{-3})	1.21	1.13
Protein-lipid ratio	3.6	1.2
% Cardiolipin	22	3
% Phosphatidyl inositol	4	14
Permeability	highly selective	lets larger molecules across
Osmotic response	+	—
Contraction after ATP	+	—

The mitochondrion is literally packed with enzymes which show a very clearly defined distribution in the various morphological compartments (Table 2.26).

The outer membrane resembles the endoplasmic reticulum of the cell, both containing cytochrome b_5, a rotenone-insensitive cytochrome c reductase and there being a phospholipid exchange between the

⟶

Fig. 2.47.   The mitochondrion. **A** An ultrathin section of a mitochondrion from mouse kidney. **O** Outer mitochondrial membrane, **I** inner mitochondrial membrane. (Courtesy of Dr. J. Ludvik, Institute of Microbiology, ČSAV, Prague.)

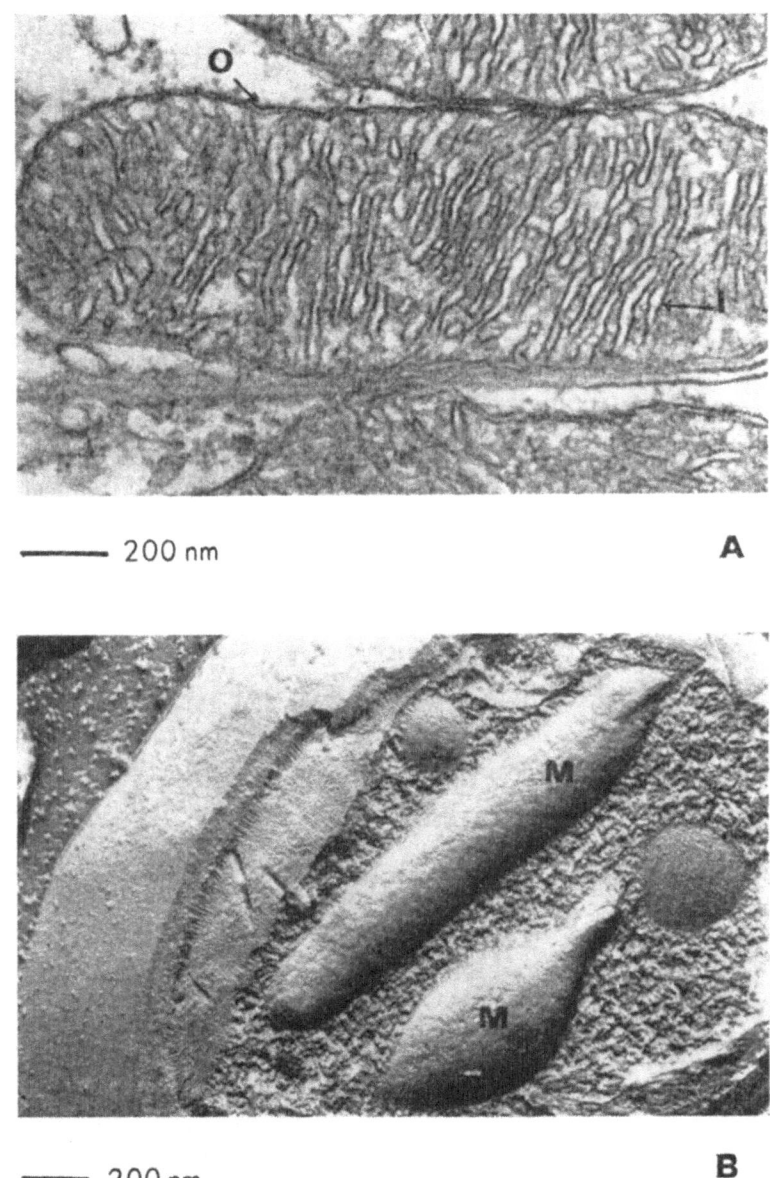

—— 200 nm   **A**

—— 200 nm   **B**

B A freeze-etched preparation of *Saccharomycodes ludwigii* showing two mitochondria (**M**) with their outer surface exposed. Parts of typical plasma membranes
with grooves are seen on the left. (Courtesy of Dr. E. Streiblová, Institute of Microbiology, ČSAV, Prague.)

TABLE 2.26. Distribution of enzymes in liver mitochondria (adapted from Ernster and Kuylenstierna, 1970; Kroger and Klingenberg, 1970)

Outer membrane	Intermembrane space	Inner membrane	Matrix
Cytochrome b_5 reductase	Adenylate kinase	Cytochrome c oxidase (1)	Malate dehydrogenase (0.4)
Amine oxidase (flavin-containing)	Nucleosidediphosphate kinase	Cytochrome a (1)	Isocitrate dehydrogenase (NAD) (0.03)
	D-Xylulose reductase		Isocitrate dehydrogenase (NADP) (0.25)
Kynureninase		Cytochrome c (1)	
Acyl-CoA synthetase		Cytochrome c_1 (0.5)	Glutamate dehydrogenase (0.01)
Glycerolphosphate acyltransferase		Cytochrome b (1)	
Lysolecithin acyltransferase		Ubiquinone (11)	Oxoglutarate dehydrogenase
Cholinephosphotransferase		Succinate dehydrogenase (0.16)	Citrate (si)-synthase
Phosphatidate phosphatase		NADH dehydrogenase (0.1)	Aconitate hydratase
Phospholipase A_1		NAD (11)	Fumarate hydratase
Nucleosidediphosphate kinase		NADP (3)	Pyruvate carboxylase
D-Xylulose reductase		3-Hydroxybutyrate dehydrogenase	Phosphoenolpyruvate carboxylase
Fatty acid elongation system		Ferrochelatase	Aspartate aminotransferase
		Carnitine palmitoyltransferase	Ornithine carbamoyltransferase
		L-Xylulose reductase	Acyl-CoA synthetase
		Fatty acid elongation chain	L-Xylulose reductase

The figures in parentheses show the relative molar amounts of enzymes and coenzymes.

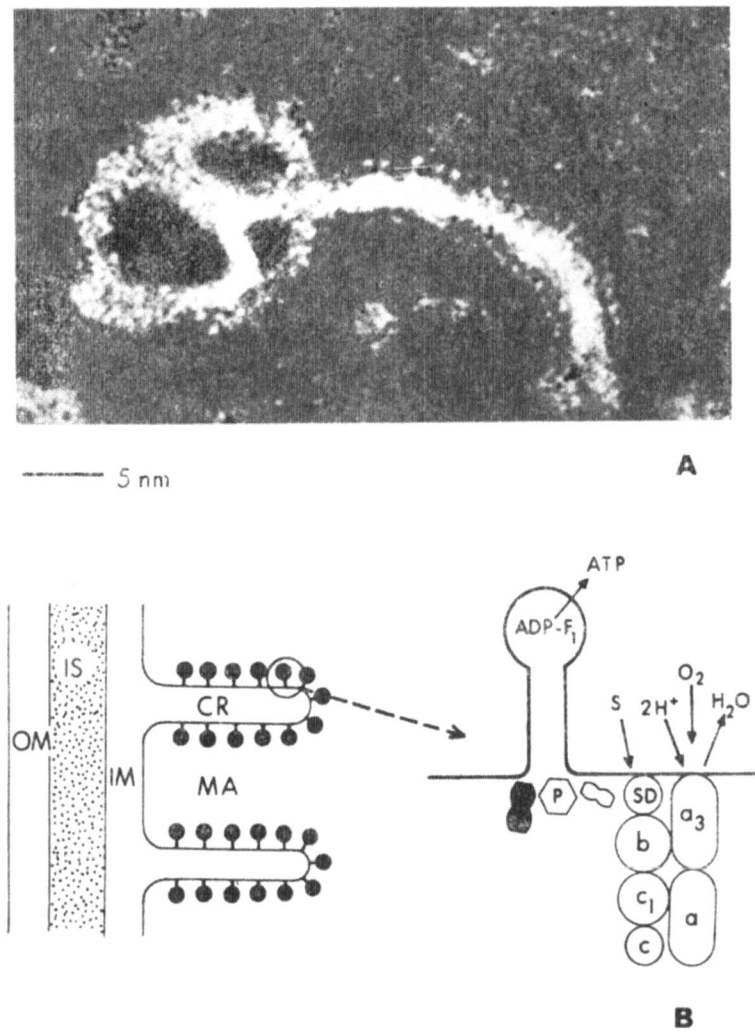

Fig. 2.48. Details of intramitochondrial architecture. **A**. A negatively stained part of the inner mitochondrial membrane of rat liver, showing the typical knobs. (Reproduced with permission from Ernster, L. and Kuylenstierna, B. (1970). In: *Membranes of Mitochondria and Chloroplasts* (ed. by Racker, E.), p. 172. Van Nostrand—Reinhold, New York.) **B**. Schematic drawing of mitochondrial membranes and an enlarged section of the vicinity of a cristal knob, together with the postulated functional components. OM Outer mitochondrial membrane, IS intermembrane space, IM inner mitochondrial membrane, CR crista, MA matrix, F_1 one of the coupling factors, P phosphate, S succinate, SD succinate dehydrogenase, b, c_1, c, a and a_3 various cytochromes.

outer membrane and the endoplasmic reticulum. The inner membrane, on the other hand, resembles very much a bacterial cytoplasmic membrane. It is the inner membrane that has been investigated by some of the best minds in the fields of biochemistry and membrane research, the most provocative feature being its ability to carry out oxidative phosphorylation, the detailed mechanisms of which is still a matter of dispute.

Three more or less antagonistic hypotheses exist, the chemical one, mainly due to Slater (e.g., 1958) and Racker (e.g., 1970), the chemiosmotic one due to Mitchell (e.g., 1966) and the electromechanochemical one due to Green and Ji (1972). The first of these envisages the function of various high-energy intermediates on the way from the oxidative chain to ATP, the second ascribes the mediating role to the translocation of protons across the inner membrane, the proton-motive force thus created generating ATP from ADP and inorganic phosphate. The third model endeavors probably most successfully to combine the merits of several previous hypotheses by introducing several configurations of the basepiece of the mitochondrial inner membrane. For thorough, healthily biased, reviews, the reader is referred to Greville (1969), Mitchell (1973) and Green (1974).

The various factors active in oxidation and oxidative phosphorylation have apparently a definite localization in the inner membrane, the latest information being summarized in Fig. 2.48.

The inner membrane is also the locale of various specific transport systems for anions, cations, organic acids (particularly those of the Krebs cycle), and nucleotides (see, e.g., Pressman, 1970).

2.7.3. Chloroplast

Like mitochondria, chloroplasts are cell organelles of considerable autonomy, containing ribosomes and a complete protein-synthesizing apparatus, starting from DNA to tRNA's. Like mitochondria, chloroplasts are suspected by some authors to represent early bacterial invaders of primitive plants with no photosynthetic apparatus.

Chloroplasts are enclosed in an outer envelope, the outer chloroplast membrane and are formed from lamellar vesicles (thylakoids) stacked in various ways, from very simple superposition in blue-green and red algae to irregular arrays several lamellae thick in green and brown algae to a more elaborate architecture in most green algae and higher plants (Fig. 2.49) where grana are seen at places where the lamellae overlap and partly fuse.

The obvious major function of chloroplast lamellae is to provide for conversion of radiation energy to chemical-bond energy, the key component of this photosynthetic process being chlorophyll of one kind or another. The postulated embedding of chlorophyll molecules

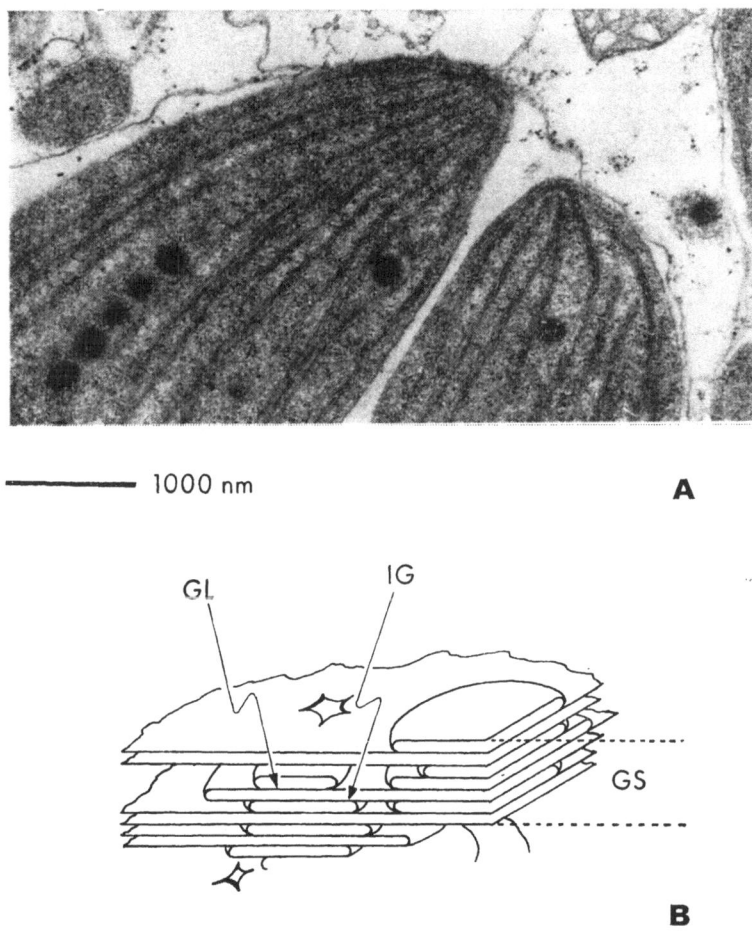

1000 nm

A

GL   IG

GS

B

Fig. 2.49. The chloroplast. **A**. An ultrathin section through a cell of the green alga *Vaucheria* with strand-like chloroplasts. (Courtesy of Dr. S. Janda and Dr. V. Pokorný, Institute of Microbiology, ČSAV, Prague.) **B**. A schematic representation of the lamellar structure of a well developed chloroplast. GL Granum lamella, IG intergranum lamella, GS stack of grana.

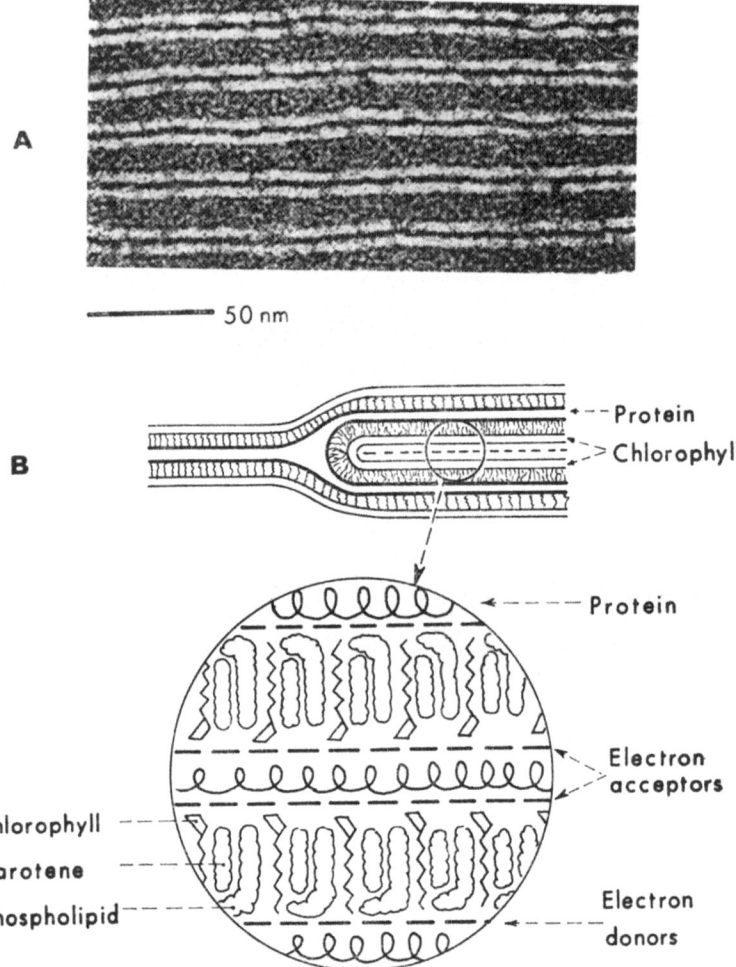

A

━━━━━━━━ 50 nm

B

Protein

Chlorophyll

Protein

Electron
acceptors

Chlorophyll

Carotene

Phospholipid

Electron
donors

Fig. 2.50. Details of intrachloroplast architecture. **A.** An ultrathin section of a spinach chloroplast granum, stained with potassium phosphotungstate. **B.** A schematic cross section through the thylakoid membrane and an enlarged view of its part. The thylakoid here is a part of the granum stack shown in Fig. 2.49.

in the lamellar structure is shown in Fig. 2.50, together with a high magnification of a chloroplast lamella.

The chlorophyll is almost exclusively associated with galactosyl-diglyceride which has a characteristic spectrum of fatty acyl residues which is taxonomically defined; in photosynthetic bacteria it contains almost exclusively 16 : 0 and 16 : 1 acids, in blue-green algae also

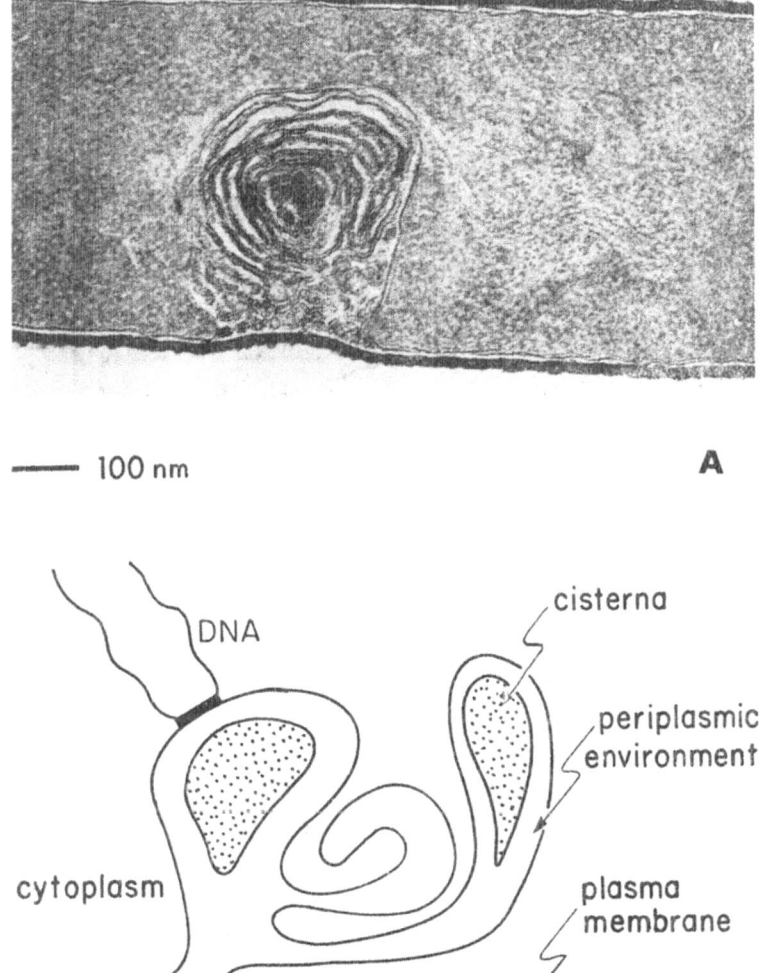

100 nm   **A**

DNA

cisterna

periplasmic environment

cytoplasm

plasma membrane

cell wall   **B**

Fig. 2.51. **A**. A well developed mesosome in *Bacillus subtilis*. (Reproduced with permission from Weibull, C. (1968). In: *Microbial Protoplasts, Spheroplasts and L-Forms* (ed. by Guze, L. B.), p. 62. The Williams and Wilkins Co., Baltimore.) **B**. A schematic drawing of the mesosome showing its interconnection with the plasma membrane.

a great proportion of 18 : 2 and 18 : 3 fatty acids while in algae it has major amounts of higher unsaturated acids, such as 20 : 4, 20 : 5 and 18 : 4. In higher plants, the practically sole component of this glycolipid is the 18 : 3 acid.

The thylakoid membrane apparently possesses certain transport functions that have been investigated using cations and anions but very little is known about the translocation of molecules into and out of chloroplasts across the outer membrane. The outer membrane, like that of mitochondria, is apparently permeable to small as well as larger molecules. In contrast with the mitochondrial inner membrane, the thylakoid membrane carries a potential which is positive inside.

2.7.4. Mesosome

A number of bacteria, both Gram-negative (*Spirillum serpens*, *Escherichia coli*) but mainly Gram-positive (*Bacillus subtilis, B. licheniformis, B. medusa, B. megaterium, B. cereus, B. thuringiensis, Micrococcus lysodeikticus, Mycobacterium phlei, M. smegmatis, Lactobacillus casei, L. corinoides, Staphylococcus aureus, Streptococcus faecalis*) have been found to contain membranous structures, apparently derived from the cytoplasmic (plasma) membrane, of characteristic appearance (Fig. 2.51) that are called mesosomes.

Of the many postulated functions (energy metabolism, cell division, cell-wall synthesis, secretion of enzymes, site of episome attachment to membrane) none has been unequivocally confirmed. It appears most likely that the mesosomes are the site of DNA penetration into bacterial cells (*cf.* p. 320) (Reusch and Burger, 1973).

2.7.5. Endoplasmic reticulum

A common feature of the internal architecture of practically all eukaryotic cells is a more or less extensive network of membrane elements studded with particles now known to be ribosomes (Fig. 2.52). The primary function of the ribosomes is the synthesis of proteins but

———————————————————————————▶

Fig. 2.52. Endoplasmic reticulum. **A.** An ultrathin section through the tubular and cisternal elements of rough endoplasmic reticulum studded with ribosomes,

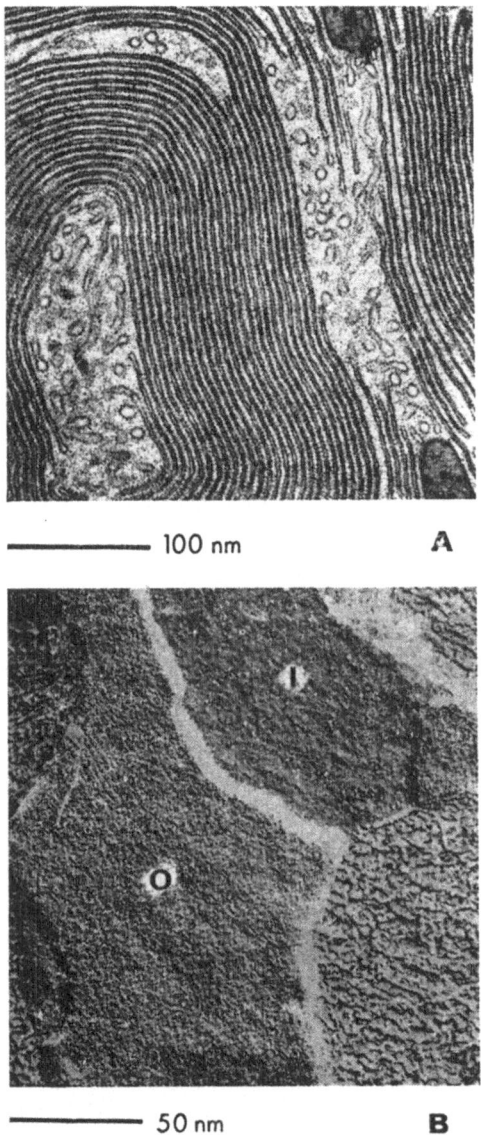

———— 100 nm   **A**

———— 50 nm   **B**

from a bat pancreas acinar cell. (Taken with permission from Fawcett, D. W. (1964). In: *Intracellular Membraneous Structure* (ed. by Seno, S. and Cowdry, E. V.), p. 15. Japan Soc. Cell Biol. Okayama.) **B**. A freeze-etched preparation of the yeast *Wickerhamia fluorescens* showing the outer surface (O) of the endoplasmic reticulum studded with 5 nm particles and the inner surface (I) of the endoplasmic reticulum with few 10 nm units. (Reproduced with permission from Bauer, H. (1970). *Can. J. Microbiol.* **16**, 219.)

this process, once the ribosomes carry the information-containing messenger RNA, probably takes place on ribosomal aggregates, the so-called polysomes whose association with the reticular membrane is not clear.

In a cell homogenate, the endoplasmic reticulum forms a fraction of "rough" microsomes which is separated from the general microsomal fraction by equilibrium density gradient centrifugation or selective aggregation in 15 mM CsCl and rate zonal sedimentation in a continuous sucrose gradient.

The membranes of the endoplasmic reticulum carry a number of enzymes, such as NADPH-cytochrome reductase, cytochrome P_{450}, cytochrome b_5, glucose-6-phosphatase, a mixed function oxidase and various esterases.

It is in the endoplasmic reticulum that a hypothetical unit, the membron, has been postulated to act in the regulation and stabilization of protein synthesis (e.g., Pitot et al., 1969).

The transport functions of endoplasmic reticulum (sometimes called the ergastoplasm) remain unexplored but it is known that even the large molecules of proteins synthesized on the ribosomes appear inside the lumen of the cisternae formed by the reticulum.

The endoplasmic reticulum is probably directly attached at places to the nuclear membrane. At the same time, it is probably functionally related to the Golgi apparatus, particularly in secretory cells.

2.7.6. Golgi apparatus

For some time, the Golgi apparatus has been included in the microsomal fraction as the "smooth" microsomes. However, it is apparently a separate system of membrane elements or cisternae (Fig. 2.53). The membranes of the apparatus or the dictyosomes are difficult to isolate without breakdown to small vesicles. The purest fraction is obtained at $1.05 - 1.09$ g cm^{-3} from a pellet sedimented at 10^6 g min after previous separation of a $2.6 . 10^5$ g min sediment.

The function of the apparatus is to provide a vehicle for transport of large proteins but also the site of synthesis of carbohydrates and lipoproteins. The proteins synthesized on the ribosomes of the endoplasmic reticulum are translocated inward and then released in the form of vesicles pinched off toward the Golgi apparatus. There they fuse with some of the Golgi membranes, may be further supplemented

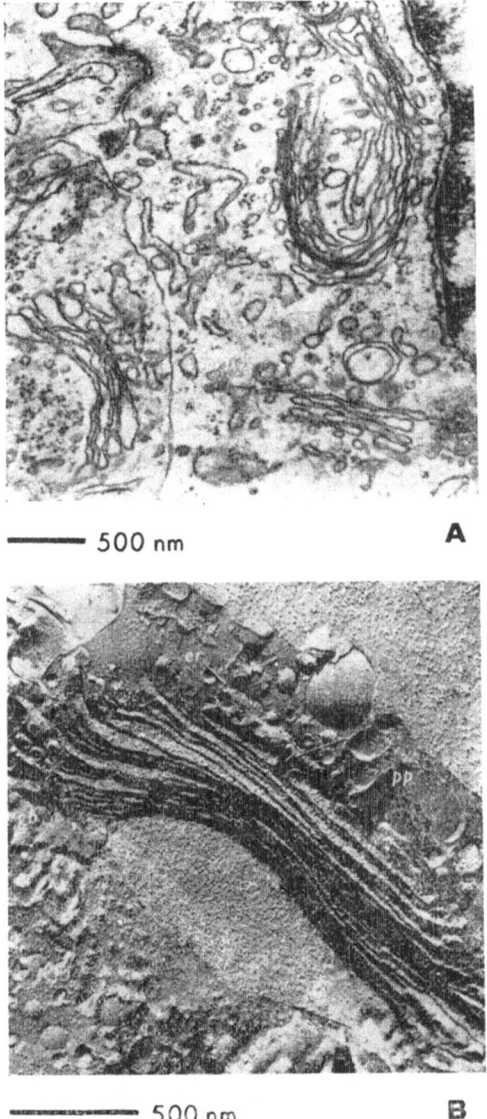

500 nm   **A**

500 nm   **B**

Fig. 2.53. The Golgi apparatus. **A**. An ultrathin section through a mouse tumor cell with several Golgi zones. A part of the nuclear membrane is also visible at right (Courtesy of Dr. J. Ludvik, Institute of Microbiology, ČSAV, Prague.) **B**. A freeze-etched Golgi apparatus (dictyosome) with individual cisternae seen in cross section (toward upper right) and in surface view (toward center and lower left). The membranes are associated with the endoplasmic reticulum at *pp*. (Reproduced with permission from Staehelin, L. A. and Kiermayer, O. (1970). *J. Cell Sci.* **7**, 787.)

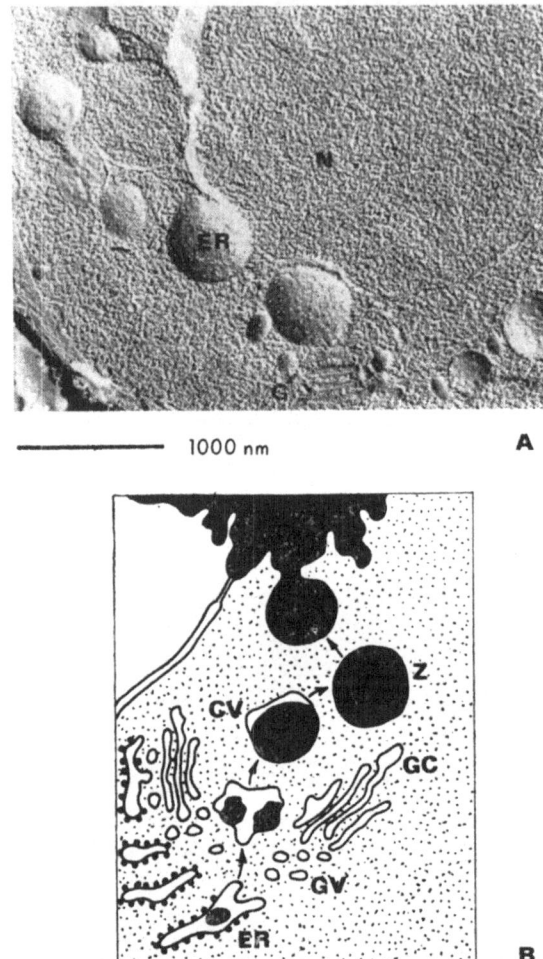

1000 nm   **A**

B

Fig. 2.54. Association of endoplasmic reticulum, Golgi lamellae and the exocrine process. **A**. A freeze-etched preparation of *Saccharomyces pombé* showing cisternae of endoplasmic reticulum (**ER**) associated with the nuclear membrane and moving toward the Golgi zone (**G**). (Courtesy of Dr. E. Streiblová, Institute of Microbiology, ČSAV, Prague.) **B**. A diagrammatic representation of the pancreatic exocrine cell function. Cisternae of the endoplasmic reticulum (**ER**) with and without ribosomes merge into the smooth vesicles of the Golgi peripheral region (**GV**) which arise from the Golgi cisternae proper (**GC**) to form the condensing vacuoles (**CV**) and finally the zymogen granules (**Z**) which are excreted into the lumen (black at top). (Compare with Fig. 7.3.) (Adapted from Jamieson, J. D. (1972.) In: *Current Topics in Membranes and Transport* (ed. by Bronner, F. and Kleinzeller, A.), vol. **3**, p. 273. Academic Press, New York—London.)

with carbohydrate moieties and stored in storage granules. These can then be expelled from cells by a process of reverse pinocytosis (p. 319). The train of events is shown schematically in Fig. 2.54.

The Golgi membranes are sometimes credited with the role of synthesizing new membrane elements.

2.7.7. Lysosome

Lysosomes are roughly spherical structure-less particles, about the size of a small mitochondrion, enclosed in a trilaminar membrane which may be variously corrugated (Fig. 2.55). Their apparent function is to wall off hydrolytic, digestive, enzymes. They may represent pinocytotic vacuoles but, more likely, they are special digestive

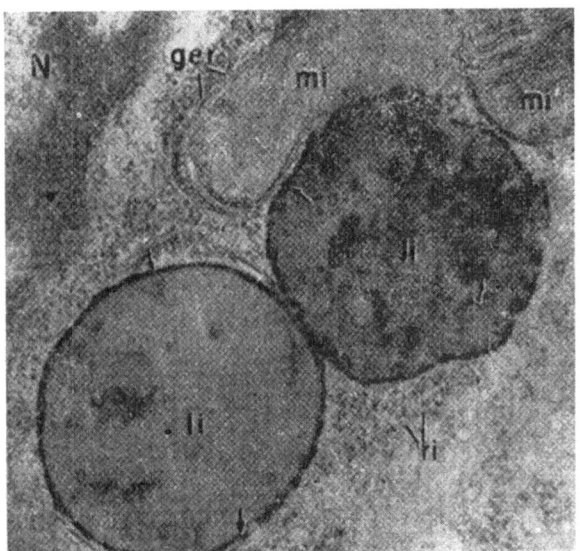

200 nm

Fig. 2.55. A thin section of a proximal convoluted tubule cell of mouse kidney, after injection of crystalline bovine hemoglobin. The most prominent features are the lysosomes (li) showing a positive acid phosphatase reaction on their surface (arrows). Other organelles to be seen include the nucleus (N), mitochondria (mi) and ribosomes (ri) as well as the rough endoplasmic reticulum (ger). (Reproduced with permission from De Robertis, E. D. P., Nowinski, W. N., and Saez, F. A. (1965). *Cell Biology*. W. B. Saunders Co., Philadelphia—London.)

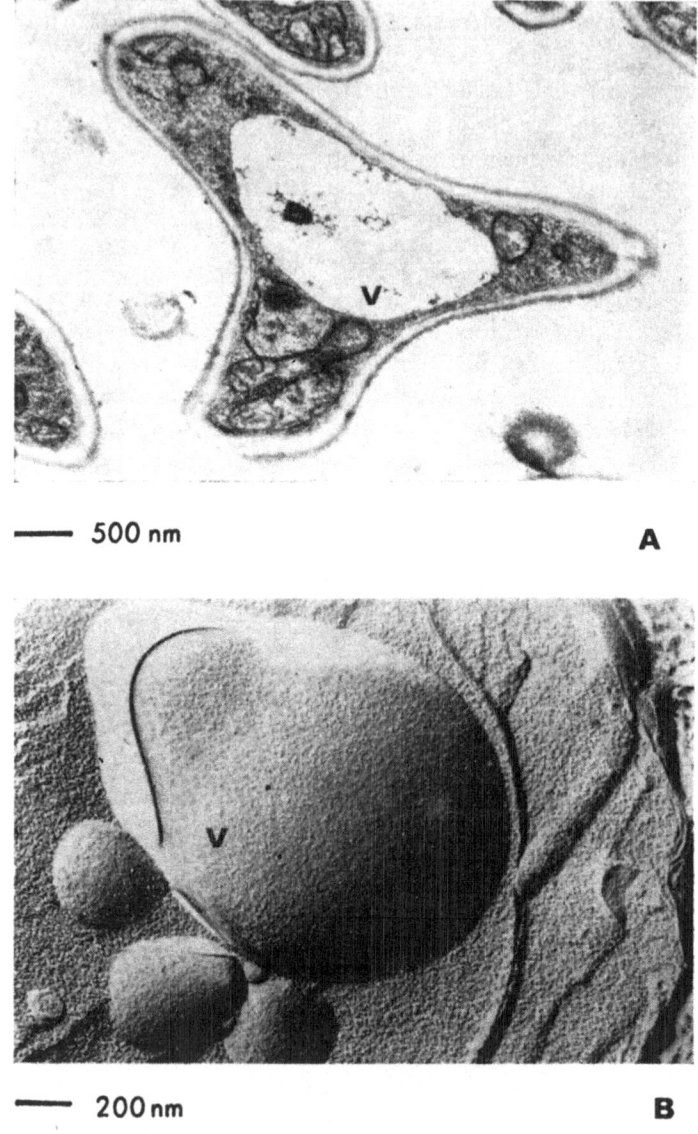

―― 500 nm   **A**

―― 200 nm   **B**

Fig. 2.56. The vacuole. **A**. A thin section through the triangular form of the dimorphic yeast *Trigonopsis variabilis* showing the vacuole (**V**) with various inclusions. (Courtesy of Dr. V. Šnejdar, Institute of Microbiology, ČSAV, Prague.) **B**. A freeze-etched preparation of the vacuole (**V**) of *Saccharomyces cerevisiae* with typical surface particles. (Courtesy of Dr. E. Streiblová, Institute of Microbiology, ČSAV, Prague.)

organelles of the cell. In damaged or dying cells they lyze and release their sanitary enzymes into the cell cytoplasm.

2.7.8. Tonoplast

Many fungal and plant cells contain large transparent structureless inclusions called vacuoles which are surrounded by a typical membrane called the tonoplast, shown in surface view in Fig. 2.56. Vacuoles can be isolated from lyzed protoplasts in the presence of 8 % Ficoll when they float on top (Matile and Wiemken, 1967). Like the lysosome of animal cells, the vacuole contains various hydrolytic enzymes (two proteases, an esterase, ribonuclease, aminopeptidase) as well as lipoamide dehydrogenase (NADH). It is, however, doubtful that these enzymes are associated with the tonoplast. The vacuole also contains metaphosphate deposits and is the principal storage pool for some amino acids (arginine, ornithine) in yeasts (Wiemken and Nurse, 1973).

If the vacuole has the lysosome function it is probably involved in protein degradation and thus in the regulation of protein synthesis but it is not known how proteins cross the tonoplast to come into contact with the proteolytic enzymes.

The transport properties of the tonoplast still require elucidation but this membrane might offer a simple model for transport studies.

2.7.9. Nucleus

The nuclear membrane is a typical structure found in all eukaryotic cells. It is actually a double membrane, the outer one being in close association with endoplasmic reticulum and covered with ribosomes. These two membranes form a continuum through large-size pores (Fig. 2.57) often plugged with electron-dense "blebs". In spite of these pores, the nuclei of cells can be isolated as such, sedimenting readily at 10^4 g min.

The two membranes are difficult to separate and usually collapse into small fragments at a density of about $1.18 - 1.22$ g cm^{-3}.

Very little is known about the enzyme equipment or permeability of nuclear membranes although it is readily appreciated that there must be channels permitting the passage of ribonucleic

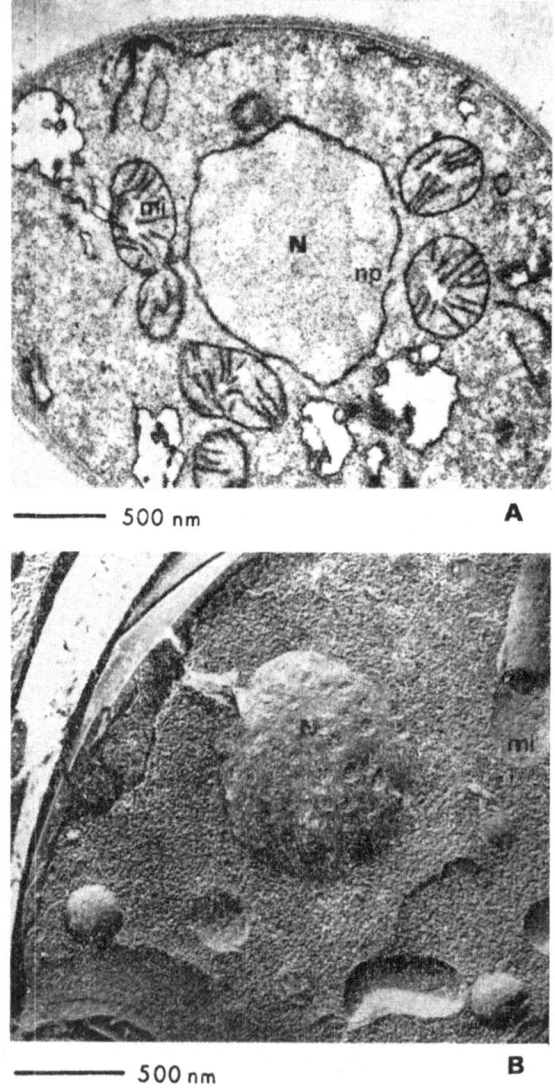

Fig. 2.57. The nucleus. **A**. A thin section through a cell of the yeast *Rhodotorula glutinis*, fixed with $KMnO_4$, showing a prominent nucleus (**N**) with several pores in the membrane (np) and several mitochondria (mi). (Courtesy of Dr. V. Šnejdar, Institute of Microbiology, ČSAV, Prague.) **B**. A freeze-etched preparation of the yeast *Saccharomyces cerevisiae* showing the nucleus (**N**) with numerous large pores. Several mitochondria are also visible (mi). (Courtesy of Dr. E. Streiblová, Institute of Microbiology, ČSAV, Prague.)

acids as well as proteins from and into the nucleus. There are some electron-microscopic indications that the transport may take place through the pores when the plugs are removed or dissolved after a suitable signal.

2.7.10. Other membranes

There is a variety of membranous organelles specific either for certain cells or certain organisms (the catalase-containing vesicular peroxisomes, the synaptic vesicles, hydrogenosomes etc.), some of their membranes possessing an informative architecture. A much studied example is the membrane of the retinal rod outer segment which is instrumental in the visual process. It is a regularly arranged membrane of globules about 4 nm in diameter, apparently the protein complexes including rhodopsin, retinal, vitamin A and the appropriate enzymes (Fig. 2.58).

SYNOPSIS

Biological membranes are recognized as trilaminar structures after negative staining in the electron microscope. They are the basic architectural framework of almost all cells and organelles (plasma membrane, nuclear membrane, mitochondria, chloroplasts, Golgi apparatus, endoplasmic reticulum, tonoplasts, lysosomes, mesosomes, etc.).

They contain mainly lipids and proteins with admixtures of carbohydrates. The lipids are either glycerol-based or sphingosine-based, phospholipids and glycolipids, plus sterols and minor lipids. Their fatty acid content is variable, the highest amount being found with palmitic and oleic acids. Membrane proteins are of two types, integral and peripheral, the integral ones being slightly more hydrophobic than the others and containing no cysteine.

The lipids are organized mainly in a bimolecular film with polar heads pointing outward, interspersed with isolated or clustered proteins, immersed in or spanning the entire thickness of the membrane. The lipids are present in several physical states, at biological temperatures in a liquid or liquid crystal form, below 12−18 °C

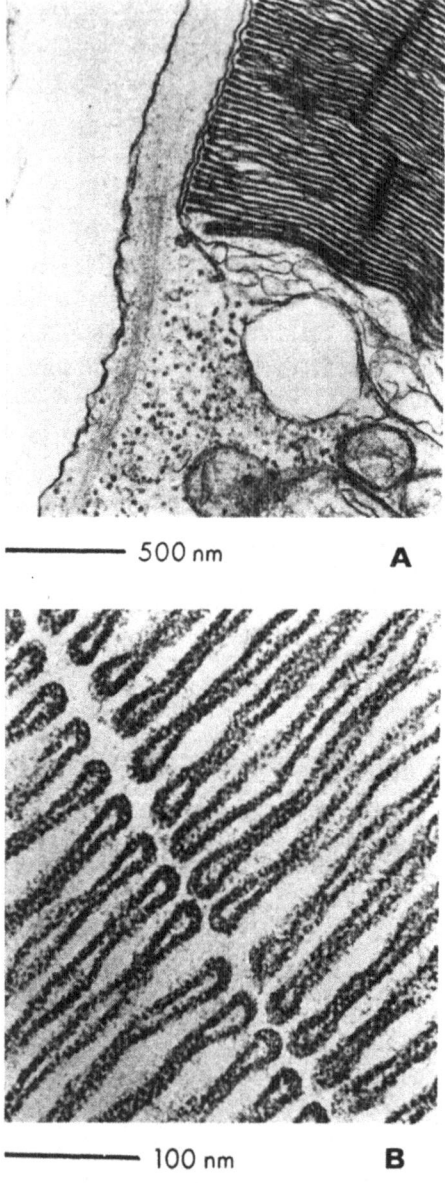

———— 500 nm   **A**

———— 100 nm   **B**

Fig. 2.58.   Retinal rod outer membranes. **A**. Part of a retinal rod of frog, stained with osmium tetroxide. The lamellae are stacked at upper right. **B**. A greater magnification of the same object showing the granular structure of the lamellae. (Taken with permission from Robertson, J. D. (1964). In: *Intracellular Membraneous Structure* (ed. by Seno, S. and Cowdry, E. V.), p. 379. Japan Soc. Cell Biol., Okayama.)

in a crystalline, less mobile state. There is considerable lateral diffusion of all membrane components.

The membranes are put together by more or less random insertion of presynthesized proteins and lipids but a distinct functional pattern is rigorously maintained.

The functions of membranes, besides transporting molecules and ions, are varied: They act as receptors for signals, antigenic and hormonal, they are the site of many enzyme activities, they aid in the synthesis of external wall components, they serve as mechanical support and as electrical insulators.

3. THERMODYNAMICS OF TRANSPORT

"For Nature is very consonant and comformable to her self."
Isaac Newton, Opticks

Since thermodynamics deals with properties of matter and laws which can be understood without knowledge of the inner structure of matter, it is of great utility in the exploration of unknown physicochemical mechanisms of transport phenomena. Although thermodynamics itself does not describe the microscopic details of transport mechanisms, it shows which of the various hypothetical mechanisms are thermodynamically permissible. An admirable "Introduction to Thermodynamics" was written by Spanner (1964) and is recommended in this connection. Only a brief summary of the most important implications of thermodynamics in the field of transport is presented here.

3.1. THERMODYNAMIC EQUILIBRIUM, PASSIVE AND ACTIVE TRANSPORT PROCESSES

Thermodynamic equilibrium of a system is characterized by the absence of spontaneous processes. Since work can be performed only by a system which tends to a spontaneous change, an equilibrium system is one that is not capable of doing work.

The typical system encountered in membrane transport is both isothermal and isobaric. The constancy of the temperature is ensured by high thermal conductivity of the surrounding tissues or media and the constancy of pressure by the ambient atmosphere. The ability of an isothermal and isobaric system to perform work over the inevitable volume work at constant pressure is expressed by its Gibbs free energy (free enthalpy according to older nomenclature and used especially in German literature). In an equilibrium no changes of the Gibbs free energy occur. It may be shown (see, e.g., Edsall and Wyman, 1958, p. 161 – 162) that the same criterion of equilibrium holds also when the system consists of two or more compartments, all at the same temperature, but each at a different constant pressure. Such situations are encountered when studying transport phenomena in microorganisms or plant cells with rigid and elastic cell walls.

Compartments separated by a membrane represent open systems and their Gibbs free energy can be changed by transport of substances across the membrane. When the transfer of a substance across the membrane brings about a decrease in the Gibbs free energy of the system, the process proceeds spontaneously. Such transport processes are termed "passive" in order to distinguish them from those which can, as a result of a decrease of free energy in the course of coupled chemical reactions or in coupled downhill transport, proceed uphill and are called "active" and "secondary active", respectively.

The change in the Gibbs free energy, G, of a system, resulting from an inflow of dn_j moles of substance j may be written as

$$dG = \left(\frac{\partial G}{\partial n_j}\right)_{T,p} dn_j \tag{3.1}$$

where the partial derivative is the partial molal Gibbs free energy of substance j at constant temperature and pressure and is called the chemical potential of the substance j, μ_j. There are other possible definitions of the chemical potential: it is also equal to the partial molal Helmholtz free energy at a constant temperature and volume, or to the partial molal internal energy at a constant volume and entropy, or to the partial molal heat content or enthalpy at a constant pressure and entropy. The latter two definitions, however, are less convenient; there are no obvious means of controlling the constancy of entropy in an experiment. With charged substances, ions, the term electrochemical potential with the symbol $\tilde{\mu}_j$ is used for the partial molal free energy.

Let us assume that a membrane separates two solutions of substance j and that its chemical potential in the outer solution is μ_{j_o} and that in the inner one μ_{ji}, Transfer of dn_j moles of substance j from the inner to the outer compartment involves a change in the Gibbs free energy of the whole system equal to

$$dG = (\mu_{j0} - \mu_{ji})\, dn_j \qquad (3.2)$$

and the process is spontaneous when the free energy change is negative, i.e., when $\mu_{j_o} < \mu_{ji}$. The spontaneous (i.e., passive) transport comes to a stop and the substance is in thermodynamic equilibrium when

$$\mu_{j_o} = \mu_{ji}. \qquad (3.3)$$

The chemical potential of a nonelectrolyte in a dilute solution may be conveniently written as

$$\mu_j = \mu_{j0} + RT \ln c_j \qquad (3.4)$$

where μ_{j0} is the standard partial molal free energy of substance j and for a given solute and solvent it may be considered as a constant, independent of concentration; c_j is the molar concentration of substance j in the solution. From eq. (3.3) and (3.4) it then follows that in thermodynamic equilibrium

$$c_{ji} = c_{jo} \qquad (3.5)$$

i.e., the concentration of a nonelectrolyte is to be the same in the two solutions. We shall see later that with ions an additional electrical term has to be included in the expression for electrochemical potential and that when transport of the abundant species (the solvent) is described, the chemical potential cannot be expressed by its concentration and the hydrostatic pressure term cannot be neglected; slightly more complex conditions of equilibria are then obtained (see the chapters on equilibria of ions and transport of water).

Equation (3.2) is of assistance when the energy requirement for active transport processes is evaluated; when the transport proceeds "uphill", from a lower to a higher chemical potential, the Gibbs free energy of the system would increase at the rate

$$\frac{dG}{dt} = (\mu_{ji} - \mu_{jo})\frac{dn_j}{dt} \qquad (3.6)$$

or, per unit membrane area,

$$\frac{1}{A} \frac{dG}{dt} = (\mu_{ji} - \mu_{jo}) J_j \qquad (3.7)$$

where $\mu_{ji} > \mu_{j0}$ and

$$J_j = \frac{1}{A} \frac{dn_j}{dt}. \qquad (3.8)$$

This is the flux of substance j, the amount in moles transferred per unit area in a unit of time. This increase in the free energy must be compensated for by an at least equal decrease in the free energy in the energy-providing chemical reactions in or at the membrane or in a coupled downhill transport of another substance. However, it is not a simple matter to evaluate experimentally the quantities appearing in eq. (3.8). When the nature of the metabolic energy-providing reactions is known – most commonly it is the splitting of ATP – the energy available may be computed with a reasonable accuracy, especially in energy-depleted cells. It may be more difficult to evaluate the actual intracellular concentration and especially to estimate the net flux through the pumping mechanism J_j, not to be mistaken with the unidirectional flux of the substance measurable with tracers (*cf.* p. 203). Pumping mechanisms are, in general, reversible; thus the reversibility of the sodium pump is obvious not only from the possibility of ATP synthesis on its reversal (Glynn and Lew, 1970) but also from the existence of a critical energy barrier which it cannot overcome (Conway and Mullaney, 1961). For this reason there is always a back flux through the mechanism and each pump, in addition to the uphill transport, performs an exchange diffusion which does not require a supply of free energy.

3.2. THERMODYNAMICS OF THE STEADY STATE

The entropy of an isolated system which exchanges neither matter nor energy increases or remains constant. On the other hand, in systems which exchange energy or both energy and matter with their surroundings, the changes of entropy may be divided into the entropy flow $d_e S$ and entropy production $d_i S$

$$dS = d_e S + d_i S \qquad (3.9)$$

The production of entropy in every macroscopic system is positive or nil

$$d_i S \geqq 0 \qquad (3.10)$$

whereas the entropy flow and hence also the entropy change dS can be either negative or positive. The production of entropy is a result of irreversible processes in the system — irreversible in the sense that their mathematical description is not invariant with respect to the sign of the time variable and that their results are never completely reversed. Even when the system itself is brought to the original state, there are necessarily changes left in other systems. Diffusion and heat conduction are typical examples of such irreversible processes.

The course of irreversible processes in the steady state is described by the phenomenological theory (i.e., a theory which does not take into account the inner structure of matter), called steady-state thermo-dynamics or thermodynamics of irreversible processes. Special forms of phenomenological relations of this thermodynamics have been known in the whole realm of physics for a long time: they describe a generalized flux J as being directly proportional to a generalized force X. Thus an electrical current is proportional to the gradient of electrical potential (Ohm's law), heat flow to the temperature gradient (Fourier's law) and diffusional flow to the concentration gradient (Fick's law). The principal aim of steady-state thermodynamics is to describe interactions between individual flows and to this end it makes a general assumption that any generalized flow J_i is proportional not only to its conjugate force X_i, but also in a higher or lesser degree to each of other generalized forces $X_j (j \neq i)$ operating in the system. Thus, e.g., temperature gradient gives rise not only to heat flow but also to electric current by the thermoelectric effect and to diffusional flow in the phenomenon of thermodiffusion. Hence the phenomeno-logical relations may be written as

$$
\begin{aligned}
J_1 &= L_{11}X_1 + L_{12}X_2 + \ldots + L_{1n}X_n \\
J_2 &= L_{21}X_1 + L_{22}X_2 + \ldots + L_{2n}X_n \qquad (3.11) \\
&\ldots\ldots\ldots\ldots\ldots\ldots\ldots\ldots\ldots\ldots \\
J_n &= L_{n1}X_1 + L_{n2}X_2 + \ldots + L_{nn}X_n
\end{aligned}
$$

where L_{ii} are straight coefficients relating the flows to their conjugate forces and $L_{ij} (i \neq j)$ are cross coefficients relating them to noncon-jugate forces. Phenomena which cannot be at least approximately described by linear relations are not considered by steady-state thermodynamics. More conscisely the phenomenological relations may be written as

$$J_i = \sum_{j=1}^{n} L_{ij}X_j \qquad (i = 1, ..., n). \qquad (3.12)$$

A proper choice of flows and forces requires the rate of entropy production in the system to be equal to the sum of flows multiplied by their conjugate forces

$$\frac{d_i S}{dt} = \sum_{i=1}^{n} J_i X_i > 0. \qquad (3.13)$$

In isothermal system such a choice of flows and forces is preferable as results in the sum of products of flows with the conjugate forces giving the so-called dissipation function which is the rate of entropy production multiplied by absolute temperature

$$T\frac{d_i S}{dt} = \sum_{i=1}^{n} J_i X_i \qquad (3.14)$$

where J_i's and X_i's now represent another set of flows.

When irreversible processes in a system are satisfactorily described by linear relations (3.11) and flows and forces are chosen in a way to satisfy eq. (3.13) or (3.14), Onsager's law

$$L_{ij} = L_{ji} \qquad (3.15)$$

is valid for the cross coefficients of the linear relations. According to Onsager's law the matrix of phenomenological coefficients is symmetrical, being unchanged when its columns are replaced by its rows. The necessary conditions of the validity of Onsager's law are usually satisfied in the proximity of thermodynamic equilibrium. The practical value of Onsager's law is considerable; it allows to predict from the measurement of the dependence of flow J_i on force X_j the dependence of flow J_j on force X_i, which may be more difficult to measure. Onsager's law can be deduced mathematically from the principle of microscopic reversibility, stating that equilibrium on a molecular scale is not achieved by a cyclic process.

Following Denbigh (1951), the equivalence of Onsager's law with the principle of microscopic reversibility can be demonstrated for a simple system of three chemical reactions:

Let there be in a solution three tautomeric forms of some substance, denoted A, B and C, their concentrations being generally

a, b and c and in equilibrium $\bar{a}$, $\bar{b}$ and $\bar{c}$. The stability of the equilibrium concentrations $\bar{a}$, $\bar{b}$ and $\bar{c}$ could be imagined to result from a cyclic process

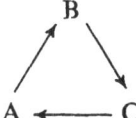

and no theorem of classical thermodynamics contradicts such scheme. The principle of microscopic reversibility, however, requires a detailed balancing of individual reactions

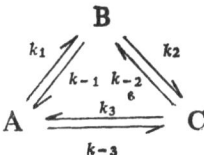

so that the following equations are valid

$$k_1 \bar{a} = k_{-1} \bar{b}, \qquad (3.16a)$$

$$k_2 \bar{b} = k_{-2} \bar{c}, \qquad (3.16b)$$

$$k_3 \bar{c} = k_{-3} \bar{a}. \qquad (3.16c)$$

Expressing now the reaction rates by phenomenological equations, it will be obvious that the validity of equations of detailed balancing (3.16) is equivalent to the validity of Onsager's law (3.15).

The general kinetic equations of the system are the following

$$\frac{da}{dt} = -(k_1 + k_{-3})a + k_{-1}b + k_3 c, \qquad (3.17a)$$

$$\frac{db}{dt} = k_1 a - (k_{-1} + k_2)b + k_{-2}c, \qquad (3.17b)$$

$$\frac{dc}{dt} = k_{-3}a + k_2 b - (k_3 + k_{-2})c. \qquad (3.17c)$$

Near equilibrium the actual concentrations deviate from the equilibrium ones only by small differences Δa, Δb and Δc, so that we can write

$$\Delta a = a - \bar{a} \qquad (3.18)$$

and the rate of change of Δa is equal to that of the actual concentration a, i.e., to the reaction velocity v_A:

$$\frac{d\,\Delta a}{dt} = \frac{da}{dt} = v_A. \tag{3.19}$$

Introducing a and da/dt from (3.18) and (3.19) into (3.17a) we obtain

$$v_A = -(k_1 + k_{-3})(\Delta a + \bar{a}) + k_{-1}(\Delta b + \bar{b}) + k_3(\Delta c + \bar{c}) =$$
$$= -(k_1 + k_{-3})\,\Delta a + k_{-1}\,\Delta b + k_3\,\Delta c - (k_{-1} + k_{-3})\,\bar{a} + k_{-1}\bar{b} + k_3\bar{c}.$$

The sum of the last three terms, however, is equal to $d\bar{a}/dt$ and hence to zero; the equilibrium concentration does not change in time. Thus

$$v_A = -(k_1 + k_{-3})\,\Delta a + k_{-1}\,\Delta b + k_3\,\Delta c. \tag{3.20}$$

Let now $\bar{\mu}_A$ be the equilibrium chemical potential of substance A. Assuming the solution to be ideal and dilute, so that the chemical potential is proportional to the logarithm of concentration of substance A, we can write

$$\Delta\mu_A = \mu_A - \bar{\mu}_A = RT \ln\frac{a}{\bar{a}} = RT \ln\left(1 + \frac{\Delta a}{\bar{a}}\right) \tag{3.21}$$

and when the deviation from the equilibrium concentration, Δa, is small

$$\Delta\mu_A = RT\,\frac{\Delta a}{\bar{a}} \tag{3.22}$$

so that

$$\Delta a = \frac{\Delta\mu_A}{RT}\,\bar{a}. \tag{3.23}$$

Introducing Δa from eq. (3.23) into eq. (3.20) and treating analogously the reaction velocities ve and v_B we obtain

$$v_A = -\frac{(k_1 + k_{-3})\,\bar{a}}{RT}\,\Delta\mu_A + \frac{k_{-1}\bar{b}}{RT}\,\Delta\mu_B + \frac{k_3\bar{c}}{RT}\,\Delta\mu_C, \tag{3.24a}$$

$$v_B = \frac{k_1\bar{a}}{RT}\,\Delta\mu_A - \frac{(k_{-1} + k_2)\,\bar{b}}{RT}\,\Delta\mu_B + \frac{k_{-2}\bar{c}}{RT}\,\Delta\mu_C, \tag{3.24b}$$

$$v_C = \frac{k_{-3}\bar{a}}{RT}\,\Delta\mu_A + \frac{k_2\bar{b}}{RT}\,\Delta\mu_B - \frac{(k_3 + k_{-2})\,\bar{c}}{RT}\,\Delta\mu_C \tag{3.24c}$$

which may be written

$$v_A = L_{AA} \Delta\mu_A + L_{AB} \Delta\mu_B + L_{AC} \Delta\mu_C, \qquad (3.25a)$$

$$v_B = L_{BA} \Delta\mu_A + L_{BB} \Delta\mu_B + L_{BC} \Delta\mu_C, \qquad (3.25b)$$

$$v_C = L_{CA} \Delta\mu_A + L_{CB} \Delta\mu_B + L_{CC} \Delta\mu_C \qquad (3.25c)$$

where $L_{AB} = \dfrac{k_{-1}\bar{b}}{RT}$, $L_{BA} = \dfrac{k_1\bar{a}}{RT}$, and so on.

We can immediately see that when the equation of detailed balancing (3.16a)

$$k_1\bar{a} = k_{-1}\bar{b}$$

is satisfied, necessarily

$$L_{AB} = L_{BA}$$

so that Onsager's law follows from the principle of microscopic reversibility.

Methods of steady-state thermodynamics will be encountered in this book when discussing the phenomenological theory of transport of water as developed by Kedem and Katchalsky (1958).

Here at least the rigorous definition of active transport given by this type of thermodynamics (Kedem, 1961) may be mentioned, viz. that the transport is active if there is a demonstrable interaction between the transmembrane flow of the substance and a metabolic reaction at the membrane, which may be expressed by a suitable nonzero cross coefficient. As a result of this coupling the active transport can and commonly does (but need not) proceed uphill. When a transport is not coupled to a metabolic reaction but rather to the passive backflow of an actively transported substance it can also proceed uphill and may be conveniently denoted by Stein's (1967) term "secondary active transport".

3.3. NETWORK THERMODYNAMICS

In the following limited space only little can be presented about the intricate field of network thermodynamics beyond the reference to a conscise explanation of the basic ideas by Kedem (1972) and a thorough treatment of the subject by Oster and co-workers (1973). Unlike steady-state thermodynamics, the network thermodynamics is capable of treating theoretically also phenomena which are nonstationary and nonlinear. The basic approach of network thermo-dynamics is reticulation: each continuous system is mentally divided into homogeneous subsystems and each subsystem subdivided into reversible parts storing energy without dissipation and irreversible parts dissipating energy without storage. The whole system is then represented by a topological graph if the system is simple and by the so-called bond graph, if the system is more complex and involves energy transduction. The graph is equivalent to a set of differential equations describing the system and may be transformed into them using a suitable algorithm and, moreover, shows how the parts of the system are interconnected, i.e., reveals its topology.

The basic state variables of network thermodynamics can be divided into two classes: *effort* of force variables, also called "across" variables, since they can be estimated by a two-point measurement, and *flow* variables which can be estimated by one-point measurement and hence are also called "through" variables. Typical flow variables are electrical current, volume flow, molar flow and reaction rate; the conjugate effort variables being voltage, pressure difference, chemical potential and chemical affinity, respectively. An effort variable multiplied by the conjugate flow variable gives the energy rate or power.

Although network graphs are not to be confused with electrical analogues, the same topological constraints are applicable to both. Thus the flow variables are subject to Kirchhoff's current laws (KCL), being conserved at each node of the network, and the effort variables obey Kirchhoff's voltage law (KVL), having at each node a unique value. For this reason the flow variables are also called KVL variables and the effort variables KVL variables.

Physical properties of the system are expressed by "constitutive relations" between individual variables. The constitutive relation between a flow variable and the conjugate effort variable is given by the resistance of the system. Thus the dissipative flow of a substance across a membrane, J_m (in moles per second per unit area), is related to the difference of the chemical potential of the substance, $\Delta\mu$, by

$$J_m = \frac{\Delta\mu}{R_m} \tag{3.26}$$

where R_m is the membrane resistance to flow. On the other hand, the constitutive relation between an integrated flow variable (electrical charge for electrical current, volume for volume flow, number of moles for molar flow and reaction advancement for reaction rate) and the effort variable is given by the capacitance of the subsystem. Thus, e.g., the differential or incremental capacitance of a membrane for a substance (all other variables being kept constant) is given by

$$C_m = \frac{dn_m}{d\mu_m} \tag{3.27}$$

where n_m is the number of moles of the substance per unit membrane area and μ_m the chemical potential of the substance in the membrane. The reversible flow of the substance, charging the membrane capacitance, J, is then given by

$$J = \frac{dn_m}{dt} = \frac{dn_m}{d\mu_m} \frac{d\mu_m}{dt} = C_m \frac{d\mu_m}{dt}. \tag{3.28}$$

If the membrane behaves as an ideal system, chemical potential may be expressed by $\mu_m = \text{const.} + RT \ln n_m$ and the capacitance is given by

$$C_m = \frac{n_m}{RT} = \frac{c_m V_m}{RT} \tag{3.29}$$

where c_m is the concentration of the substance in the membrane and V the volume of the membrane per unit area, i.e., membrane thickness.

In order to illustrate the representation of a simple system by topological graph we shall now consider the permeation of a nonelectrolyte across a homogeneous membrane by simple diffusion. In the topological graph in Fig. 3.1, drawn after Oster and co-workers (1973), J_m^1 and J_m^2 represent the dissipative flows of the nonelectrolyte into and out of the membrane, respectively. The constitutive relations involving these flows are

$$J_m^1 = \frac{\mu_1 - \mu_m}{R_1}, \tag{3.30}$$

$$J_m^2 = \frac{\mu_m - \mu_2}{R_2}$$

where

$$R_1 = R_2 = R_m/2 \tag{3.31}$$

the membrane being mentally subdivided into two dissipative elements and a capacitative element with capacitance C_m. There is a common reference potential μ_{ref} for the chemical potentials in the two solutions and in the membrane, the broken lines denoting the capacitative

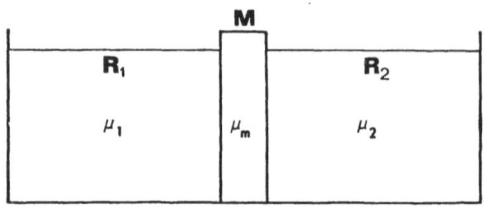

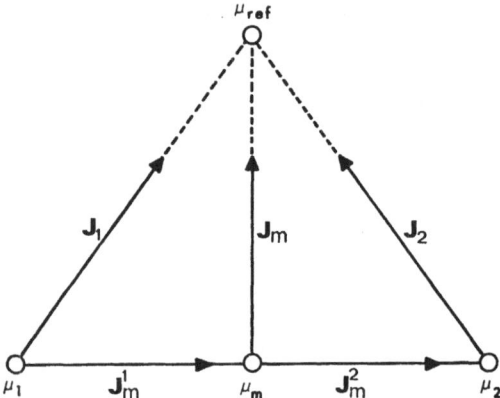

Fig. 3.1.   Nonelectrolyte permeation across a homogeneous membrane represented by a topological graph (adapted from Oster et al., 1973). R_1, R_2 Reservoirs; M membrane.

elements and expressing the fact that there is no actual flow into the reference node.

J_1 is the "reservoir flow" from the first reservoir, $J_1(\mu_1 - \mu_{ref})$ being the power delivered by the reservoir. An analogous relation is valid for J_2. J_m, on the other hand, is the rate of nonelectrolyte accumulation in the membrane. All these are reversible flows.

Kirchhoff's current law written for the central node representing the membrane yields

$$J_m = J_m^1 - J_m^2. \tag{3.32}$$

Expressing the reversible flow J_m by eq. (3.28) and introducing for the two dissipative flows the explicit expressions (3.30) we obtain

$$C_m \frac{d\mu_m}{dt} = \frac{\mu_1 - \mu_m}{R_1} - \frac{\mu_m - \mu_2}{R_2} = \frac{\mu_1}{R_1} + \frac{\mu_2}{R_2} - \mu_m \left(\frac{1}{R_1} + \frac{1}{R_2} \right) \tag{3.33}$$

and since $R_1 = R_2 = R_m/2$ (eq. 3.31)

$$\frac{d\mu_m}{dt} = 2\frac{\mu_1 + \mu_2}{R_m C_m} - \frac{4}{R_m C_m}\mu_m. \tag{3.34}$$

On introducing the relaxation time τ_m of charging or discharging the membrane with the permeant,

$$\tau_m = \frac{R_m C_m}{4} \tag{3.35}$$

eq. (3.34) may be written

$$\tau_m\frac{d\mu_m}{dt} = \frac{\mu_1 + \mu_2}{2} - \mu_m. \tag{3.36}$$

This is easily integrated to describe the time course of charging or discharging the membrane with the permeant. Thus, when the membrane initially free of the permeant is exposed at time zero to two reservoirs with the values of the permeant chemical potential equal to μ_1 and μ_2, respectively, the integration yields

$$\mu_m = \frac{\mu_1 + \mu_2}{2}\left(1 - e^{-\frac{t}{\tau_m}}\right). \tag{3.37}$$

The relaxation time for a simple membrane is easily interpreted in terms of membrane thickness Δx and the diffusion coefficient of the permeant in the membrane, D_m. Introduction of an expression for membrane resistance

$$R_m = RT\,\Delta x/c_m D_m \tag{3.38}$$

and of an expression for membrane capacitance (cf. eq. 3.29)

$$C_m = c_m\,\Delta x/RT \tag{3.39}$$

(where Δx is equal to the membrane volume per unit of its area) into eq. (3.35) yields (Oster et al., 1973):

$$\tau_m = \frac{R_m C_m}{4} = \frac{RT\,\Delta x}{4c_m D_m}\cdot\frac{c_m\,\Delta x}{RT} = \frac{(\Delta x/2)^2}{D_m}. \tag{3.40}$$

SYNOPSIS

The phenomenological and hence general laws of thermodynamics are of manifold assistance in the classification and description of transport phenomena. Criteria of equilibrium of classical thermodynamics allow to discriminate between two types of stationary distribution of substances across membranes: equilibrium distribution and steady-state distribution resulting from active transport. Further, classical thermodynamics makes it possible to evaluate the rate of energy supply required by active transport transferring a substance to a higher free-energy level. Thermodynamics of the steady state ("irreversible thermodynamics") is suited for description of interactions between various steady flows and, by Onsager's law, it reduces the necessary number of experimental measurements. Finally, network thermodynamics is capable of treating nonstationary and nonlinear phenomena. In network thermodynamics each system is reticulated into homogeneous subsystems and reversible and irreversible parts and represented by a graph. The graph is equivalent to a set of differential equations describing the system and displays, at the same time, the topology of the system.

4. TRANSPORT
OF NONELECTROLYTES

*"And therefore I scruple not to propose
the Principles of Motion above-men-
tion'd, they being of very general
Extent, and leave their Causes to be
found out."*

Isaac Newton, Opticks

Depending on their size, degree of hydrophobicity and specific structural features, molecules can cross biological membranes by three principally distinct mechanisms: (1) Nonspecific "diffusion" through the lipid and, in a limited degree, through the polar parts of the membrane; (2) specific "carrier" transport – this category includes, as a subgroup, systems where the transported substance is altered chemically; (3) mechanisms involving profound, even if transient, changes in membrane architecture, such as pinocytosis, penetration of biopolymers.

The first two categories will be dealt with in the present chapter, the third will be discussed separately later (chapter 7). Likewise, the movement of water, actually belonging to the first category, will be treated in the special chapter 6.

Although our primary concern here is with the movement across membranes, it will aid in understanding the more advanced kinetic treatment if we review briefly the features of diffusion in a continuous (homogeneous) phase.

4.1. PRINCIPLES OF DIFFUSION

Particles dissolved or suspended in liquids are subject to permanent thermal movement. In the absence of external fields, such as electrical field which influences considerably the movement of charged particles, or high gravitational fields obtainable in an ultracentrifuge, which raise the gravity of heavier particles to appreciable values, there is no preferential direction in particle movement. As a result of the laws of probability, more particles leave the volume element in which they are concentrated while a smaller number of particles flow during the same time into the element from its dilute surroundings. For this reason, concentration gradients in solutions tend to disappear with time and the process of equalizing of particle concentrations, called diffusion, is commonly observed.

The rate of diffusion follows an important phenomenological law, called Fick's first law. When diffusion proceeds or is observed in one direction only, it has the form

$$\frac{1}{A}\frac{\mathrm{d}n}{\mathrm{d}t} = -D\frac{\partial c}{\partial x}. \tag{4.1}$$

The left-hand side represents the number of particles diffusing in a unit of time per unit area normal to the direction of the x-axis, along which the diffusion proceeds; in transport studies this quantity is commonly called flux and is usually denoted by J. According to Fick's first law it is directly proportional, by the diffusion coefficient D, to the concentration gradient $\partial c/\partial x$, prevailing at a given time at given x. The minus sign expresses the fact that when the concentration gradient is a positive number, i.e. when concentration increases in the direction of the positive x-axis, the diffusion flow proceeds in the opposite direction.

It will be seen later (p. 193) that Fick's first law may serve as a convenient approximation when diffusion across a thin layer separating two mixed reservoirs is to be described. However, when describing diffusion proceeding across greater distances in continuous systems, the same equation contains four variables even in the simplest, one-dimensional case. For this purpose Fick's first law is transformed into a partial differential equation, called simply the diffusion equation or Fick's second law. In the one-dimensional case the procedure gives

$$\frac{\partial c}{\partial t} = D\frac{\partial^2 c}{\partial x^2} \tag{4.2}$$

so that the rate of concentration change is proportional to the second derivative of the concentration in space. Phenomenologically, Fick's first law may be considered as an approximation of a specific case of the general Teorell's (1953) formula for the flux of a substance:

$$\text{Flux} = \text{Mobility} \cdot \text{Concentration} \cdot \text{Total driving force} \qquad (4.3)$$

where the flux J is the amount of substance in moles which in unit time penetrates per unit area normal to the direction of the transport:

$$J = \frac{1}{A} \frac{dn}{dt}. \qquad (4.4)$$

In the absence of external fields the system tends to an equilibrium in which the chemical potential of solute is the same at each point of the available space (see the Thermodynamics of Transport, p. 169). Hence the suitable total driving force to be introduced into Teorell's equation to describe simple diffusion is the negative of the space gradient of the chemical potential; in the one-dimensional case equal to the partial derivate of the chemical potential μ with respect to x, $\partial\mu/\partial x$. Teorell's equation (4.3) then takes the form

$$\frac{1}{A} \frac{dn}{dt} = -Uc \frac{\partial\mu}{\partial x} \qquad (4.5)$$

where U is the mobility and c the concentration at x at time t. Using the expression for chemical potential of solute in a dilute solution of ideal behaviour

$$\mu = \mu_0 + RT \ln c \qquad (4.6)$$

eq. (4.5) may be written

$$\frac{1}{A} \frac{dn}{dt} = -UcRT \frac{\partial \ln c}{\partial x} \qquad (4.7)$$

and, since $d \ln y = dy/y$,

$$\frac{1}{A} \frac{dn}{dt} = -RTU \frac{\partial c}{\partial x}. \qquad (4.8)$$

Denoting the quantity RTU by D we obtain

$$\frac{1}{A} \frac{dn}{dt} = -D \frac{\partial c}{\partial x} \qquad (4.9)$$

which is Fick's first law, derived by its author as an analogy to Fourier's law, describing the flow of heat.

Fick's second law (the partial differential equation of diffusion) may be derived from Fick's first law as follows.

Let there be two infinitesimally distant parallel planes normal to the x-axis, one at x, another at $x + dx$. The concentration gradient being $\partial c/\partial x$ at the former plane, it is equal to

$$\frac{\partial c}{\partial x} + \frac{\partial(\partial c/\partial x)}{\partial x}\,dx = \frac{\partial c}{\partial x} + \frac{\partial^2 c}{\partial x^2}\,dx$$

at the latter. Hence also the diffusional fluxes across the two planes differ, being

$$\frac{1}{A}\frac{dn}{dt} = -D\frac{\partial c}{\partial x}$$

and

$$\frac{1}{A}\frac{dn'}{dt} = -D\left(\frac{\partial c}{\partial x} + \frac{\partial^2 c}{\partial x^2}\,dx\right)$$

respectively. Their difference is the rate of change of the number of moles of solute between the two planes per unit area:

$$\frac{1}{A}\frac{dn - dn'}{dt} = D\frac{\partial^2 c}{\partial x^2}\,dx \tag{4.10}$$

and, since the volume included between the planes per unit area is numerically equal to dx, the rate of concentration change between the planes is

$$\frac{\partial c}{\partial t} = \frac{1}{A\,dx}\frac{dn - dn'}{dt}. \tag{4.11}$$

Combining equations (4.10) and (4.11) we obtain

$$\frac{\partial c}{\partial t} = D\frac{\partial^2 c}{\partial x^2}$$

which is Fick's second law (eq. 4.2). When observable diffusional flow is not limited to one dimension of space, a more general form of the diffusion equation must be used:

$$\frac{\partial c}{\partial t} = D\nabla^2 c \tag{4.12}$$

where the symbol ∇^2 is the operator nabla squared (Laplacean) and takes a form according to the choice of space coordinates, which in their turn depend on the geometry of the system observed.

Solutions of the diffusion equation (4.2) or (4.12) are of interest for a biologist studying diffusion into or out of tissue samples of simple geometrical forms, provided that the process is not rate-limited by membranes (otherwise the problem of diffusion would be reduced to the problem of permeation, see p. 192). Problems of this kind may be encountered when diffusion in extracellular spaces is followed, or when the solute permeates across cell membranes extremely rapidly. Solution of eq. (4.2) for a plane sheet of thickness d, exposed at $t = 0$ on both sides to concentration c_0 of the solute of which the initial concentration inside the sheet is zero (so that $t = 0$, $0 \leq x \leq d$, $c = 0$) is (Hill, 1928):

$$c = c_0 \left[1 - \frac{4}{\pi} \left(e^{-D\pi^2 t/d^2} \sin \frac{\pi x}{d} + \frac{1}{3} e^{-9D\pi^2 t/d^2} \sin \frac{3\pi x}{d} + \right.\right.$$
$$\left.\left. + \frac{1}{5} e^{-25 D\pi^2 t/d^2} \sin \frac{5}{d} \frac{x}{d} + \ldots \right) \right] \tag{4.13}$$

which represents a rapidly convergent series and gives the concentration c at any x (distance from one of the surface planes) inside the sheet, at any time t. Especially useful are formulas giving the fractional equilibration m_t/m_∞ of simple geometric shapes, where m_t is the amount of the solute present in the body at time t and m_∞ the amount present after full equilibrium has been reached. The fractional equilibration of the above plane sheet may be calculated as

$$\frac{m_t}{m_\infty} = \frac{\int_0^d c \, dx}{c_0 d}$$

and is equal to (Hill, 1928):

$$\frac{m_t}{m_\infty} = 1 - \frac{8}{\pi^2} \left(e^{-D\pi^2 t/d^2} + \frac{1}{9} e^{-9D\pi^2 t/d^2} + \frac{1}{25} e^{-25 D\pi^2 t/d^2} + \ldots \right) \tag{4.14}$$

Fractional equilibration of a sphere of radius r is given by

$$\frac{m_t}{m_\infty} = 1 - \frac{6}{\pi^2} \left(e^{-D\pi^2 t/r^2} + \frac{1}{4} e^{-4D\pi^2 t/r^2} + \frac{1}{9} e^{-9D\pi^2 t/r^2} + \ldots \right) \tag{4.15}$$

and fractional equilibration of a cylinder of radius r by

$$\frac{m_t}{m_\infty} = 1 - 4\left(\frac{1}{\mu_1^2} e^{-\mu_1^2 Dt/r^2} - \frac{1}{\mu_2^2} e^{-\mu_2^2 Dt/r^2} + \ldots\right) \quad (4.16)$$

where μ's are the zeros of the Bessel function J_0, $\mu_1 = 2.4048$, $\mu_2 = 5.5201$, $\mu_3 = 8.6537$, $\mu_4 = 11.7915$, etc.

The diffusion coefficients D of substances of low molecular weight are of the order of 10^{-5} cm^2 s^{-1}, ranging from $2.5 \cdot 10^{-5}$ for water to $0.5 \cdot 10^{-5}$ for sucrose. It should be added that, as shown by Hartley and Crank (1949), the rate of diffusion is determined by the intrinsic diffusion coefficient of an individual substance only in the so-called self-diffusion experiments, where no net concentration change occurs and the process of diffusion is followed with tracers. The reason for this is that only pure diffusion can be observed under such conditions, whereas a net transfer of a substance is always accompanied by the bulk flow of the solution. Since the intrinsic rates of pure diffusion of interdiffusing substances (solute and solvent) differ, there is a tendency to build up a hydrostatic pressure difference in the course of a net transfer, relieved by a bulk flow which enhances the flow of the less mobile solute and slows down the flow of the more mobile solvent. The rate of interdiffusion is then determined by a single mutual diffusion coefficient, depending on the conditions of the experiment. The effect of solvent diffusion, however, is likely to be small in the case of diffusion of minor components of the solution.

As shown by Einstein (1905) the diffusion equation (4.2) can be derived not only from the phenomenological Fick's first law but also directly from the very basic principles. In his famous derivation of the theory of Brownian movement, Einstein (1905) considers n particles in a liquid, each of them being displaced during a short time interval τ by some individual distance Δ in the direction of the x-axis. A probability law describes the extent of the displacement; the number of particles dn being displaced during the time interval τ by a distance between Δ and $\Delta + d\Delta$ is given by

$$dn = n\Phi(\Delta)\, d\Delta \quad (4.17)$$

where

$$\int_{-\infty}^{\infty} \Phi(\Delta)\, d\Delta = 1$$

since each particle is subject to some displacement between minus and plus infinity, zero displacements included. Moreover

$$\Phi(\Delta) = \Phi(-\Delta) \tag{4.18}$$

movements of particles in positive and negative direction being equally probable. (Actually, the function Φ deviates from zero only for very small values of Δ). The concentration of particles is a function of the coordinate x and the time t, $c(x, t)$. The concentration distribution at time $t + \tau$ is equal to the distribution at time t with the x-coordinate of each particle changed by its proper displacement Δ:

$$c(x, t + \tau) = \int_{\Delta = -\infty}^{\Delta = \infty} c(x + \Delta, t)\, \Phi(\Delta)\, d\Delta. \tag{4.19}$$

Since τ is very small, the left-hand side of the equation may be expressed by

$$c(x, t + \tau) = c(x, t) + \tau \frac{\partial c}{\partial t} \tag{4.20}$$

and the function $c(x + \Delta, t)$ under the integration sign may be expanded in a Taylor series:

$$c(x + \Delta, t) = c(x, t) + \Delta \frac{\partial c(x, t)}{\partial x} + \frac{\Delta^2}{2} \frac{\partial^2 c(x, t)}{\partial x^2} \dots . \tag{4.21}$$

Eq. (4.19) then has the form

$$c + \tau \frac{\partial c}{\partial t} = c \int_{-\infty}^{\infty} \Phi(\Delta)\, d\Delta + \frac{\partial c}{\partial x} \int_{-\infty}^{\infty} \Delta \Phi(\Delta)\, d\Delta +$$
$$+ \frac{\partial^2 c}{\partial x^2} \int_{-\infty}^{\infty} \frac{\Delta^2}{2} \Phi(\Delta)\, d\Delta \dots . \tag{4.22}$$

As a result of (4.18) the second, fourth, etc., terms on the right-hand side vanish. The integral in the first right-hand side term is equal to 1 by eq. (4.17). The integral in the third term is a constant value which may be put equal to the constant time τ multiplied by another constant D:

$$\int_{-\infty}^{\infty} \frac{\Delta^2}{2} \Phi(\Delta)\, d\Delta = \tau D. \tag{4.23}$$

The other odd terms are negligibly small. Eq. (4.22) then can be written as

$$\frac{\partial c}{\partial t} = D \frac{\partial^2 c}{\partial x^2}$$

which is the partial differential diffusion equation (4.2).

Einstein (1905) then solves the diffusion equation to describe the time-dependent space distribution of particles which, at time zero, are concentrated in the immediate vicinity of the plane at $x = 0$. Thus, for all x different from zero, $c(x, t) = 0$ for $t = 0$ and

$$\int_{-\infty}^{\infty} c(x, t)\, dx = n \tag{4.24}$$

where n is the total number of particles, conserved during the diffusional process. With these conditions the solution is

$$c(x, t) = \frac{n}{\sqrt{4\pi D}} \frac{e^{-\frac{x^2}{4Dt}}}{\sqrt{t}}. \tag{4.25}$$

It may be easily shown, by taking the appropriate derivatives, that the function (4.25) satisfies the diffusion equation. Thus, since

$$\frac{d(u(t)/v(t))}{dt} = \frac{1}{v^2}\left(v\frac{du}{dt} - u\frac{dv}{dt}\right),$$

$$\frac{\partial c(x, t)}{\partial t} = \frac{n}{\sqrt{4\pi D}} \frac{\left(\sqrt{t}\,\frac{x^2}{4Dt^2} - \frac{1}{2\sqrt{t}}\right)e^{-\frac{x^2}{4Dt}}}{t},$$

whereas

$$\frac{\partial c(x, t)}{\partial x} = \frac{n}{\sqrt{4\pi D}} \frac{-\frac{x}{2Dt}e^{-\frac{x^2}{4Dt}}}{\sqrt{t}};$$

and since

$$\frac{d(u(x) \cdot v(x))}{dx} = u\frac{dv}{dx} + v\frac{du}{dx},$$

$$\frac{\partial^2 c(x, t)}{\partial x^2} = \frac{n}{\sqrt{4\pi D}} \frac{\left(\frac{x^2}{4D^2t^2} - \frac{1}{2Dt}\right)e^{-\frac{x^2}{4Dt}}}{\sqrt{t}} \frac{\sqrt{t}}{\sqrt{t}} =$$

$$= \frac{n}{\sqrt{4\pi D}}\frac{1}{D} \frac{\left(\sqrt{t}\,\frac{x^2}{4Dt^2} - \frac{1}{2\sqrt{t}}\right)e^{-\frac{x^2}{4Dt}}}{t}$$

so that

$$\frac{\partial c(x, t)}{\partial t} = D\frac{\partial^2 c(x, t)}{\partial x^2}.$$

The concentration change at distance x and time t (equal for zero initial concentration to the function $c(x, t)$ itself) divided by the total number of particles is equal to the probability of an individual particle being displaced by a distance x during time t. Hence the arithmetic means of the squares of displacements may be evaluated as (Fürth, 1956)

$$\overline{x^2} = \frac{1}{n} \int_{-\infty}^{\infty} c(x, t)\, x^2 \, dx = \frac{1}{\sqrt{4\pi Dt}} \int_{-\infty}^{\infty} e^{-\frac{x^2}{4Dt}}\, x^2 \, dx = 2Dt. \quad (4.26)$$

For those who might be interested in the procedure we describe this integration in details. First, the substitution $y = x^2/4Dt$ is done so that $x^2 = 4Dty$, $x = 2\sqrt{Dt}\sqrt{y}$ and $dx = (\sqrt{Dt}/\sqrt{y})\, dy$. The integral (4.26) thus becomes

$$\overline{x^2} = \frac{1}{\sqrt{4\pi Dt}} \int_{-\infty}^{\infty} e^{-y} \cdot 4Dty \frac{\sqrt{Dt}}{\sqrt{y}} \, dy =$$

$$= \frac{2Dt}{\sqrt{\pi}} \int_{-\infty}^{\infty} e^{-y}\sqrt{y}\, dy = \frac{4Dt}{\sqrt{\pi}} \int_{0}^{\infty} e^{-y}\sqrt{y}\, dy.$$

The integration can now be carried out per partes according to

$$\int u \, dv = uv - \int v \, du$$

with $u = \sqrt{y}$ and $dv = e^{-y}\, dy$, so that $du = dy/(2\sqrt{y})$ and $v = -e^{-y}$. The integral becomes

$$\overline{x^2} = \frac{4Dt}{\sqrt{\pi}} \left\{ [-e^{-y}\sqrt{y}]_0^{\infty} + \frac{1}{2} \int_0^{\infty} \frac{e^{-y}}{\sqrt{y}} \, dy \right\}.$$

The first term on the right-hand side vanishes at both boundaries, the second is simplified by the substitution $y = u^2$, $\sqrt{y} = u$, and $dy = 2u\, du$, so that

$$\frac{1}{2} \int_0^{\infty} \frac{e^{-y}}{\sqrt{y}} \, dy = \int_0^{\infty} e^{-u^2} \, du.$$

The value of this integral has been known since Euler's time, being $\sqrt{\pi}/2$. Thus

$$\overline{x^2} = 2Dt$$

as indicated by eq. (4.26).

This relationship, often written as $\sqrt{\overline{x^2}} = \sqrt{2Dt}$, gives a vivid picture of the rate at which diffusion proceeds. Since the diffusion coefficients of common small molecules are of the order of

10^{-5} cm^2 s^{-1}, the mean displacement of these molecules is approximately 4.4 μm s^{-1}, 34.6 μm min^{-1}, 0.27 cm h^{-1}, 1.31 cm d^{-1}, so that diffusion is justly described as a rapid process for short and slow one for longer distances.

The excellent book on Diffusion Processes by Jacobs (1967) is recommended to the reader interested in a clear and complete treatment of the theory of diffusion.

4.2. DIFFUSION ACROSS MEMBRANES

Unlike diffusion in space which is properly described by solutions of the partial differential equation (Fick's second law), permeation across a thin membrane by a diffusional process may be described to a good approximation by an ordinary differential equation, *viz.* Fick's first law as well as its modifications taking into account the saturability of mediated diffusion and/or volume changes of the compartment into which the diffusion across the membrane proceeds. This results from the fact that the equilibration process in the media at the two sides of the membrane is mostly much faster than the process of permeation across the membrane; even with rapidly permeating substances it is usually sufficient to introduce a correction for the diffusion in the unstirred layers adjacent to the membrane and behaving like high permeability membranes in series with the membrane proper.

The intricacies of permeation by saturable processes are the subject of all the following sections of this chapter and we shall see that in most cases it is not even necessary to integrate the modified laws of Fick to determine the characteristics of a saturable process. Estimation of the so-called initial velocities is based on the fact that measurable quantities of a substance crossing the membrane do not change the concentrations in the compartments in question appreciably and hence nonintegrated modified laws of diffusion in which differentials are replaced by small finite differences can be used to describe the process quantitatively.

The aim of the present section is to discuss the differential equations governing the permeation of a nonelectrolyte across a membrane by a nonsaturable process (simple diffusion) and to show how they can be integrated in special cases. It will be seen that analogous integrated equations describe the tracer exchange in steady state over a much wider range, from nonsaturable equilibrating transport of a nonelectrolyte to saturable active transport of ions.

Fick's first law relates the flux of a substance, J (amount of substance in moles crossing a unit surface perpendicular to the direction of the movement in unit time) to the concentration gradient of the substance

$$J = -D \frac{dc}{dx} \qquad (4.27)$$

where D is the diffusion coefficient. If the flux is expressed in $mol\ cm^{-2}\ s^{-1}$, the length x in cm and concentration in $mol\ cm^{-3}$, the dimensions of D are $cm^2\ s^{-1}$.* For a thin membrane, in a steady state of transport, the derivative of concentration may be replaced with the quotient of a finite concentration difference across the membrane thickness

$$J = -D \frac{\Delta c}{\Delta x} = -D \frac{c_{II} - c_I}{l} = P(c_I - c_{II}) \qquad (4.28)$$

where $P(= D/l)$ is the permeability coefficient across the membrane of a substance permeating by simple diffusion according to a diffusion coefficient across the given membrane equal to D. The units of P are $cm\ s^{-1}$.

Equation (4.28) indicates that the rate of transport will depend on the concentration of the transported solute. If the initial rate is followed (or if the unidirectional flux inward is measured) the dependence is between J and c_I; if the net flux is examined, the dependence will be between J and Δc (Fig. 4.1).

The value of P includes an important constant for every given solute and every given membrane, viz. the membrane : water partition coefficient** defined as

$$K = c_{Im}/c_I = c_{IIm}/c_{II} \qquad (4.29)$$

(the equation assumes identical solvent properties of both bulk solutions and the same lipid composition near the two membrane faces). The indexed values refer to concentrations just within the membrane where it holds that

$$J = P_m(c_{Im} - c_{IIm}). \qquad (4.30)$$

---

* The dimension of D does not change if concentration is expressed in $g\ cm^{-3}$ and flux in $g\ cm^{-2}s^{-1}$.

** This is experimentally approximated satisfactorily by using olive oil or olive oil plus oleic acid (for basic solutions) as membrane substitute.

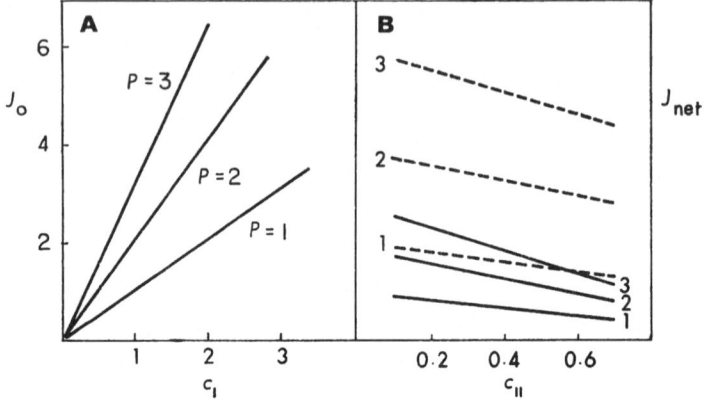

Fig. 4.1. Concentration dependence of diffusional flow across a membrane. In **A**, the initial (or unidirectional) flux is plotted against concentration at the starting (*cis*) side; in **B**, the net flow is plotted against concentration at the *trans* side for $c_I = 1$ (full lines) and $c_I = 2$ (dashed lines) at three values of P (1, 2, 3).

Hence

$$J = P_m K(c_1 - c_{II}) = \frac{D_m}{l} K(c_1 - c_{II}). \qquad (4.31)$$

It should be appreciated that the transmembrane permeability is thus comprised of two factors, the distribution between water and membrane and the "diffusion" across the membrane.

The diffusion coefficient included in the fundamental eq. (4.27) is itself a function of the size and shape of the molecule. For spherical molecules, where the Stokes–Einstein equation* holds, it is seen (Davson and Danielli, 1952) that

$$DM^{1/3} = \text{constant}$$

where M is the molecular weight. In other words, the relationship between $\log D$ and $\log M$ should give a straight line of slope equal to -0.33. This is actually found for diffusion in water of large molecules of proteins. For small molecules, ranging from hydrogen to a trisaccharide, the empirical slope is steeper, corresponding to

$$DM^{1/2} = \text{constant}$$

* $D = RT/6N\pi\eta r$, N being the Avogadro constant, η the viscosity of the medium and r the particle radius.

apparently because the size of the molecules is such that the surrounding water medium is not "seen" by them as a continuum.

These relationships have also been applied with greater or lesser success to diffusion across membranes, the usual way of plotting being $\log (PM^{1/2})$ vs. $\log K$, when points lying within a broad straight band are obtained (cf. Collander, 1949; Stein, 1967) (Fig. 4.2).

For a long time, these data have been taken as indicative that molecules pass through the lipid parts of the membrane merely by

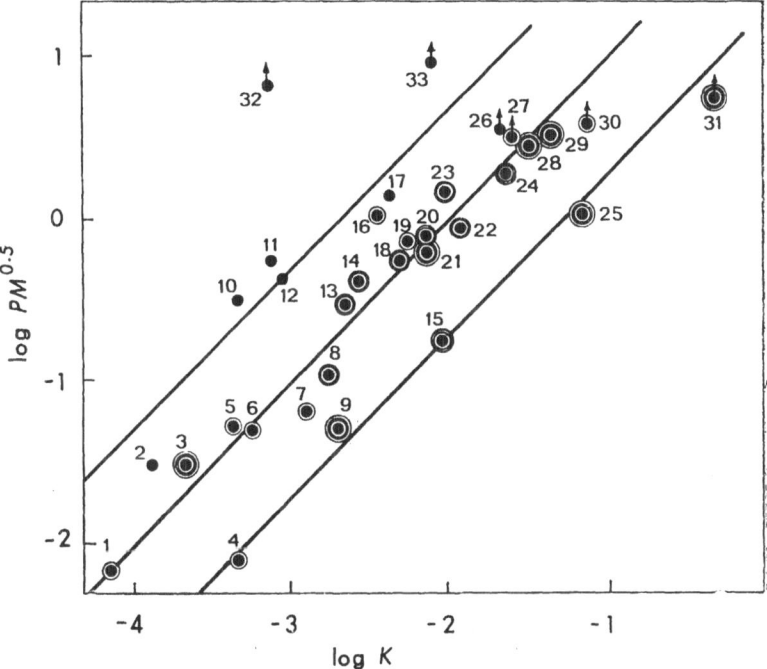

Fig. 4.2. The permeability of *Chara ceratophylla* to nonelectrolytes. P is the permeability constant, M the molecular weight, K the partition coefficient between olive oil and water. The size of the circles is related to the molar refraction, ranging from less than 15 for simple dots, to 15—22, 22—30 and more than 30 for the points with increasing number of concentric circles. 1 Glycerol, 2 urea, 3 hexamethylenetetramine, 4 dicyanodiamide, 5 methylurea, 6 lactamide, 7 thiourea, 8 ethylurea, 9 diethylmalonamide, 10 ethylene glycol, 11 formamide, 12 acetamide, 13 dimethylurea, 14 glycerol methyl ether, 15 monoacetin, 16 propionamide, 17 cyanamide, 18 succinimide, 19 propylene glycol, 20 glycerol ethyl ether, 21 diethylurea, 22 chlorhydrin, 23 butyramide, 24 valeramide, 25 diacetin, 26 ethanol, 27 urethylan, 28 antipyrin, 29 trimethyl citrate, 30 urethane, 31 triethyl citrate, 32 water, 33 methanol.

virtue of their lipid solubility (and, of course, the more slowly the greater the molecule). However, the deviating data on rather lipo-philic molecules, on the one hand, and on very small hydrophilic molecules, such as water, on the other, have led to several developments of the model.

Thus, if compounds with a log K of 0 and higher are considered, the Collander graph of Fig. 4.2 ceases to be linear but actually follows a parabola as shown in Fig. 4.3. The behaviour has been explained

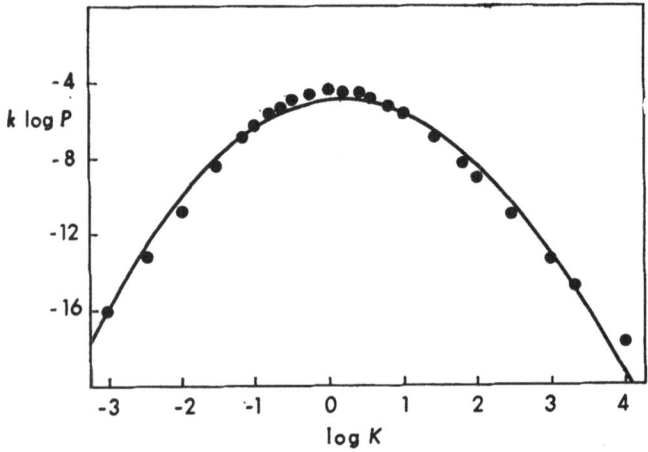

Fig. 4.3. Dependence of concentration in a distant compartment (approximated by permeability) on the partition coefficient of compounds between ether and water. The points are fitted by a parabola using the least-squares method. (According to Penniston *et al.*, 1969).

in analogy with equations derived by Hansch and Fujita (1964) who showed that a standard pharmacological response relates to the hydrophobicity of a solute in a rather complicated way, implying that for highly hydrophobic substances (so as to be practically insoluble in water) the rate of transport across the membrane is diminished:

$$\log (1/c) = -k(\log K)^2 + k' \log K + k'' + \varrho\sigma \qquad (4.32)$$

where c is the concentration resulting in equal flux, k, k', and k'' are empirical constants and ϱ and σ are Hammett constants, defined as $\log (k_d/k_0) = \sigma\varrho$, k_d and k_0 being reaction rate constants of a derivative and of the parent compound, respectively; σ depends only on the character and position of a substituent, ϱ on the type of reaction involved. Thus, to elicit the same effect (or flux in an approximation) a compound must be present at a higher concentration if its partition coefficient is too high (as K increases, so does c, the concentration required to produce the same response).

If c is replaced with the permeability constant to which it may be proportional, relationships of the type of eq. (4.32) are seen to fit experimental data very well indeed (*cf.* Penniston *et al.*, 1969).

The deviations concerning small hydrophilic molecules have been treated recently by Lieb and Stein (1969). The concept views the membrane as a polymer sheet which undergoes thermal movement whereby "holes" are transiently opened so as to permit the passage of molecules across the membrane. This is conceptually related to the still later theory of "kinks" moving through the membrane with a "diffusion coefficient" of 10^{-5} cm^2 s^{-1}. The kinks are mobile structural defects in the hydrocarbon phase of the membrane (Träuble, 1971).

In this development, the D_m is defined as kM_r^{-s} where M_r is the relative molecular weight referred to methanol as unity, k is a constant and s is the so-called differential mass selectivity coefficient, its value being $0.3-0.5$ in water but as high as 3.5 in a biological membrane. This results in a steeper differentiation according to size than expected on the basis of diffusion in a continuous aqueous medium. Thus, the diffusion coefficients in the molecular weight range from 45 (formamide) to 122 (erythritol) lie between $1.6 . 10^{-5}$ and $8 . 10^{-6}$ cm^2 s^{-1} in water (a factor of 2) but between $1.4 . 10^{-8}$ and $2 . 10^{-10}$ cm^2 s^{-1} in *Chara ceratophylla* (a factor of 70).

Another, conceptually different, extension of the sieving-plus-lipophilicity hypothesis, was the introduction of the pore concept, envisaging small permanent hydrophilic pores that would permit the passage of small water-soluble molecules across the membrane. This concept evolved from measuring the so-called reflection coefficient σ (discussed on p. 312). Using the development by Kedem and Katchalsky (1958) and Kedem (1961) one can show that

$$P_m = (1 - \sigma) L_p RT / K\overline{V}_s \qquad (4.33)$$

where L_p is the hydraulic permeability coefficient (*cf.* eq. 6.41) and $\overline{V}_s$ is the partial molal volume of the solute. Thus, the reflection coefficient σ is related to the permeability coefficient in a straightforward manner.

The use of σ for computing the equivalent pore radius has been found to be somewhat more objectionable, the principal flaw being in the fact that even in a membrane with no pores, $\sigma < 1$ for a permeant solute which contributes to the volume flow (see p. 314). Still, reproducible values have been obtained for various cells, using the relationship

$$1 - \sigma = \frac{[2(1 - a/r)^2 - (1 - a/r)^4] \times}{\times [1 - 2.104a/r + 2.09(a/r)^3 - 0.95(a/r)^5]}{[2(1 - a_W/r)^2 - (1 - a_W/r)^4] \times}{\times [1 - 2.104a_W/r + 2.09(a_W/r)^3 - 0.95(a_W/r)^5]} \tag{4.34}$$

where r is the equivalent pore radius, a is the radius of the permeating molecule and a_W the radius of a water molecule (Renkin, 1954; Goldstein and Solomon, 1960). The equivalent pore radii thus assessed range from 0.42 nm in human erythrocytes to 0.56 nm in *Necturus* kidney to 2.3 nm in a Visking dialysis tubing.

Although the existence of permanent hydrophilic pores in the membrane is still contested their function at least in bulk flow of solutions appears to be generally accepted.

The magnitude of the permeability constant estimated from the flux and bulk concentrations of solute according to eq. (4.28) is usually in error because of the fact that layers of solution adjacent to the membrane surface are not stirred vigorously enough, with the consequence that the solute concentration in this unstirred layer may differ substantially from that in the bulk phase. The unstirred layers (also called Nernst diffusion layers) can be both morphological (such as the cell wall of bacteria or yeasts or the glycocalyx of epithelial cells) as well as (as layers of slow laminar flow) located directly in the solution itself, the thickness of the former being perhaps 1 μm, that of the latter ranging from 20 to 500 μm, depending on the efficiency of mechanical stirring (Dainty, 1963).

Due to the presence of the unstirred layer, the concentration difference across the membrane proper is less than that between the bulks of the solutions. The situation is depicted schematically in Fig. 4.4, in which the so-called effective thickness of the unstirred layer at each side of the membrane is shown. Layers of this thickness would exert the observed effects if the concentration gradients in them were constant. In fact, the concentration profiles are not linear but have the form shown approximately by the dot-and-dash lines in the figure.

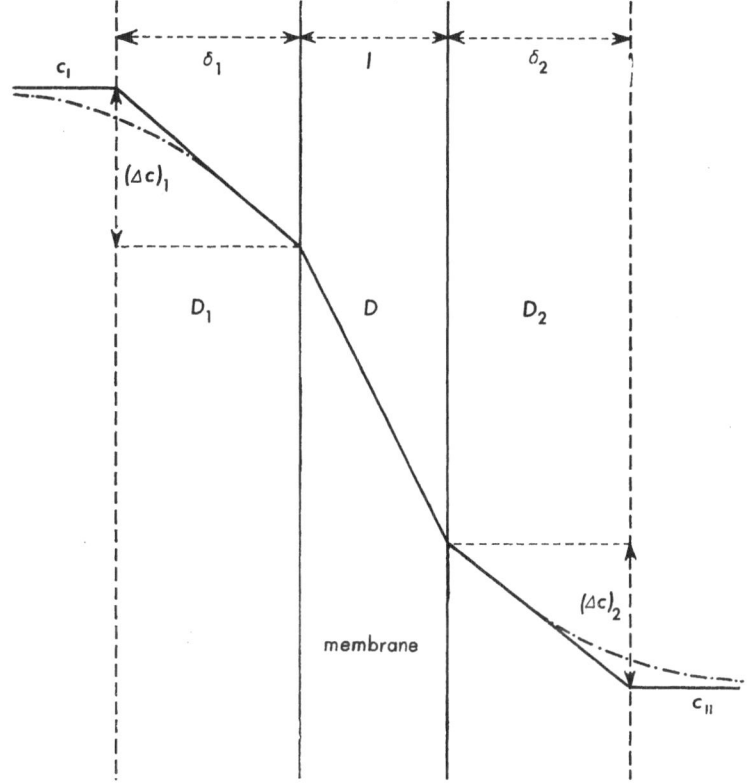

Fig. 4.4.   Concentration profile in a membrane and the adjacent effective unstirred layers. The diffusion coefficients are shown in capital D's.

The true permeability properties are expressed by the permeability coefficient

$$P_{\text{true}} = \frac{J}{c_{\text{I}} - c_{\text{II}} - (\Delta c)_1 - (\Delta c)_2} \tag{4.35}$$

rather than by eq. (4.28). Since

$$(\Delta c)_1 = \frac{J}{\dfrac{D_1}{\delta_1}} \qquad \text{and} \qquad (\Delta c)_2 = \frac{J}{\dfrac{D_2}{\delta_2}}$$

it may be shown that

$$P_{\text{true}} = \frac{1}{\dfrac{1}{P_{\text{measured}}} - \dfrac{\delta_1}{D_1} - \dfrac{\delta_2}{D_2}}. \tag{4.36}$$

Most diffusion coefficients are of the order of magnitude of 10^{-5} $cm^2\ s^{-1}$ and since the thickness of the unstirred layers is of the order of 100 μm (10^{-2} cm), the "permeability" of an unstirred layer D/δ is of the order of 10^{-3} cm s^{-1}. Hence measured permeability coefficients as high as 10^{-3} cm s^{-1} show that the process is controlled by the diffusion in the unstirred layers and give little information about the properties of the membrane. Those of the order of 10^{-4} cm s^{-1} require a considerable correction.

The permeability constants are markedly influenced by temperature, something that rules out diffusion through large aqueous pores but does not exclude the possibility of passage through narrow pores where the hydration molecules would have to be stripped before the permeation. Values of activation energy of $40-80$ kJ mol^{-1} are not uncommon, the main contribution being represented by the translocation across the membrane.

Fick's law for simple diffusion across a thin membrane (eq. 4.28) may be used to express concentration changes in a membrane-surrounded compartment of surface A and volume V, which are brought about by the permeation of the diffusing substance. The total inflow of the substance per unit time being $J \cdot A$, the rate of the concentration change due to permeation is

$$\frac{dc_{II}}{dt} = \frac{PA}{V}(c_I - c_{II}). \tag{4.37}$$

It should be observed that c_{II} can change even in the absence of any permeation process, due to changes in compartment volume. If we assume that there is no permeation of the substance in question across the membrane the total amount of substance in compartment II will be constant

$$c_{II}V = \text{constant}. \tag{4.38}$$

By taking derivatives with respect to time we obtain

$$V\frac{dc_{II}}{dt} + c_{II}\frac{dV}{dt} = 0$$

or

$$\frac{dc_{II}}{dt} = -\frac{c_{II}}{V}\frac{dV}{dt} \tag{4.39}$$

which is the rate of concentration change due to changing volume of the compartment (e.g., a cell). Superposition of the concentration changes due to the permeation process (4.37) and to the changes of compartment volume yields

$$\frac{dc_{II}}{dt} = \frac{PA}{V}(c_I - c_{II}) - \frac{c_{II}}{V}\frac{dV}{dt}. \tag{4.40}$$

Although volume changes of the compartment may be caused by various factors, those which interest us most are brought about by osmotic effects of the permeating substance itself. As far as the behavior of the compartment may be approximated by that of an ideal osmometer, the rate of volume change is proportional to the difference between the internal (II) and the external (I) osmolarity:

$$\frac{dV}{dt} = k\left(c_{II} + \frac{m}{V} - c_I - \frac{m_0}{V_0}\right) \tag{4.41}$$

where m is the amount of other osmotically active substances present in the osmometer and m_0/V_0 is the initial osmolarity in the osmometer, presumably equal to that of the outer medium before addition of the permeating substance. Equations (4.40) and (4.41) are equivalent to those given by Jacobs (1952) and are not easy to solve. One possibility is to calculate (under reasonable assumptions about the value of m, such as $m = m_0$, the permeation of other substances being negligible during the experiment) small changes of c_{II} and V in turns and, from the small finite differences obtained, to construct curves approximating the time course of concentration and volume (Sigler and Janáček, 1971). Another way is to determine the course of volume changes experimentally. It should be added that equations (4.40) and (4.41) may represent a good approximation only in the case of reflection coefficient σ for the solute practically equal to 1 (i.e., when there is in the membrane no appreciable interaction between the flow of a slowly permeating solute and the flow of water), otherwise equations (4.40) and (4.41) have to be corrected in accordance with equations (6.34) and (6.35); see p. 111. Equations so modified were developed and successfully used by Johnson and Wilson (1967).

It is of great advantage if the volume changes of the compartment (cell) during permeability measurements are negligibly small, as can be always achieved (unless the saturability of the process is tested) by using a minute concentration of radioactively labelled permeant, or when the volume changes are minimized by the presence of a rigid cell wall. The same objective is accomplished by applying a small amount of tracer to the system which is in a steady state with respect to the unlabelled substance. Then the rate of the concentration change of the permeant is given by eq. (4.37), where P is a constant, in the case of simple diffusion across the membrane. In laboratory experiments it is often feasible to employ a large outer compartment, a reservoir, so that c_I is constant to a good approximation. Eq. (4.37) can then be easily integrated to obtain the time course of concentration:

$$\int_{c_{II}=0}^{c_{II}} \frac{dc_{II}}{c_I - c_{II}} = \frac{PA}{V} \int_{t=0}^{t} dt \tag{4.42}$$

yielding

$$\ln \frac{c_I - c_{II}}{c_I} = -\frac{PA}{V} t \tag{4.43}$$

or

$$c_{II} = c_I \left(1 - e^{-\frac{PA}{V}t}\right) \tag{4.44}$$

which describes the exponential equilibration of concentrations. When the outflow of a substance from a preloaded compartment into a reservoir with practically zero concentration is measured the situation is even simpler, the decay of the internal concentration being described by

$$c_{II(t)} = c_{II(0)} \, e^{-\frac{PA}{V}t} \tag{4.45}$$

which, in the logarithmic, linear form can be conveniently used to determine the coefficient PA/V from the slope of a straight line (cf. Fig. 4.5)

$$\ln \frac{c_{II(t)}}{c_{II(0)}} = 2.303 \log \frac{c_{II(t)}}{c_{II(0)}} = -\frac{PA}{V} t. \tag{4.46}$$

An expression frequently used in this connection is the half-time of equilibration defined as

$$\ln 0.5 = -\ln 2 = -\frac{PA}{V} t_{0,5} \quad \text{or} \quad t_{0,5} = 0.693 V/PA. \tag{4.46'}$$

When the outer volume V_I is of a size comparable to that of the inner compartment, V_{II}, the concentration c_I in eq. (4.37) cannot be any more considered as a constant, but must be expressed from the condition

$$c_I V_I + c_{II} V_{II} = c_{I(0)} V_I \tag{4.47}$$

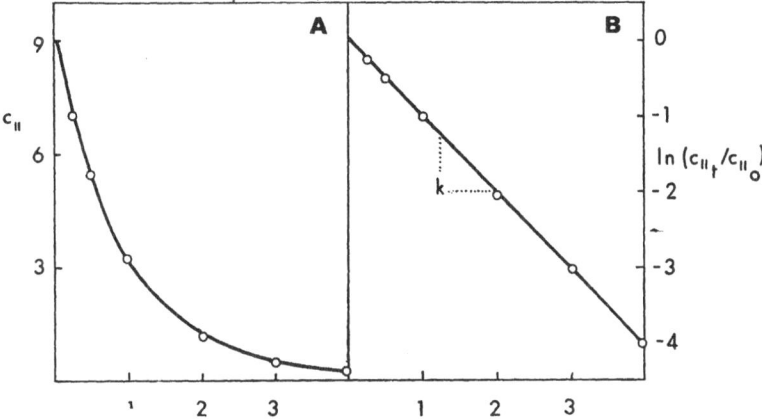

Fig. 4.5. Diffusional outflow of solute from cells. In **A**, the residual cellular concentration is plotted against time; in **B**, the logarithm of the ratio of initial and actual concentrations is plotted against time. The rate constant k is given by the slope of the straight line.

and the equilibration of compartment II is described by

$$c_{II} = c_{I(0)} \frac{V_I}{V_I + V_{II}} \left(1 - e^{-\frac{V_I + V_{II}}{V_I V_{II}} P A t} \right). \tag{4.48}$$

In the case of mediated diffusion the above equations do not apply, since P is then not a constant, but rather a function of concentration; in the case of an active transport, moreover, the stationary concentration in the two compartments is not the same. But even there the equilibration of a tracer, resulting in the same specific activity of the labelled substance in the two compartments, is described by analogous equations provided that the tracer is added in a small amount to the system in a steady state. The appropriate equations are solutions of the differential equation

$$\frac{dc_{II}^*}{dt} = \frac{J_s}{V_{II}} A \frac{c_I^*}{c_I} - \frac{J_s}{V_{II}} A \frac{c_{II}^*}{c_{II}} \tag{4.49}$$

where c^*'s are concentrations of the tracer and J_s is the steady-state unidirectional flux, i.e., a constant under the given experimental conditions.

The equation for tracer equilibration, equivalent to eq. (4.44) then will be

$$c_{II}^* = c_I^* \frac{c_{II}}{c_I} \left(1 - e^{-\frac{A}{V_{II}} \frac{J_s}{c_{II}} t} \right) \tag{4.50}$$

that equivalent to eq. (4.45) is

$$c_{II}^* = c_{II(0)}^* \, e^{-\frac{A}{V_{II}} \frac{J_s}{c_{II}} t} \tag{4.51}$$

and finally that equivalent to (4.48)

$$c_{II}^* = c_{I(0)}^* \frac{c_{II} V_I}{c_I V_I + c_{II} V_{II}} \left(1 - e^{-\frac{(c_I V_I + c_{II} V_{II}) J_s}{c_I V_I c_{II} V_{II}} At} \right) \tag{4.52}$$

the validity of which in a steady state is general and which enable us to estimate the value of the steady-state unidirectional flux of substance from the data on tracer exchange.

When there are several compartments present in a parallel or a series arrangement the situation is more complex, but still it may be described by a sum of several exponential terms and the experimental curves may be analyzed graphically (e.g., Kotyk and Janáček, 1975) or fitted by a sum of exponential functions using a computer (e.g., Berman *et al.*, 1962).

Compounds transported across biological membranes by some kind of "simple", non-carrier diffusion are of many types. Besides the smallest hydrophilic molecules, such as water, dimethylsulfoxide formamide, etc., there is a vast variety of lipid-soluble compounds, ranging from methanol to complicated drugs which apparently cross membranes via the lipid domains; of naturally produced compounds of chemotherapeutic importance, one may name the penicillins, macrolides, rifamycins, actinomycin D, puromycin and chloramphenicol (*cf.* Franklin, 1974). Streptomycin, bacitracin and several other antibiotics penetrate across the cell membrane only after disrupting the external architecture of the cell.

On the other hand, even relatively small polar substances, such as glycols, monosaccharides, amino acids, cannot cross the membranes of most cells to any considerable degree by simple diffusion. There are usually specific carriers for such substances present in the cell membrane.

Apparent diffusion of a number of compounds can be induced by the application of various membrane-active substances, typically the polyene antibiotics, e.g., filipin, nystatin, etruscomycin, pimaricin, *N*-acetylcandidin, amphotericin B, and azalomycin (e.g., van Zupthen *et al.*, 1971).

Filipin

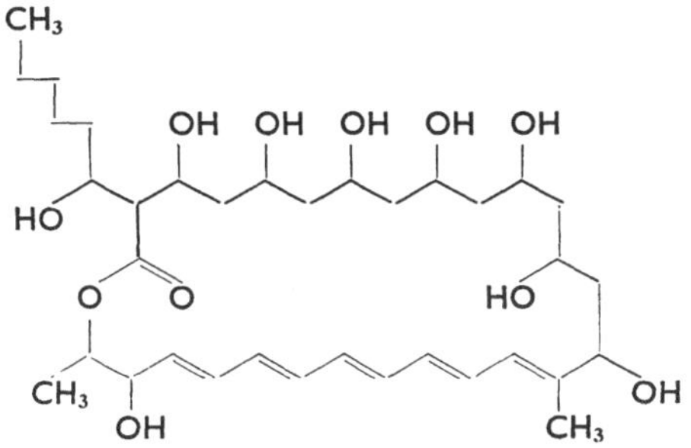

Amphotericin B

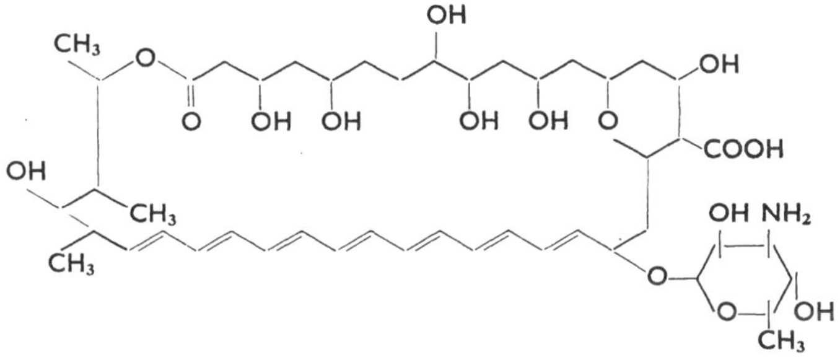

The mechanism of action these antibiotics appears to consist in binding to the sterol component of cells, preferring ergosterol to cholesterol, and thus forming aqueous "pores" which are in a continuous process of forming and breaking up. The holes thus formed range in size from a diameter of some 0.4 nm after nystatin to 12.5 nm after filipin, depending also on the concentration of the antibiotic. Table 4.1 shows some examples of the permeability changes involved.

TABLE 4.1. Effect of amphotericin B on the permeability properties of artificial sterol membranes (adapted from Dennis et al., 1970)

Molecule	P cm s^{-1} . 10^6	
	control	with 10^{-7} M amphotericin B
Water	630	1310
Urea	< 5	900
Glycerol	< 5	220
Ribose	< 5	67

Another group of antibiotics known to cause the leakage (= diffusion) of substances across the membrane are the tyrocidines, colistins and polymyxins which probably bind to cardiolipin, phosphatidyl choline and phosphatidyl serine through an electrostatic interaction. Cyclic peptides known to make nonselective holes in membranes are the surfactin from *Bacillus subtilis* and monamycin from *Streptomyces* species.

Polymyxin B$_1$

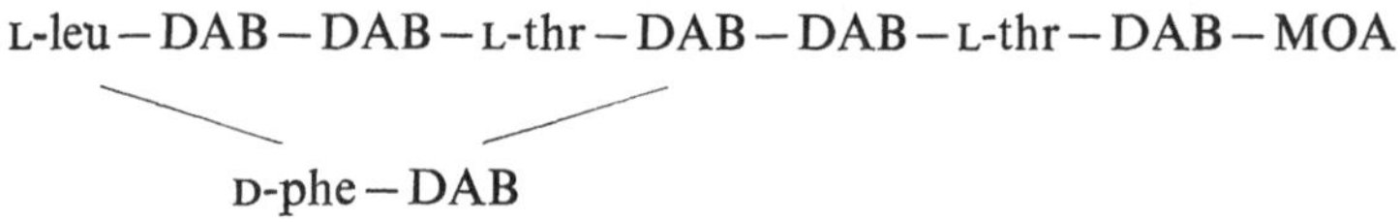

$$\text{L-leu} - \text{DAB} - \text{DAB} - \text{L-thr} - \text{DAB} - \text{DAB} - \text{L-thr} - \text{DAB} - \text{MOA}$$

$$\text{D-phe} - \text{DAB}$$

(MOA = 6-methyloctanoic acid, DAB = diaminobutyric acid)

Tyrocidine A

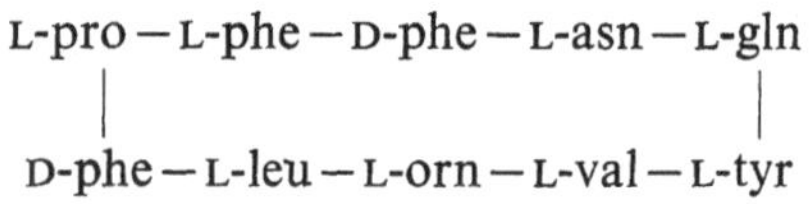

$$\text{L-pro} - \text{L-phe} - \text{D-phe} - \text{L-asn} - \text{L-gln}$$
$$\text{D-phe} - \text{L-leu} - \text{L-orn} - \text{L-val} - \text{L-tyr}$$

Surfactin

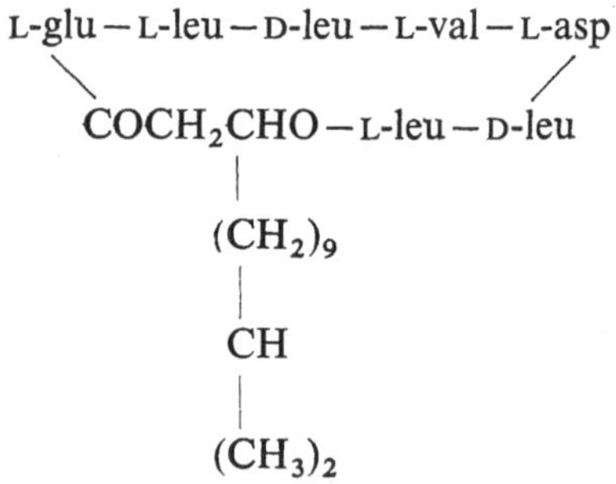

$$\text{L-glu} - \text{L-leu} - \text{D-leu} - \text{L-val} - \text{L-asp}$$
$$\text{COCH}_2\text{CHO} - \text{L-leu} - \text{D-leu}$$
$$(\text{CH}_2)_9$$
$$\text{CH}$$
$$(\text{CH}_3)_2$$

There are a number of various other naturally produced compounds which decrease the resistance to permeation of biological as well as artificial membranes. However, their effects being specific for various ions (and usually not associated with actual pore formation), they will be dealt with in the chapter on the transport of ions.

4.3. KINETICS OF MEDIATED TRANSPORT

The plasma membranes, mitochondrial and chloroplast membranes and possibly (but not certainly) other cell membranes contain mechanisms which allow for the specific transport of many substances, both of uncharged and ionic character, from one side of the membrane to the other. All the systems of this type are characterized besides their specificity by the fact that the rate of transport through their mediation rises with concentration only to a certain limiting value: they are saturable. The most straightforward (and probably justified in most cases) explanation for this phenomenon is that the transported solute binds transiently to a specific receptor on the membrane surface and only after this binding can it be translocated to the other side. It is of no consequence whether the receptor itself with the bound solute moves across the membrane or whether it transmits it to another membrane molecule which moves from one side to the other or, finally, whether it releases the solute into an internal "pool" from which it is attached by an analogous receptor at the other membrane face. The only property that such saturable transports must possess is that a given bidning site for substrate must never be exposed simultaneously to both external media. If this should happen, some of the typical properties of saturable transport, such as countertransport, would be lost.

4.3.1. Mediated or facilitated diffusion

The simplest of all the mechanisms so far proposed (but not necessarily the one universally valid) is the concept of a mobile carrier, the kinetics of which will be developed first.

If the carrier binds a single substrate we may depict the situation as follows:

$$
\begin{array}{ccc}
\text{I} & & \text{II} \\[4pt]
\mathbf{CS_I} & \underset{k_4}{\overset{k_3}{\rightleftharpoons}} & \mathbf{CS_{II}} \\[6pt]
k_1 s_I \Big\updownarrow k_2 & & k_6 s_{II} \Big\updownarrow k_5 \\[6pt]
\mathbf{C_I} & \underset{k_7}{\overset{k_8}{\rightleftharpoons}} & \mathbf{C_{II}}
\end{array}
$$

4.3.1.1. Steady-state approach

If no assumption is made as to the rate limitation of the system the expression for the rate of flow J_s of substrate S is derived either rom the set of steady-state equations

$$J_s = k_1 c_I s_I - k_2 c s_I = k_3 c s_I - k_4 c s_{II} =$$
$$= k_5 c s_{II} - k_6 c_{II} s_{II} = k_7 c_{II} - k_8 c_I \qquad (4.53)$$

and the carrier conservation equation

$$2c_t = c_I + c s_I + c_{II} + c s_{II} \qquad (4.54)$$

or by one of the abbreviated techniques described by King and Altman (1956), Wong and Hanes (1962) or Fromm (1970) (*cf.* Kotyk, 1974)

$$J_s = c_t \frac{k_1 k_3 k_5 k_7 s_I - k_2 k_4 k_6 k_8 s_{II}}{\begin{aligned}(k_7 + k_8)(k_2 k_4 + k_2 k_5 + k_3 k_5) + \\ + k_1[k_5(k_3 + k_7) + k_7(k_3 + k_4)] s_I + \\ + k_6[k_4(k_2 + k_8) + k_8(k_2 + k_3)] s_{II} + \\ + k_1 k_6(k_3 + k_4) s_I s_{II}\end{aligned}} \qquad (4.55)$$

If the system has no energy input we may speak of mediated or facilitated diffusion which is characterized by the fact that net flow ceases ($J_s = 0$) when $s_I = s_{II}$. From this it follows that $k_1 k_3 k_5 k_7 = k_2 k_4 k_6 k_8$ so that the numerator of eq. (4.55) may be written as $c_t k_1 k_3 k_5 k_7(s_I - s_{II})$. (In systems of active transport to be dealt with later, $k_1 k_3 k_5 k_7 \neq k_2 k_4 k_6 k_8$.)

The initial rate of uptake (when $s_{II} = 0$) is defined by a much simpler expression, *viz.*

$$J_{s(0)} = c_t \frac{k_1 k_3 k_5 k_7 s_I}{\begin{aligned}(k_7 + k_8)(k_2 k_4 + k_2 k_5 + k_3 k_5) + \\ + k_1[k_5(k_3 + k_7) + k_7(k_3 + k_4)] s_I\end{aligned}} \qquad (4.56)$$

which, like all expressions relating to transport by a univalent carrier, is formally identical with the Michaelis–Menten equation of enzyme kinetics $v = Vs/(K_m + s)$ or $J_s = J_{max} s/(K_T + s)$. Here

$$J_{max} = c_t \frac{k_3 k_5 k_7}{k_5(k_3 + k_7) + k_7(k_3 + k_4)}$$

and

$$K_T = \frac{(k_7 + k_8)(k_2 k_4 + k_2 k_5 + k_3 k_5)}{k_1[k_5(k_3 + k_7) + k_7(k_3 + k_4)]}.$$

If we work with very low concentrations of substrate (well below the apparent K_T) eq. (4.56) reduces to a diffusion-like process

$$J_{s(0)} = c_t \frac{k_1 k_3 k_5 k_7 s_1}{(k_7 + k_8)(k_2 k_4 + k_2 k_5 + k_3 k_5)} \qquad (4.57a)$$

while at very high concentrations (well above the apparent K_T) a zero-order process is obtained so that

$$J_{s(0)} = c_t \frac{k_3 k_5 k_7}{k_3 k_5 + k_5 k_7 + k_3 k_7 + k_4 k_7}. \qquad (4.57b)$$

Equation (4.55) does not show the unidirectional fluxes. These can be derived either by a somewhat lengthy iterative procedure (cf. Britton, 1966) or by the King and Altman or Fromm method proceeding from the following model:

$$
\begin{array}{ccc}
\mathbf{I} & & \mathbf{II} \\[4pt]
\mathbf{CS}_I & \underset{k_4}{\overset{k_3}{\rightleftharpoons}} & \mathbf{CS}_{II} \\[4pt]
k_1 s_I \Big\updownarrow k_2 & & k_6 s_{II} \Big\updownarrow k_5 \\[4pt]
\mathbf{C}_I & \underset{k_7}{\overset{k_8}{\rightleftharpoons}} & \mathbf{C}_{II} \\[4pt]
k_2 \Big\updownarrow k_1 s_I^* & & k_5 \Big\updownarrow k_6 s_{II}^* \\[4pt]
\mathbf{CS}_I^* & \underset{k_4}{\overset{k_3}{\rightleftharpoons}} & \mathbf{CS}_{II}^*
\end{array}
$$

where S* is he labelled form of S.

Because of the chemical identity of S and S* both the dissociation constants and the translocation constants are the same for S and S*. The expression for the flux from left to right is then

$$\overrightarrow{J}_{s*} = c_t \frac{k_1 k_3 k_5 s_I^*[k_7 + k_2 k_4 k_6 s_{II}/(k_2 k_4 + k_2 k_5 + k_3 k_5)]}{\text{denominator}} \qquad (4.58a)$$

the denominator being as in eq. (4.55). The flux from right to left is given by

$$\overleftarrow{J}_{s*} = c_t \frac{k_2 k_4 k_6 s_{II}^*[k_8 + k_1 k_3 k_5 s_I/(k_2 k_4 + k_2 k_5 + k_3 k_5)]}{\text{denominator}} \qquad (4.58b)$$

Both fluxes are thus seen to depend not only on the *cis* but also on the *trans* concentrations of the substrate ($\overrightarrow{J}_{s*}$ on s_{II} and $\overleftarrow{J}_{s*}$ on s_I). It is a question of the magnitude of the translocation constants (k_3, k_4, k_7, k_8) whether the effect will result in stimulation or inhibition.

If the system is intrinsically symmetrical, i.e. the translocation constants in both directions are equal ($k_3 = k_4$; $k_7 = k_8$) and the rate constants of association and dissociation, respectively, are equal at the two sides ($k_1 = k_6$; $k_2 = k_5$), eq. (4.55) for the overall rate of flow becomes

$$J_s = c_t \frac{k_1 k_3 k_5 k_7 (s_I - s_{II})}{2k_5 k_7 (k_5 + 2k_3) + k_1 (k_3 k_5 + k_5 k_7 + 2k_3 k_7)(s_I + s_{II}) + 2k_1^2 k_3 s_I s_{II}} \tag{4.59}$$

and the unidirectional fluxes will be

$$\overrightarrow{J}_{s*} = c_t \frac{k_1 k_3 k_5 s_I^*[k_7 + k_1 k_3 s_{II}/(2k_3 + k_5)]}{\text{denominator}} \tag{4.60a}$$

and

$$\overleftarrow{J}_{s*} = c_t \frac{k_1 k_3 k_5 s_{II}^*[k_7 + k_1 k_3 s_I/(2k_3 + k_5)]}{\text{denominator}} \tag{4.60b}$$

the denominator being the same as in eq. (4.59). In this case, the transport rate expression may be written as

$$J_s = \frac{J_{\max} K_T (s_I - s_{II})}{(K_T + s_I)(K_T + s_{II}) - x^2 K_T^2} \tag{4.61}$$

with $K_T = \alpha k_5/k_7$, $J_{\max} = c_t k_7/2\alpha$, $x = (1 - \alpha)/\alpha$, where $\alpha = (k_3 k_5 + k_5 k_7 + 2k_3 k_7)/2k_3 k_5$ (cf. Lieb and Stein, 1971).

For very low concentrations of S, the initial rate derived from eq. (4.59) reduces to

$$J_{s(0)} = c_t/2 \frac{k_1 k_3 s_I}{2k_3 + k_5} \tag{4.62a}$$

while for very high concentrations to

$$J_{s(0)} = c_t \frac{k_3 k_5 k_7}{k_3 k_5 + k_5 k_7 + 2k_3 k_7} \tag{4.62b}$$

Macroscopically, this system cannot be distinguished from one where no two constants are identical but where $K_{CS_I} k_8/k_3 = K_{CS_{II}} k_7/k_4$, K_{CS_I} being k_2/k_1, $K_{CS_{II}}$ being k_5/k_6 (cf. Geck, 1971).

An expression which is identical with eq. (4.55) but which allows a simple calculation of flows when $s_I = s_{II}$ was derived by Regen and Morgan (1964)

$$J_s = c_t \left[\frac{As_I}{1 + Bs_I + \dfrac{(s_I - s_{II})\,C}{1 + s_{II}/D}} - \frac{As_{II}}{1 + Bs_{II} + \dfrac{(s_{II} - s_I)/C'}{1 + s_I/D}} \right] \quad (4.63)$$

where $A = k_1 k_3 k_5 k_7 (= k_2 k_4 k_6 k_8)/\alpha\beta$;

$B = k_1 k_7 \gamma / k_2 k_4 \alpha = k_6 k_8 \gamma / k_3 k_5 \alpha$;

$C = k_1 k_3 k_5 [k_2 k_4 - k_7(k_2 + k_3 + k_4)]/\alpha\beta k_2 k_4$;

$C' = k_2 k_4 k_6 [k_3 k_5 - k_8(k_3 + k_4 + k_5)]/\alpha\beta k_3 k_5$;

$D = k_7 \beta / k_2 k_4 k_6$; and

$\alpha = k_7 + k_8$;

$\beta = k_2 k_4 + k_2 k_5 + k_3 k_5$;

$\gamma = k_3 + k_4$.

4.3.1.2. *Equilibrium approach*

While the above expressions are more universal than those that will follow, they are bulky and in many cases a rather substantial simplification is justified. One makes use of the assumption that the translocation constants are much smaller than the dissociation constants at the membrane surface, whereupon*

$$J_s = 2k_c k_{cs} c_t \frac{s_I' - s_{II}'}{2(k_c + k_{cs} s_I' s_{II}') + (k_c + k_{cs})(s_I' + s_{II}')} \quad (4.64)$$

where $k_c = k_7 = k_8$, $k_{cs} = k_3 = k_4$; $s_I' = s_I/K_{CS} = s_I k_1/k_2$; $s_{II}' = s_{II}/K_{CS} = s_{II} k_6/k_5$.
The initial rate ($s_{II} = 0$) is then

$$J_{s(0)} = 2k_c k_{cs} c_t \frac{s_I'}{2k_c + s_I'(k_c + k_{cs})} \quad (4.65)$$

so that $J_{max} = 2k_c k_{cs} c_t/(k_c + k_{cs})$ and $K_T = 2k_c K_{CS}/(k_c + k_{cs})$. Like with eq. (4.60), one can derive simple formulae for uptake at very low or very high concentrations of substrate.

The sum of unidirectional fluxes is

$$J_{s*} = 2k_{cs} c_t \frac{s_I^{*\prime}(k_c + k_{cs} s_{II}') - s_{II}^{*\prime}(k_c + k_{cs} s_I')}{2(k_c + k_{cs} s_I' s_{II}') + k_c(s_I' + s_{II}')}. \quad (4.66)$$

---

* The formula can be derived either by introducing the simplifying assumptions into eq. (4.55) or from the assumption of equilibria at the surfaces and setting up carrier flux equations (for J_C and J_{CS}), the sum of which must be equal to zero (*cf.* Kotyk and Janáček, 1975).

The ratio of k_{cs}/k_c $(= \varrho)$ can be determined experimentally by comparing the initial rate of efflux of S into an equilibrium concentration of S $(J_=)$ with that into a substrate-free medium (J_0).

$$J_0 = -2c_t k_c k_{cs} s_{II}^* / [2k_c K_{CS} + s_{II}(k_c + k_{cs})] \qquad (4.67a)$$

and

$$J_= = -c_t k_{cs} s_{II}^* / (K_{CS} + s) \qquad (4.67b)$$

(here $s = s_1 = s_{II}$).

By dividing these two equations we obtain

$$\varrho = \frac{2K_{CS}(1 - x) + s(2 - x)}{sx} \qquad (4.68a)$$

where $x = J_0/J_=$.

For very high values of s, moreover,

$$\varrho = (2 - x)/x. \qquad (4.68b)$$

In this equilibrium model, like in the previous steady-state one, if $k_{cs} > k_c$, one will observe the phenomenon of trans acceleration or the preloading effect, i.e., that the initial rate of uptake of substrate will be increased by preincubation with the same or a related substrate.

We may proceed further in simplifying the rate expression by assuming ϱ to be equal to one $(k_c = k_{cs} = k)$ whereupon

$$J_s \doteq kc_t/2 \left(\frac{s_1}{s_1 + K_{CS}} - \frac{s_{II}}{s_{II} + K_{CS}} \right) \qquad (4.69)$$

this being applicable probably only in rare cases (cf. Regen and Morgan, 1964). For the initial rate of uptake then

$$J_{s(0)} = kc_t/2 \frac{s_1}{s_{1} + K_{CS}} \qquad (4.70)$$

so that $J_{max} = kc_t/2$ and $K_T = K_{CS}$.

The unidirectional fluxes here are determined by the individual terms in brackets, thus

$$\overrightarrow{J_{s*}} = kc_t \frac{s_1^*}{s_1^* + K_{CS}} \qquad (4.71a)$$

and

$$\overleftarrow{J_{s*}} = kc_t \frac{s_{II}^*}{s_{II}^* + K_{CS}}. \qquad (4.71b)$$

Thus, in this case, the unidirectional fluxes are independent of each other and of the trans concentration.

. If two substrates are present which share the same carrier, the simplest model yields the following equation:

$$J_{s(r)} = J_{max}\left(\frac{s_1}{s_1 + K_{CS}(1 + r_1/K_{CR})} - \frac{s_{11}}{s_{11} + K_{CS}(1 + r_{11}/K_{CR})}\right) \quad (4.72)$$

the expression (for $s_{11}, r_{11} = 0$) being identical with that for competitive inhibition in enzyme kinetics.

The three cases discussed so far (eq. 4.56, 4.65 and 4.70) are thus seen to be indistinguishable if only the apparent J_{max} and K_T are evaluated. All of them, moreover, will obey first-order kinetics at very low concentrations of substrate when

$$J_s = J_{max}(s_I - s_{II}) \quad (4.73)$$

This presents a danger when studying systems with a high K_T (when the values of s used are relatively small), when one is inclined to define the situation as simple, rather than mediated, diffusion. Other properties should always be examined concurrently, such as inhibition by heavy metal ions (UO_2^{2+}, Th^{4+}, La^{3+}) or competition with a high-affinity analogue, etc.

At high values of s, all the systems studied yield

$$J_s = J_{max}K'(1/s_{II} - 1/s_I) \quad (4.74)$$

where K' is a composite constant, depending on the degree of complexity of the system (it is equal to K_{CS} according to eq. 4.69).

At higher levels of substrate, the actual J_s is thus very substantially dependent on the magnitude of s_{II}, particularly at its low levels (beginning of uptake experiments).

The similarity between the three expressions mentioned above extends further. One can derive various parameters related to J_{max} and K_T by the application of tracers (usually the radioactively labeled form of substrate examined). The movement of a tracer in a steady state of the system (when $s_I = s_{II}$ or, more generally, when $J_s = 0$) always obeys the laws of diffusion (*cf.* the cases of various compartments on p. 203).

Taking the case described by eq. (4.71) we see for $s_I = s_{II} = s$ that

$$\overrightarrow{J_{s*}} \equiv ds_{II}^*/dt = J_{max}(s_I^* - s_{II}^*)/(s + K_{CS}). \quad (4.75)$$

Integration from 0 to s_{II} yields

$$J_{max}t = (s + K_{CS}) \ln \left[s_I^*/(s_I^* - s_{II}^*) \right] \tag{4.76}$$

the half-time of tracer equilibration (when $s_{II} = 0.5s_I$) being

$$t_{0.5} = 0.693(s + K_{CS})/J_{max}. \tag{4.77}$$

Plotting $t_{0.5}$ against different concentrations of s yields a straight line, intersecting the abscissa at $-K_{CS}$ ($= -K_T$) and having a slope of $0.693/J_{max}$ ($= 1.386/c_t k$).

In the medium-complex case of eq. (4.66) an analogous plot will intersect the abscissa at $-K_{CS}$ (here, however, $K_T = 2k_c K_{CS}/(k_c + k_{cs})$!) and the slope will be $0.693/k_{cs}c_t$ (J_{max} is equal to $2k_c k_{cs}c_t/(k_c + k_{cs})$).

Unfortunately, in the steady-state case of eq. (4.58a, b) an analogous plot will yield a complex J_{max} and a constant related to the K_T in a complicated manner.

An interesting feature of carrier-mediated transports is the difference in the rate of equilibration of a tracer in the presence of a larger amount of unlabeled substance and that of the analytical substance itself.

Taking the case when cells do not change volume on taking up solute and when the size of the external medium is much greater than that of cells, the half-time of equilibration of a tracer in simple diffusion is equal to the half-time of equilibration of an analytical substance, viz. $t_{0.5} = 0.693/k$ (cf. eq. 4.43 and 4.46'). On the other hand, in carrier-mediated diffusion, the tracer half-equilibration time (taking the simplest model) is $t_{0.5} = 0.693(s + K_{CS})/J_{max}$ (cf. eq. 4.77) while the half-equilibration of the total concentration is given by

$$t_{0.5} = (s + K_{CS}) (0.693 + 0.193s/K_{CS})/J_{max}. \tag{4.78}$$

Hence, for a high-affinity process, where $K_{CS} = 10^{-4} M$ and for $s = 0.1 M$ we have for the tracer equilibration time $0.0694 M/J_{max}$, while for the analytical half-equilibration time $19.39 M/J_{max}$. (For the situation where the cells respond to osmotic changes in the medium and when their amount is comparable with that of the medium, a much more complicated case, consult LeFevre and McGinnis (1960).)

The above consideration is directly related to the well-known Ussing (1949) flux ratio which states that only in simple diffusion will the ratio of fluxes $\overrightarrow{J_s}/\overleftarrow{J_s}$ be equal to the ratio of concentrations s_I/s_{II}. In carrier-mediated transport (taking again the simplest case) we have, indeed, $\overrightarrow{J_s}/\overleftarrow{J_s} = (s_I/s_{II})\,(K_{CS} + s_{II})/(K_{CS} + s_I) < 1$, unless, of course, $s_I = s_{II}$.

The ratio shows the interesting property of carrier fluxes that at high concentrations (when $s > K_{CS}$) the ratio will be practically equal to unity even if s_I is very different from s_{II}.

It may be of importance to decide whether the translocation across the membrane is rate-limiting (then the "equilibrium" equations 4.64, etc., apply) or whether none of the rate constants is pronouncedly smaller or, finally, whether the surface dissociation is much slower than the translocation. While the last-named situation is very unlikely to occur (uphill transport most of countertransport and could not be observed), distinction between the first two cases deserves attention.

The simplest and oldest approach is due to Wilbrandt (1954) who suggested plotting J_s against s_I for various s_{II} (Fig. 4.6).

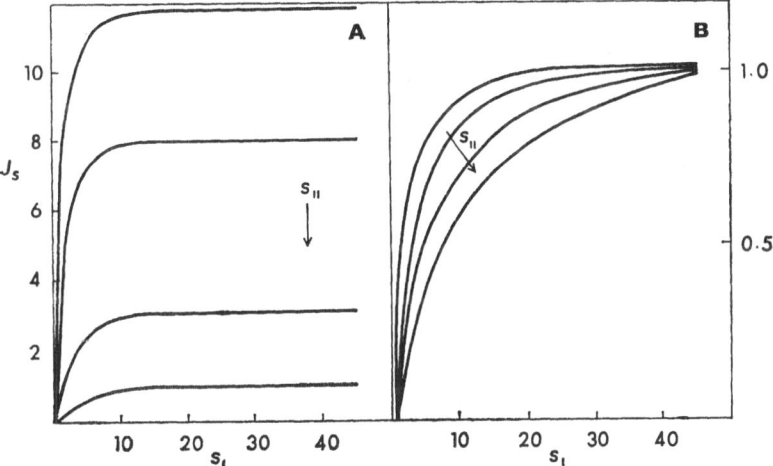

Fig. 4.6. Differentiation between a carrier transport with rate limitation by the translocation across the membrane (**A**) and with no pronounced rate limitation (**B**). The values of s_{II} in this particular example (taken from Wilbrandt, 1954) range from 0.08 to 0.13, 0.3 and 0.8.

The other approach was developed by Hoare and is founded on comparing the maximum rate of influx into preequilibrated cells with the maximum rate of influx into empty cells. The latter is derived from eq. (4.56) as

$$\overrightarrow{J}_{max} = \frac{c_t k_3 k_5 k_7}{k_7(k_3 + k_4) + k_5(k_3 + k_7)} \qquad (4.79a)$$

while the former (from eq. 4.63) is

$$\overleftrightarrow{J}_{max} = \frac{c_t k_2 k_3 k_4 k_5}{(k_3 + k_4)[k_5(k_2 + k_3) + k_2 k_4]}. \tag{4.79b}$$

If the translocation is rate-limiting ($k_1, k_2, k_5, k_6 \gg k_3, k_4, k_7, k_8$)

$$\overrightarrow{J}_{max}/\overleftrightarrow{J}_{max} = k_7(k_3 + k_4)/k_4(k_3 + k_7) \tag{4.80a}$$

which can have values both smaller and greater than unity, depending on the relative mobilities k_4 and k_7. If the translocation is not rate-limiting (all the constants are of the same order of magnitude) we have

$$\overrightarrow{J}_{max}/\overleftrightarrow{J}_{max} = \frac{k_7(k_3 + k_4)[k_5(k_2 + k_3) + k_2 k_4]}{k_2 k_4 [k_7(k_3 + k_4) + k_5(k_3 + k_7)]} \tag{4.80b}$$

which is always greater than unity, particularly if the surface reaction between substrate and carrier should become rate-limiting. Thus, whenever the above ratio is found to be less than one we can postulate a translocation-limited mechanism.

All the above carrier mechanisms (as well as any other saturable mechanism with spatially or functionally separated events at the two membrane surfaces) display the phenomenon of countertransport which consists in an uphill movement of one carrier substrate at the expense of the downhill movement of another substrate, no metabolic energy being required.

There are two basic ways of demonstrating countertransport (it will be shown on the simple case of eq. 4.72).

1. If cells are preloaded with substrate R so that $r_I = r_{II}$ and then S is added (at that moment $s_{II} = 0$) the movement of R will be described by

$$J_r = J_{max}\left(\frac{r'}{1 + s' + r'} - \frac{r'}{1 + r'}\right) \tag{4.81}$$

where $r' = r/K_{CR}$; $s' = s/K_{CS}$). This is less than zero indicating that R will move out of cells and hence against its concentration gradient (initially $r_I = r_{II}$!).

2. The rate of uptake of a minute concentration of labelled substrate into empty cells is compared with that into cells preloaded with the substrate (Fig. 4.7) when an overshoot is observed. This demonstration of countertransport (due to Miller, 1965) is superior

to the first case because it involves no osmotic effects. In the first case, a countertransport-like movement of substrate out of cells might be produced by osmotic pressure increase and hence cell shrinkage due to adding the high concentration of substrate S.

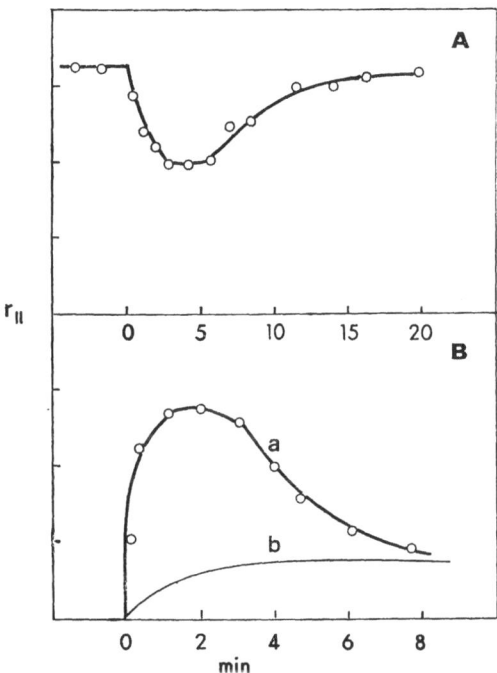

Fig. 4.7. Countertransport according to Rosenberg and Wilbrandt (1957) using the example of D-xylose in baker's yeast (**A**), and according to Miller (1965) using the example of galactose in human erythrocytes. (**B**) Curve a refers to cells preincubated with unlabeled galactose, curve b to a control without preincubation.

Unstirred layers (*cf.* p. 198) will affect the apparent K_T of a saturable transport process (Winne, 1973). The transport rate of S across the unstirred layer (in mol s^{-1} cm^{-2}) may be written as

$$J_s = (D/\delta)(s_B - s_M) \qquad (4.82)$$

where D is the diffusion coefficient in the unstirred layer (cm^2 s^{-1}), δ the layer thickness (cm), and s_B and s_M are the molarities of substrate in the mixed bulk phase and at the membrane surface, respectively. Combining eq. (4.82) with the simple eq. (4.70) ($s_M \equiv s_I$) we can solve for s_M and introduce the suitable root into eq. (4.82). The bulky expression

$$J_s =$$
$$= (D/\delta)[0.5(K_T + s_B + J_{max}\delta/D) \pm \sqrt{0.25(K_T - s_B + J_{max}\delta/D)^2 + s_B K_T}]$$
$$(4.83)$$

simplifies for $s \ll K_T$ to

$$J_s = s_B/(\delta/D + K_T/J_{max}) \qquad (4.84a)$$

and for $s \gg K_T$ to

$$J_s = J_{max} \qquad (4.84b)$$

and permits to show that for $J_s = 0.5 J_{max}$ (when s_B is the concentration giving rise to half maximum transport) we have

$$s_B = K_T + 0.5 J_{max}\delta/D \qquad (4.85)$$

The value of K_T obtained experimentally is thus greater than the true one (cf. Fig. 4.8). The presence of interfering unstirred layers is

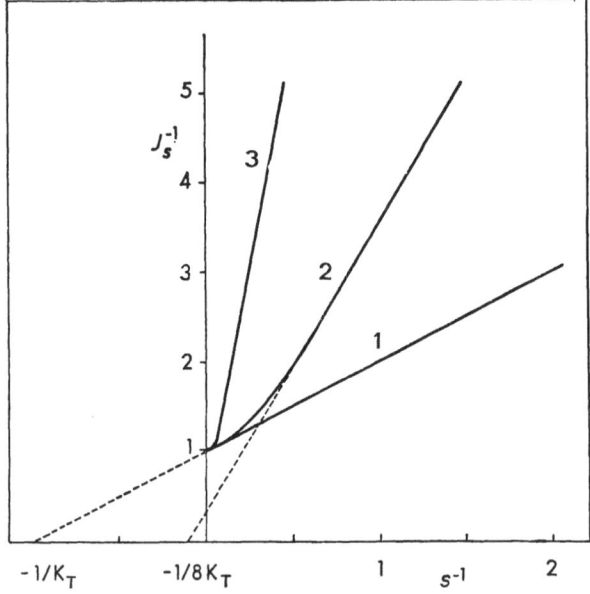

Fig. 4.8. Influence of unstirred layer on a Lineweaver—Burk plot of carrier-mediated transport. Curve 1, no unstirred layer; curve 2, $K_T = 0.5 J_{max}\delta/D$; curve 3, $5K_T = 0.5 J_{max}\delta/D$. Values used: $J_{max} = 1$ nmol s^{-1} cm^{-2}, $K_T = 1$ mM, $D = 5 . 10^{-6}$ cm^2 s^{-1}, $\delta = 0.1$ mm (curve 2) or 0.5 mm (curve 3). Even with the lower thickness of the unstirred layer the error in the estimation of K_T amounts to eight-fold. (Adapted from Winne, 1973.)

indicated if, on increasing the rate of stirring, the experimental value of K_T decreases.

If one sets $D/\delta = P$ and rearranges eq. (4.82) one obtains for the unstirred layer at side I

$$s_\mathrm{I} = s_{\mathrm{B,I}} - J_s/P_1 \qquad (4.86a)$$

and similarly for side II

$$s_\mathrm{II} = s_{\mathrm{B,II}} + J_s/P_2. \qquad (4.86b)$$

Manipulation of these equations in combination with eq. (4.55) (for an extremely lucid treatment see Lieb and Stein, 1974) permits to determine the effective permeability of the unstirred layer by plotting $1/J_s$ against $s_\mathrm{B}/(J_\infty - J_s)$ (Fig. 4.9) where J_s is the net flow rate from

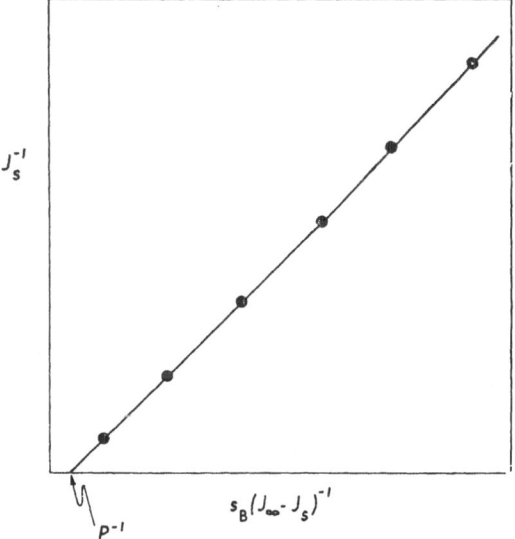

Fig. 4.9. Determination of unstirred layer permeabilities.
(According to Lieb and Stein, 1974.)

an extremely high *cis*-concentration against a concentration s_B, J_∞ is the net flow rate from the same high *cis*-concentration when $s_\mathrm{B} = 0$. Starting from a very high $s_{\mathrm{B,I}}$ against different values of $s_{\mathrm{B,II}}$, the intercept of the line in Fig. 4.9 denotes $1/P_2$, i.e., the unstirred layer permeability inside the cell in the usual arrangement.

Underestimation of the role of unstirred layers on transport kinetics has in fact led to attempts to supplant the carrier model with more sophisticated versions, e.g., that by Lieb and Stein (1970) where tetrameric proteins were assumed to span the membrane, possessing high- and low-affinity subunits equally distributed on both sides. Another model was that of introvertive hemiports lining the membrane faces (LeFevre, 1973) and another still that by Naftalin (1970) who postulated the existence of transmembrane pores lined with binding sites, the substrate having restricted chances of transfer from one site to the other.

However, a recent report by Regen and Tarpley (1974) demonstrates rather convincingly that a carrier of the type described by equation (4.55) and unstirred layers at both sides of the membrane provide an explanation for all the existing experimental findings. Thus, although the polymeric models would provide greater versatility particularly if cooperative interactions were allowed, there is at present no need to replace the conceptually simpler mobile carrier model. An extremely clear review of the situation can be found in Le Fevre (1975).

4.3.1.3. Two-site carrier

In principle, multi-site carriers can exist and at least a few cases have been described in active or coupled transports (adenosinetriphosphatase transport of Na^+ and K^+: Repke and Schön, 1973; symport of two sodium ions with amino acid in pigeon erythrocytes: Vidaver and Shepherd, 1968). They may be kinetically indistinguishable from single-site carriers if the sites are mutually independent or they may show some kind of cooperativity. If it is negative (corresponding to a Hill coefficient of enzyme kinetics less than one), high concentrations of substrate would decrease the actual rate of uptake (a type of excess-substrate inhibition). If it is positive (corresponding to a Hill coefficient greater than one) it may be due either to a greater affinity of any subsequent site for its substrate, or to a greater translocation constant of complexes with more than one molecule of bound substrate.

A rate equation proceeding from the equilibrium assumption (translocation constants being rate-limiting) for two competing substrates S and R and two binding sites* has the form

$$J_{s(r)} = 2c_t \frac{x_{II}y_I - x_I y_{II}}{x_{II}w_I + x_I w_{II}} \qquad (4.87)$$

* Any greater number of cooperating binding sites will produce qualitatively the same effects.

where $x = s'(k_{cs} + k_{css}s'' + k_{csr}r''') +$
$\qquad + r'(k_{cr} + k_{crr}r'' + k_{crs}s''') + k_c$
$\quad y = s'(k_{cs} + k_{csr}r'' + 2k_{css}s'') + k_{crs}s'''r'$
$\quad w = 1 + s'(1 + s'' + r''') + r'(1 + r'' + s''').$

Here $s' = s/K_{CS}$, $s'' = s/K_{CSS}$, $s''' = s/K_{CRS}$ and analogously for r', r'' and r'''.

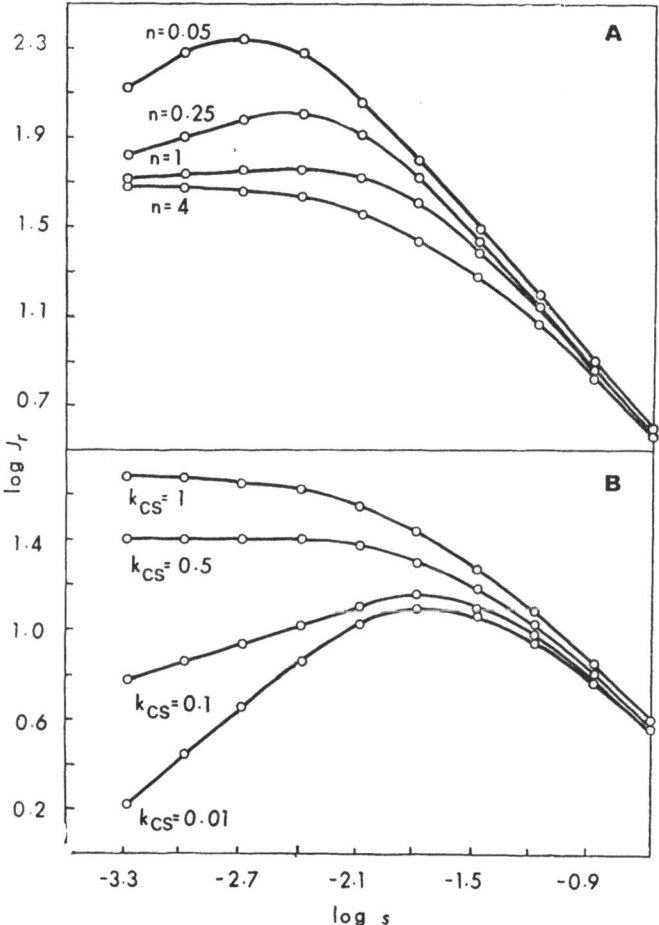

Fig. 4.10. Dependence of the rate of uptake of labelled solute (J_r) added to various concentrations of unlabelled solute S on the concentration of the latter (s), in a two-site carrier system with positive cooperativity. In **A**, n denotes the ratio of K_{CSS}/K_{CS}; in **B**, the translocation constant of the singly charged carrier is varied. (Adapted from Wilbrandt, W. and Kotyk, A. (1964). *Naunyn—Schmiedebergs Arch. exp. Path. u. Pharmak.* **249**, 279.)

The equation predicts several interesting phenomena.

1. The dependence of initial rate on concentration will be sigmoid.

2. If cells are prequilibrated with R and the S is added and its rate of uptake plotted against r (preferably in logarithmic coordinates), the curve will pass through a maximum rather than have a zero slope at low and a slope of -1 at high concentrations (Fig. 4.10).

3. In an experiment with countertransport of 1st type (p. 216) using two forms of the same substrate, the movement of preequilibrated R at the moment of adding S will be given by

$$J_r = sr(1 - r's'' - r'r'')\, a/D$$

where a is a complex positive constant and D the denominator containing only positive terms. Thus, for relatively high concentrations of S and R, $J_r < 0$ (*countertransport*) while for very low concentrations of S and R, $J_r > 0$ (the so-called *cotransport*). Here the addition of the new substrate will actually cause more of the originally equilibrated substrate to move into cells.

It should be noted that the same phenomena are to be observed in a system with no pronounced rate limitation by transmembrane translocation although the appropriate equations are prohibitively complicated.

4.3.2. Primary active transport

4.3.2.1. Kinetics

The only difference between the kinetics of mediated diffusion and that of active carrier transport is in the fact that for active transport energy is "fed" into one of the steps in such a way that for J_s to become zero we have $s_{II}/s_I \neq 1$. In the case described by eq. (4.55) the zero net flow condition is satisfied for $s_{II}/s_I = k_1 k_3 k_5 k_7 / k_2 k_4 k_6 k_8$. Thus, an increase (by metabolic energy) of any of the odd-numbered constants or a decrease of any of the even-numbered ones will result in accumulation of substrate S.

Using a stratagem of enzyme kinetics (*cf.* Cleland, 1963) one can convert the unwieldy groups of rate constants of eq. (4.55) into more meaningful kinetic constants. The equation can be rewritten as

$$J_s = \frac{N_1 s_I - N_2 s_{II}}{\text{const.} + D_1 s_I + D_2 s_{II} + D_{12} s_I s_{II}}$$

or

$$J_s/J_{\max}^{\rightarrow} = \frac{s_{\mathrm{I}} - s_{\mathrm{II}}/K_{\mathrm{eq}}}{K_{T_1}(1 + s_{\mathrm{II}}/K_{T_2}) + s_{\mathrm{I}}(1 + s_{\mathrm{II}}/K_i)}$$

where $K_{\mathrm{eq}} = N_1/N_2 = K_{T_2}J_{\max}^{\rightarrow}/K_{T_1}J_{\max}^{\leftarrow}$; $K_{T_1} = \mathrm{const.}/D_1$; $K_{T_2} = \mathrm{const.}/D_2$; $K_i = D_1/D_{12}$; $J_{\max}^{\rightarrow} = N_1/D_1$. On setting $\Delta s = s_{\mathrm{I}} - s_{\mathrm{II}}/K_{\mathrm{eq}}$ and inverting the last equation we obtain

$$1/J_s = \left[\frac{K_{T_1}}{J_{\max}^{\rightarrow}}(1 + s_{\mathrm{II}}/K_{T_2}) + \frac{s_{\mathrm{II}}}{J_{\max}^{\rightarrow}K_{\mathrm{eq}}}(1 + s_{\mathrm{II}}/K_i)\right](1/\Delta s) +$$

$$+ \frac{1}{J_{\max}^{\rightarrow}}(1 + s_{\mathrm{II}}/K_i)$$

On plotting $1/J_s$ against $1/\Delta s$ for different values of s_{II} we arrive at a set of $J_{\max_{\mathrm{app}}}$'s which are related to true $J_{\max}^{\rightarrow}$ by

$$1/J_{\max_{\mathrm{app}}} = s_{\mathrm{II}}/J_{\max}^{\rightarrow}K_t + 1/J_{\max}^{\rightarrow}$$

A plot of $1/J_{\max_{\mathrm{app}}}$ against s_{II} will yield $1/J_{\max}^{\rightarrow}$ as the intercept with the ordinate and $-K_i$ as the intercept with the abscissa. Similarly, the intercepts with the abscissa in the $1/J_s$ vs. $1/\Delta s$ plot will yield a set of $K_{T_{\mathrm{app}}}$'s which are related to the true K_T's by

$$K_{T_{\mathrm{app}}} = K_{T_1}\frac{1 + s_{\mathrm{II}}/K_{T_2}}{1 + s_{\mathrm{II}}/K_i} + s_{\mathrm{II}}/K_{\mathrm{eq}}$$

Since K_{T_1} is known from measuring the initial velocity of uptake at different concentrations of S_1 and K_i was determined in the previous replot and since K_{eq} can be determined readily from the steady-state accumulation ratio (for mediated diffusion it is equal to 1) we can calculate the value of K_{T_2} and hence $J_{\max}^{\leftarrow}$ can be derived from the definition of K_{eq} (cf. Cuppoletti and Segel, 1975; Betz et al., 1975).

The above mechanism has a serious drawback in that it predicts constant $s_{\mathrm{II}}/s_{\mathrm{I}}$ ratios irrespective of concentration. It is generally observed, however, that these ratios decrease with increasing concentrations, either tending toward unity or decreasing further, the value of s_{II} being then seemingly constant (Fig. 4.11).

Hence somewhat more elaborate models have to be developed, viz. for the first case a model where the carrier exists in two forms with different affinities at each of the membrane sides, the conversion between them being relatively slow; for the second case a model where the movement' across the membrane of the association-

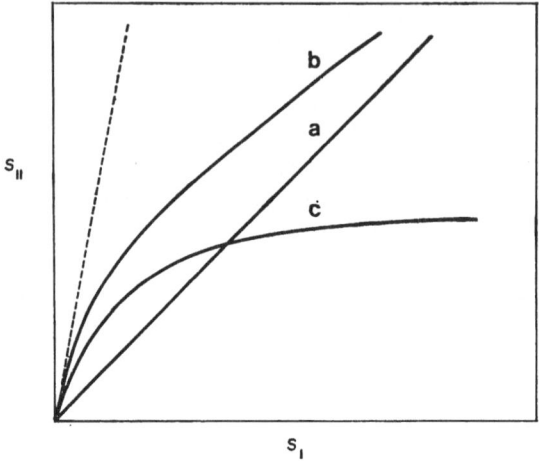

Fig. 4.11. Dependence of steady-state intracellular concentration (s_{II}) on the external concentration (s_I). Curve a (e.g. D-xylose in *Saccharomyces cerevisiae*) reflects a mediated diffusion; curve b (e.g. D-xylose in *Rhodotorula glutinis*) reflects an active transport which resembles mediated diffusion at high s_I (then $s_{II}/s_I = 1$); curve c (e.g. D-xylose in *Candida beverwijkii*) reflects an active transport which is strictly dependent on a source of energy which becomes less available at high values of s_I (then s_{II} rises very slowly). The dashed line depicts the case of a constant ratio over the entire concentration range as predicted by the simple models.

dissociation reaction is obligatorily dependent on the presence of a source of energy which is only present in a limited amount.

The first case may be depicted as follows:

$$
\begin{array}{ccc}
\mathrm{I} & & \mathrm{II} \\
& \mathbf{CS}_I \underset{k_4}{\overset{k_3}{\rightleftharpoons}} \mathbf{CS}_{II} & \\
k_1 s_I \big\updownarrow k_2 & & k_6 s_{II} \big\updownarrow k_5 \\
& \mathbf{C}_I \underset{k_{18}}{\overset{k_{17}}{\rightleftharpoons}} \mathbf{C}_{II} & \\
k_{15} \big\updownarrow k_{16} & & k_8 \big\updownarrow k_7 \\
& \mathbf{Z}_I \underset{k_{20}}{\overset{k_{19}}{\rightleftharpoons}} \mathbf{Z}_{II} & \\
k_{13} \big\updownarrow k_{14} s_I & & k_{10} \big\updownarrow k_9 s_{II} \\
& \mathbf{ZS}_I \underset{k_{11}}{\overset{k_{12}}{\rightleftharpoons}} \mathbf{ZS}_{II} &
\end{array}
$$

C represents the carrier with high affinity for substrate, Z the carrier with low affinity. Metabolic energy may in this case stimulate either k_7 or k_{15}. Even if the association-dissociation reactions between C, Z and S are considered to be in equilibrium and symmetrical at the two faces and if $k_3 = k_4 = k_{11} = k_{12} = k_{cs}$, $k_{17} = k_{18} = k_{19} = = k_{20} = k_c$, the rate equation is quite bulky (here $s' = s/K_{CS}$ and $s'' = s/K_{ZS}$)

$$J_s = J_{CS} + J_{ZS} = 2c_t k_{cs} \frac{X_{II}(s'_I + s''_I \alpha\beta) - X_I(s'_{II} + s''_{II}\beta\gamma)}{A_I X_{II} + A_{II} X_I} \quad (4.88)$$

with $A_I = 1 + s'_I + \alpha\beta + s''_I \alpha\beta$

$A_{II} = 1 + s'_{II} + \beta\gamma + s'_{II}\beta\gamma$

$X_I = k_c(1 + \alpha\beta) + k_{cs}(s'_I + s''_{II}\alpha\beta)$

$X_{II} = k_c(1 + \beta\gamma) + k_{cs}(s'_{II} + s''_{.I}\beta\gamma)$

where $\alpha = (k_c + k_{cs}s'_I)/(k_c + k_7 + k_{cs}s'_{II})$

$\beta = k_7/k_{15}$

$\gamma = (k_c + k_{15} + k_{cs}s''_I)/(k_c + k_{cs}s''_{II})$.

The accumulation ratio s_{II}/s_I may be shown to be equal to

$$\frac{[K_{ZS}k_{15}(k_c + k_7) + K_{CS}k_c k_7]}{[K_{CS}k_7(k_c + k_{15}) + K_{CS}k_c k_{15}]} \quad (4.89)$$

for low concentrations of substrate (i.e., a constant) but equal to one for high concentrations of substrate.

To account for the values of s_{II}/s_I decreasing below unity, the most expedient model appears to be

$$
\begin{array}{ccc}
\text{I} & \overset{cq}{\underset{d}{\rightleftharpoons}} & \text{II} \\
\mathbf{CS} & & \mathbf{CS'} \\
as_I \Big\updownarrow b & \quad fs_{II} \Big\updownarrow e & \\
\mathbf{C} & \overset{h}{\underset{g}{\rightleftharpoons}} & \mathbf{C'}
\end{array}
$$

where Q is the source of energy required for the movement of S from side I to side II.

The steady-state concentration ratio of substrate S is

$$s_{II}/s_I = acegq/bdfh. \quad (4.90)$$

However, since $q_t = q + cs'$, q will decrease with increasing substrate concentration (limited, obviously, by the amount of c_t present). Since the relative concentration of CS' can be computed to be equal to

$$(acfqs_I s_{II} + acgqs_I + cfhqs_{II} + bfhs_{II})/c_t$$

a cubic equation for s_{II} can be set up:

$$b^2 d^2 f^3 h^2 (as_I + h) s_{II}^3 +$$
$$+ bdf^2 h \{b\ddot{u}eh(g + h) + as_I [bde(g + h) + dgh(b + e) +$$
$$+ egh(b + cc_t - cq_t)] + a^2 egs_I^2 (d + cc_t - cq_t)\} s_{II}^2 +$$
$$+ aefgs_I \{bdh(bd + be - ceq_t)(g + h) + as_I [bcdgh(c_t - q_t) +$$
$$+ bdgh(d + e) + bceghc_t - ceq_t(bdh + dgh + bgh + bdg)] -$$
$$- s_I^2 a^2 cdegq_t\} s_{II} - a^2 ce^2 g^2 s_I^2 q_t (d + e)(ags_I + bg + bh) = 0 \quad (4.91)$$

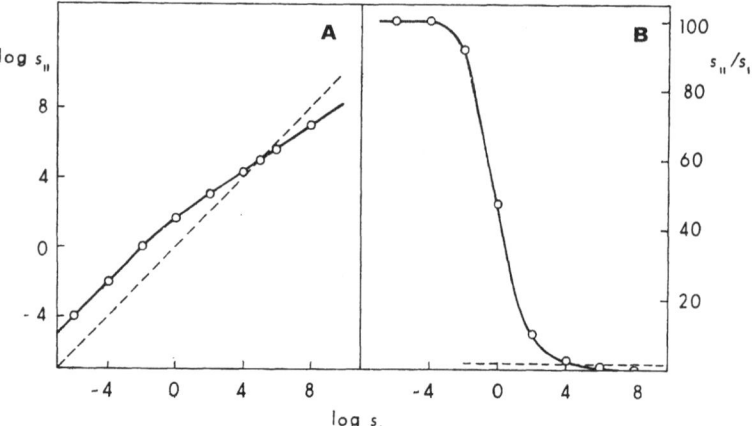

Fig. 4.12. Dependence of steady-state intracellular concentration s_{II} (**A**) and of the s_{II}/s_I ratio (**B**) on s_I, in a system with limited energy supply and with only the energized form of carrier mobile (see scheme on p. 224). The dashed lines denote distribution according to diffusion equilibrium.

An example of a numerical solution is shown in Fig. 4.12. Other models can explain this type of observation, their common feature being that the source of energy is of limited availability and that only the "energized" form can cross the membrane.

4.3.2.2. Combined systems

The case may arise that a compound is transported both by a saturable process and by simple diffusion as described on p. 192.

The situation is readily recognized by several characteristic features. 1. The inhibitory effect of a competing substrate is only partial even at extreme concentrations of inhibitor. 2. The effect of energy poisons (in the case of active transport) is only partial. 3. A plot of $J_{s(0)}$ against s or of $J_{s(0)}^{-1}$ against s^{-1} has the shape shown in Fig. 4.13.

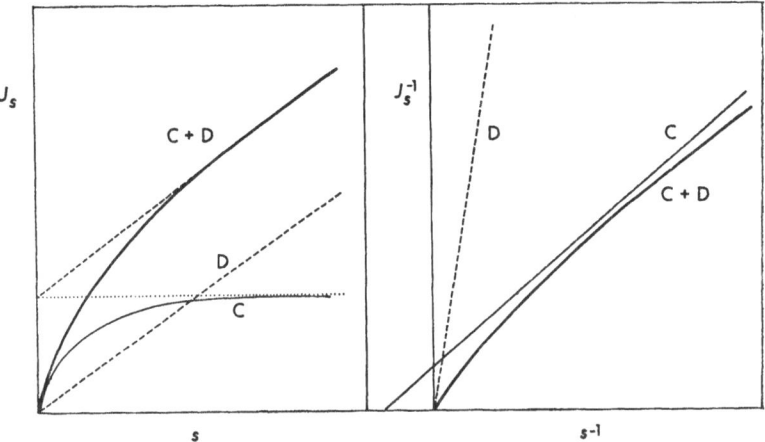

Fig. 4.13. A transport system combining diffusion with carrier transport. The flow-versus-concentration plot as well as the Lineweaver-Burk plot using reciprocal values is broken down to the diffusion (D) and carrier (C) components. (Adapted freely from Neame, K. D. and Richards, T. G. (1972). *Elementary Kinetics of Membrane Carrier Transport*. Blackwell Scientific Publications, Oxford.)

An analysis of such an event is fairly simple and follows from the figure.

A slightly more complicated case is the rather frequent transport of one substrate by two carrier systems. A plot of $J_{s(0)}^{-1}$ against s^{-1} is amenable to extracting from it all the pertinent data according to Neal (1972); *cf.* Kotyk and Janáček (1975).

A system of active transport operating simultaneously with a simple diffusion (a pump-and-leak model) will plot just as shown in Fig. 4.13 but the steady-state accumulation ratio will be generally given (from eq. 4.55 combined with 4.28) by

$$s_{II}/s_I = \frac{P(\alpha + \beta s_I + \gamma s_{II} + \delta s_I s_{II}) + k_1 k_3 k_5 k_7}{P(\alpha + \beta s_I + \gamma s_{II} + \delta s_I s_{II}) + k_2 k_4 k_6 k_8} \qquad (4.92)$$

α, β, γ, and δ being positive constants. Thus the ratio is nearly constant for very low concentrations and tends toward unity at high ones.

A qualitatively identical result is obtained from any of the simplified versions of the uptake equation.

4.3.2.3. Energetics

Calculation of the energy (in $J s^{-1}$) required to maintain a given concentration ratio proceeds from the modified Nernst formula, viz.

$$\text{Energy/time} = J_s RT \ln (s_{II}/s_I) \tag{4.93}$$

J_s being as usual the unidirectional flux inward. While the ratio of s_{II}/s_I is determined with greater or lesser accuracy without much difficulty, an exact expression for the flux is uncommonly difficult and can be generally obtained only for a system in surface equilibrium (cf. eq. 4.88). The most expedient way is to split eq. (4.93) into expressions relating to the fluxes of the individual carrier species, these being of first order, e.g.,

$$\text{Energy/time} = k_c[(c_I - c_{II}) RT \ln(c_I/c_{II}) +$$
$$+ (z_I - z_{II}) RT \ln (z_I/z_{II})] + k_{cs}[(cs_I - cs_{II}) RT \ln (cs_I/cs_{II}) +$$
$$+ (zs_I - zs_{II}) RT \ln (zs_I/zs_{II})]. \tag{4.94}$$

This equation predicts among other things that for low values of s the energy consumption will be greater than for high values of s, this being a consequence of the s_{II}/s_I approaching unity at very high concentrations (when $\ln (s_{II}/s_I) = 0$). However, a straightforward calculation is hardly ever possible, there being systems requiring energy only for translocation but not for maintaining the gradient (cf. Kotyk and Říhová, 1972), as well as others where s_{II} falls below s_I at very high concentrations and still energy is required for the transport inward (e.g., Deák and Kotyk, 1968). On the other hand, if a leak is present in parallel with an active transport, the energy required for maintaining a concentration ratio may be more than calculated from the above formula.

Moreover, as shown by Rosenberg and Wilbrandt (1963), the economy of the system may differ, depending on the ratio of affinities at the two membrane sides as well as on the efficiency with which one of the carrier forms is changed to the other.

4.3.3. Coupled transport

Among the numerous uphill-transporting systems now known to operate in cell membranes, the secondary mechanisms have come into clear prominence so that some authors will even doubt the existence of true primary active transports of nonelectrolytes. It is indeed of a distinct advantage to the cell if it can use as a source of energy the gradient of an ion that it expels into the medium perhaps for other reasons; thus H^+ must be removed from cells to maintain the intracellular pH at a reasonable level and Na^+ is undesirable in cells because it counteracts the activating effect of K^+ on many enzymes. Still, some overzealous supporters of ubiquitous ion-nonelectrolyte coupling may overlook the fact that even if, say sodium ions will drive sugars, amino acids, as well as other ions into cells, each of the substrate types requires its specific carrier or at least recognition protein so that the number of components of a coupled transport system is probably the same as that of a primary active transport.

Kinetically speaking, a ternary complex of carrier-nonelectrolyte-ion is the form commonly assumed to mediate this type of transport. The ternary complex may be formed in one of two ways:
(1) carrier + ion + substrate (C + A + S), or
(2) carrier + substrate + ion (C + S + A) as shown schematically below:

If the system is in surface equilibrium so that dissociation constants $K_1 - K_4$ may be used to describe the partial steps, the overall expression for Path I is

$$J_s = \frac{2c_t k_{as}}{\alpha_I \beta_{II} + \alpha_{II} \beta_I} (\beta_{II} s_I a_I / K_1 K_2 - \beta_I s_{II} a_{II} / K_1 K_2) \qquad (4.95)$$

with $\quad \alpha = 1 + a/K_1 + s \cdot a/K_1K_2;$

$\quad\quad \beta = k_c + k_a a/K_1 + k_{as} s \cdot a/K_1K_2.$

The initial rate $J_{s(0)}$ (for $s_{II} = 0$) is given by Michaelis – Menten type expressions with

$$J_{max} = \frac{2c_t(k_cK_1 + k_a a_{II})\,k_{as}}{K_1(k_{as} + k_c) + a_{II}(k_{as} + k_a)} \tag{4.96a}$$

and

$$K_T = \frac{K_2[(K_1 + a_{II})(k_cK_1 + k_a a_1) + (K_1 + a_1)(k_cK_1 + k_a a_{II})]}{a_1[K_1(k_{as} + k_c) + a_{II}(k_{as} + k_a)]}. \tag{4.96b}$$

The steady-state accumulation ratio (for $J_s = 0$) is given by

$$s_{II}/s_I = \frac{a_1(k_cK_1 + k_a a_{II})}{a_{II}(k_cK_1 + k_a a_1)}. \tag{4.96c}$$

If the ion-carrier complex is immobile ($k_a = 0$) we have

$$J_{max} = \frac{2c_t k_c k_{as}K_1}{K_1(k_{as} + k_c) + a_{II}k_{as}}, \tag{4.97a}$$

$$K_T = \frac{k_cK_1K_2(2K_1 + a_1 + a_{II})}{a_1[K_1(k_{as} + k_c) + k_{as}a_{II}]} \tag{4.97b}$$

and

$$s_{II}/s_I = a_1/a_{II}. \tag{4.97c}$$

If the formation of the ternary complex proceeds via path II the corresponding rate is expression is given by

$$J_s = \frac{2c_t}{\alpha_1\beta_{II} + \alpha_{II}\beta_1}\left[\beta_{II}\left(\frac{k_s s_1}{K_3} + \frac{k_{sa} s_1 a_1}{K_3K_4}\right) - \beta_1\left(\frac{k_s s_{II}}{K_3} + \frac{k_{sa} s_{II} a_{II}}{K_3K_4}\right)\right] \tag{4.98}$$

where $\alpha = 1 + s/K_3 + s \cdot a/K_3K_4$; $\beta = k_c + k_s s/K_3 + k_{sa} s \cdot a/K_3K_4$. The parameters of the initial rate expression are

$$J_{max} = \frac{2c_t k_c(k_s K_4 + k_{sa} a_1)}{K_4(k_s + k_c) + a_1(k_{sa} + k_c)} \tag{4.99a}$$

and

$$K_T = \frac{2k_c K_3 K_4}{K_4(k_s + k_c) + a_1(k_{sa} + k_c)} \tag{4.99b}$$

while

$$s_{II}/s_I = \frac{k_s K_4 + k_{sa} a_1}{k_s K_4 + k_{sa} a_{11}}.$$ (4.99c)

If the activation is essential ($k_s = 0$) we have

$$J_{max} = \frac{2c_t k_c k_{sa} a_1}{k_c K_4 + a_1(k_{sa} + k_c)},$$ (4.100a)

$$K_T = \frac{2k_c K_3 K_4}{k_c K_4 + a_1(k_{sa} + k_c)}$$ (4.100b)

and

$$s_{II}/s_I = a_I/a_{II}.$$ (4.100c)

The above expressions predict several possibilities of influencing the various parameters by the presence of A at the *cis* or at the *trans* side (Table 4.2). They simplify greatly if one assumes equal translocation constants for all the carrier forms.

TABLE 4.2. Effect of activating ion on transport parameters in a coupled transport in surface equilibrium

Mechanism	Ion at *cis* side	Ion at *trans* side
Path I	K_T	K_T, J_{max}
for $ks = 0$	K_T	K_T, J_{max}
for $ks = 0$		
and K_1 large	K_T	K_T
Path II	K_T, J_{max}	—
for $k_s = 0$	K_T, J_{max}	—
for $k_s = 0$		
and K_4 large	J_{max}	—
Random formation of CSA complex	K_T, J_{max}	K_T, J_{max}

A special case arises when the system contains C, CA, CS, CAS and when the dissociation constant of C−S is the same as that of CA−S. Then no accumulation is predicted and $s_{II}/s_I = 1$.

The membrane potential obviously plays a role in the accumulation of a nonelectrolyte by a coupled process. Since it is usually oriented with its negative side inward it will result in an attraction for the positively charged carrier complex toward the inside face of the membrane. Whatever the charge of the free carrier, the binding of a cation will increase it by one positive unit; in path II which is simpler to treat kinetically, the ternary complex CSA will have a positive charge with respect to C and CS. If this fact is reflected in $\overrightarrow{k_{sa}} >$ $> \overleftarrow{k_{sa}} (\equiv k' > k)$ the accumulation ratio for the path II mechanism is

$$s_{II}/s_I = \frac{k_s K_4 + k'a_I}{k_s K_4 + ka_{II}} \tag{4.101a}$$

and, for the case of immobile CS,

$$s_{II}/s_I = k'a_I/ka_{II}. \tag{4.101b}$$

This expression has the important consequence that even for $a_I/a_{II} < 1$ there may be a concentrative transport of S as long as $k'/k > a_I/a_{II}$.

Expressed quantitatively, the ratio of ternary complex fluxes may be written, in analogy to Ussing's flux ratio (1949)

$$\frac{J_{CSAI}}{J_{CSAII}} \equiv \frac{k'csa_I}{kcsa_{II}} = \frac{csa_I}{csa_{II}} e^{-nF\varphi/RT} \text{ and } \frac{k'}{k} = e^{-z} \tag{4.102}$$

where φ is the membrane potential in mV, F is the Faraday (96 500 coulombs/equivalent), n is the charge of the ion. Thus a ratio of fluxes equal to 100 is to be expected for a potential of 120 mV (negative inside) even if $a_I = a_{II}$.

Like in primary active transport, the accumulation ratios appear to be constant over the whole range of concentrations unless further assumptions are introduced. If, say, the values of a_I and a_{II} are very small or if K_4 is very large, or if $k_s \gg k_{sa}$, the accumulation ratio will tend toward unity. Likewise, in a steady-state approach (pertaining to the symmetrical path II of scheme on p. 229) the accumulation ratio is given by

$$s_{II}/s_I = \frac{k_s(d + 2k_{sa}) + ck_{sa}a_I}{k_s(d + 2k_{sa}) + ck_{sa}a_{II}} \tag{4.103}$$

and hence for $k_s \gg k_{sa}$ or for d/c very large, s_{II}/s_I will approach unity.

However, to express the fact that the ratio approaches unity only at high concentrations, and to account for values of the accumulation ratio less than one, one has to assume a "dissipation" of the ion gradient by the carrier-substrate flux just as derived at eq. (4.91). For a compulsory coupling ($k_{cs} = 0$) it will always hold that $s_{II}/s_I = $ $= a_I/a_{II}$ but a_I will be indirectly proportional to the amount of CS_I and hence S_I.

Somewhat more complex models of coupled transport have been designed to accommodate the simultaneous effects of two cations (e.g., Na^+ and H^+) by Curran and associates (1970) and by Frizzell and Schultz (1970).

4.4. CHEMICAL NATURE OF NONELECTROLYTE TRANSPORT SYSTEMS

Kinetic evidence has provided invaluable information about the properties of various transport systems but, like in the corresponding stage of enzymology, the time has come to isolate and define the chemical infividuals participating in transport. In contrast with enzymology, however, the work was hampered by the lack of any specific reaction of the transport components except binding their substrate. This led to the necessity of working with rather large quantities of material but helped in developing refined techniques of measuring ligand binding to macromolecules.

4.4.1. Group-translocation systems

In the field of nonelectrolytes the greatest amount of detailed information is available on a system that may be designated as group translocation rather than carrier-mediated transport, *viz.* the phosphotransferase system. Kundig and co-workers reported its existence in 1964 in *Escherichia coli* and since then it was discovered in *Salmonella typhimurium, Bacillus subtilis, Aerobacter aerogenes, Lactobacillus plantarum, Streptococcus lactis, Staphylococcus aureus, Rhodospirillum rubrum* and several *Mycoplasma* species. It is thus widely distributed among bacteria but it is not present in eukaryotic cells.

The function of the system is to phosphorylate sugars while translocating them from the medium into the cell. In Gram-negative species, the sugars thus phosphorylated are D-glucose, D-fructose D-mannose, mannitol, sorbitol, D-glucosamine, *N*-acetyl-D-glucosamine, 2-deoxy-D-glucose and β-glucosides. In Gram-positive species, the list probably includes also D-galactose, various pentoses, lactose, sucrose, trehalose, melibiose, maltose, melezitose, and glycerol. As far as is known, the position phosphorylated in monosaccharides is usually at the terminal carbon (5 or 6) but fructose is phosphorylated at carbon 1 and so is lactose.

The overall process common to all species is the following:

$$\text{phosphoenolpyruvate} + \text{enzyme I} \underset{}{\overset{Mg^{2+}}{\rightleftharpoons}} \text{P-enzyme I} + \text{pyruvate}$$

$$\text{P-enzyme I} + \text{HPr} \rightleftharpoons \text{HPr-P} + \text{enzyme I}$$

$$\text{HPr-P} + \text{sugar} \underset{}{\overset{E\ II,\ factor\ III,\ Mg^{2+}}{\rightleftharpoons}} \text{sugar-P} + \text{HPr}$$

The third reaction has been split up in the case of lactose in *Staphylococcus aureus* into

$$3\ \text{HPr-P} + \text{factor III}^{lac} \underset{}{\overset{EIIA}{\rightleftharpoons}} \text{factor III}^{lac}\text{-P}_3 + 3\ \text{HPr}$$

$$\text{factor III}^{lac}\text{-P}_3 + 3\ \text{lactose} \underset{}{\overset{EIIB^{lac}}{\rightleftharpoons}} 3\ \text{lactose-P} + \text{factor III}^{lac}$$

Hence one high-energy phosphate is required for the transport of one sugar molecule. Of the protein components involved, HPr (the heat-stable protein of molecular weight of about 9 000), enzyme I (molecular weight about 80 000) and factor III (molecular weight about 35 000) are soluble while enzyme II (molecular weight about 40 000) is membrane bound. There is a pronounced lipid requirement (for phospatidyl glycerol) of the phosphotransferase system in *Escherichia coli*.

The HPr protein appears to have an identical primary structure in the Gram-negative species and to be phosphorylated at a histidine residue.

There are at least three other group-translocating systems assumed to operate in nonelectrolyte transport.

1. The γ-glutamyltranspeptidase described in greatest detail in kidney cortex cells (Bodnaryk, 1972; Meister, 1973) uses intracellular glutathione as one substrate and extracellular amino acid (all natural amino acids except proline) as the second substrate, giving rise to γ-glutamylamino acid and cysteinylglycine. The γ-glutamyl derivative is then hydrolyzed to the originally transported amino acid and 5-oxoproline which is then used for the resynthesis of glutathione. The system operates rather inefficiently, requiring three molecules of ATP for the resynthesis of glutathione and hence for the transport of a single amino acid.

2. In *Escherichia coli*, adenine is transported with concomitant phosphoribosylation to adenylic acid (Hochstadt-Ozer and Stadtman, 1971).

3. In the brush-border membranes of the intestine, sucrose is transported only after previous splitting to its constituent monosaccharides. The system has been highly purified and incorporated in a functional state into an artificial phospholipid membrane (Storelli *et al.*, 1972). It remains to be explored whether any of the numerous disaccharidases present in the intestine are capable of a similar feat.

4.2.2. Oxidoreductive systems

Substantial evidence has accumulated on the intimate coupling between the oxidation of D-lactate or the artificial substrate ascorbate-phenazine methosulfate and the transport of sugars, amino acids and some ions in specially prepared membrane vesicles of *Escherichia coli, Salmonella typhimurium, Pseudomonas putida, Proteus mirabilis, Bacillus megaterium, Bacillus subtilis, Micrococcus denitrificans, Mycobacterium phlei* and *Staphylococcus aureus* (see, e.g., Kaback, 1972; Konings *et al.*, 1971).

Other substrates that may be used with different degrees of efficiency are α-glycerol phosphate and much less L-lactate, DL-α-hydroxybutyrate, and even formate.

The sugars thus involved are β-galactosides, galactose, arabinose, glucose-6-phosphate, gluconate and glucuronate; amino acids include all natural ones except glutamine (and asparagine ?), arginine, methionine and ornithine; the cations thus transported are possibly K^+ and Rb^+ (after treatment of the vesicles with valinomycin).

The postulated mechanism is depicted in an abridged form as follows:

$$
\begin{array}{ccc}
 & \text{lac} \quad \text{pyr} & \\
 & \searrow \nearrow & \\
C_{ox}S_{I} & \longrightarrow & C_{red}S_{II} \\
\uparrow & & \uparrow \\
S_{I} \nwarrow \big\downarrow & \text{lac} \quad \text{pyr} & \big\downarrow \searrow S_{II} \\
 & \searrow \nearrow & \\
C_{ox_{I}} & \longleftarrow & C_{red_{II}} \\
 & \swarrow \searrow & \\
 & cyt^{2+} \quad cyt^{3+} &
\end{array}
$$

The theory of oxidoreductive transport is not without problems, particularly in that it requires different oxidative pathways for different transported solutes and in that it has an even more attractive alternative for most of the arguments rallied in its support. This other possibility is affiliated with the proton-motive force generated by a proton-extruding ATPase or a proton-generating oxidative system

TABLE 4.3. Binding proteins for nonelectrolytes isolated from various cell plasma membranes

Transported substrate	Source	Molecular weight	K_D in vitro (M)	K_T in vivo (M)	Reference	Note
L-Leucine	Escherichia coli	36 000	$7 \cdot 10^{-7}$	10^{-6}	Furlong and Weiner (1970)	
L-Leucine	Escherichia coli	36 000	10^{-6}	10^{-6}	Anraku (1968a, b, c)	Binds also isoleucine and valine
L-Histidine	Salmonella typhimurium	26 000	$1.4 \cdot 10^{-7}$	$3 \cdot 10^{-8}$	Ames and Lever (1970, 1972)	The J protein
L-Histidine	Salmonella typhimurium	25 000	$1.5 \cdot 10^{-6}$	$3 \cdot 10^{-8}$	Rosen and Vasington (1971)	
L-Lysine	Escherichia coli	28 000	$1.5 \cdot 10^{-6}$	$5 \cdot 10^{-7}$	Rosen (1971)	Binds also arginine and ornithine
L-Arginine	Escherichia coli	27 700	$3 \cdot 10^{-8}$	$4 \cdot 10^{-7}$	Rosen (1973)	Specific
L-Arginine	Saccharomyces cerevisiae	5 000	$4 \cdot 10^{-4}$	10^{-5}	Opekarová et al. (1975)	
L-Glutamine	Escherichia coli	26 000	$3 \cdot 10^{-7}$	$8 \cdot 10^{-8}$	Weiner et al. (1970)	
L-Glutamate	Escherichia coli	30 000	10^{-6}	10^{-6} (Na$^+$) 10^{-5} (no Na$^+$)	Willis and Furlong (1974)	Contains an exposed S—S bond
L-Cystine	Escherichia coli	27 000	10^{-7}	10^{-8}	Berger and Heppel (1972)	
L-Phenylalanine	Comamonas	26 000	10^{-8}—10^{-7}	$2 \cdot 10^{-5}$	Kuzuya et al. (1971)	
L-Tryptophan	Neurospora crassa	200 000	$8 \cdot 10^{-5}$	$5 \cdot 10^{-5}$	Wiley (1970)	Binds also phenylalanine and leucine
D-Galactose	Escherichia coli	35 000	10^{-7} and 10^{-4}	$4 \cdot 10^{-7}$	Boos et al. (1972)	Exists in two conformations
D-Glucose	Saccharomyces cerevisiae	40 000	$1.1 \cdot 10^{-3}$	$7 \cdot 10^{-3}$	Horák and Kotyk (1973)	Lipoprotein
D-Glucose	Rat kidney cortex	60 000	$6.7 \cdot 10^{-7}$	$7 \cdot 10^{-5}$	Thomas (1973)	Binds also NEM and phlorizin

Transported substrate	Source	Molecular weight	K_D in vitro (M)	K_T in vivo (M)	Reference	Note
D-Glucose	Rabbit jejunum	$2 \times 120\,000$	10^{-3}	10^{-3}	Storelli et al. (1972)	Sucrase-isomaltase
D-Glucose	Aspergillus nidulans	29 200	$5.3 \cdot 10^{-4}$	10^{-4}	Desai and Modi (1975)	
L-Arabinose	Escherichia coli	35 000	$2.2 \cdot 10^{-6}$	10^{-6}	Hogg and Englesberg (1969); Schleif (1969)	High-affinity system
D-Ribose	Salmonella typhimurium	31 000	$3.3 \cdot 10^{-7}$	—	Aksamit and Koshland (1972)	
D-Ribose	Escherichia coli	—	$2 \cdot 10^{-6}$	—	Hazelbauer and Adler (1971)	Involved in chemotaxis
β-Galactosides	Escherichia coli	30 000	$7 \cdot 10^{-5}$	—	Jones and Kennedy (1968)	The M protein
Maltose	Escherichia coli	40 000	$1.5 \cdot 10^{-6}$ and 10^{-5}	—	Kellermann and Szmelcman (1974)	Possibly in two forms
Glucose-1-phosphate	Agrobacterium tumefaciens	35 000 (I) 43 000 (II)	$8 \cdot 10^{-7}$ $1.3 \cdot 10^{-6}$	$4.5 \cdot 10^{-6}$	Fukui and Isobe (1973)	
D-Glucose and L-histidine	Intestinal brush border	55 000	—	—	Faust and Shearin (1974)	Na$^+$-activated; separate binding of glucose and histidine
Thiamine	Escherichia coli	42 000	$3 \cdot 10^{-7}$	—	Griffith et al. (1971)	
Thiamine	Escherichia coli	36 000	$2 \cdot 10^{-8}$	—	Nishimune and Hayashi (1973)	
Riboflavin	Escherichia coli	48 000	$3 \cdot 10^{-5}$	—	Griffith et al. (1971)	
Cyanocobalamine	Escherichia coli	22 000 or 200 000	$5 \cdot 10^{-7}$	10^{-8}	Taylor et al. (1972)	

and the accompanying membrane potential. One can easily visualize that the oxidation of various substrates simply gives rise to either a pH gradient across the membrane or to a membrane potential (negative inside) which can serve as driving forces for the translocation of nonelectrolytes.

4.4.3. Binding proteins

Several years ago the most promising way to extracting "molecular-level" information on the transport of solutes across membranes appeared to be the solubilization of membrane components that would bind the substrate without altering it and the extraction of which would reduce the transport capacity of the cell *in vivo*. Particularly due to the elegant method of osmotic shock (*cf.* p. 57) it has been possible to solubilize and purify a number of such binding proteins from bacteria. Chemical methods of extraction had to be used for obtaining such binding proteins from eukaryotic cells and tissues. The present state of affairs is summarized in Table 4.3.

The role of the binding proteins is by no means unequivocally accepted. It is very likely that most of them represent indeed essential components of the corresponding transport systems — in many cases there is evidence from gene mutations that this is so. In a most remarkable review of the subject, Slayman (1973) lists no less than 199 mutations affecting transport in microorganisms plus several score in multicellular organisms. Of these, at least seven involve the structural genes for the various binding proteins.

However, what the binding proteins do in transport remains unclear. It is possible, although not widely believed, that some of them might actually act as carriers (particularly those of large molecular weight from nonbacterial sources) but it is rather likely that the function is to recognize the substrate and transmit it further to one of the "effector" molecules which may include: (*1*) chemotactic proteins bringing about movement toward a suitable substrate concentration (in motile bacteria) (Hazelbauer and Adler, 1971); (*2*) carrier proteins which actually translocate the substrate across the membrane (with or without requirement for energy), (*3*) effectors of pinocytotic invaginations in protozoan as well as other cells.

More information on binding proteins can be found on p. 294 and in Kotyk and Janáček (1975).

4.5. DISTRIBUTION AND ROLE
OF NONELECTROLYTE TRANSPORTS

There are two main groups of compounds that have been studied in considerable detail, *viz.* the sugars and the amino acids. In addition to these, some data are available on the transport of purines and pyrimidines, polyols, vitamins, organic acids and small organic molecules (*cf.* Kotyk and Janáček, 1975).

All the specific, saturable, systems appear to aid in providing nutrition for the cells while the low-molecular waste products probably diffuse from cells by nonsaturable processes.

In all prokaryotic cells, the major groups are transported by energy-requiring systems against gradients of concentration of up to 10 000 : 1, depending on species and substrate. In the case of sugars, the energy may come from ATP but it does so usually indirectly through coupling with a gradient of H^+ ions which drive the solute inward. It may also derive from an oxidoreductive system, either by direct participation of the system in the transport of the nonelectrolyte or by the formation of a suitable membrane potential. Finally, a phosphotransferase system can be of importance.

With amino acids, the situation is not as varied but not any clearer for that. Oxidative energy is involved with some amino acids but ATP is required for others. Moreover, some amino acid transports even in prokaryotes require Na^+ as a symport cation, the specificity of the requirement differing from species to species.

While monosaccharides use carriers with a relatively broad specificity, some disaccharides and particularly amino acids are transported by systems with specificities that may be restricted to a single or to two chemical individuals. Still, there is a great deal of overlapping and one amino acid may be transported by as many as four different systems in a given species.

In lower eukaryotes (yeasts and fungi) the specificity of uptake is rather similar to that in bacteria but several major differences may be discerned in the involvement of energy. Thus, monosaccharides are transported in some species only by mediated diffusion while in others an active transport is involved. The energy required for this transport may be derived from H^+ gradients but possibly a high-molecular weight polyphosphate may have this function is some fungi (Kulaev, 1975).

In yeasts, amino acids are transported by multiple carriers against high intracellular concentrations but they cannot return to the outside even in exchange for an amino acid present in the medium. Moreover, a pronounced trans inhibition (like in *Streptomyces*) is found both in yeasts and fungi, resulting in a decreased uptake rate when high intracellular concentrations are reached.

There is relatively little information on nonelectrolyte transport in plants with the exception of algae where proton symports have been described for monosaccharides, energy being derived either from ATP or from the photosynthetic cycle. In higher plants, transmembrane movement appears to be usually by mediated diffusion but active transports are also known.

Animal tissues can be divided into several groups in this context. In epithelia, such as intestinal or kidney brush border, most monosaccharides and amino acids are transported in symport with Na^+, although probably primary active transport may also exist there. In muscle, Ehrlich ascites cells and erythrocytes, monosaccharides use mediated diffusion systems while amino acids are generally transported in conjunction with one or more sodium ions. In the extreme case of mammalian anucleate erythrocytes, both monosaccharides and amino acids are translocated by a mediated diffusion process.

The time is hardly ripe to make generalizations for lack of comparative data but it appears justified to conclude that active transports (either primary or secondary) have developed (or survived, as the case may be) in cells that depend on variable external environment to keep alive, or in cells that specialize in such transport within a higher-order framework of a differentiated organism. On the contrary, cells or organs that are not being exposed to substantial changes in the external concentration of nutrients (cultivated yeasts as well as sarcosomes) have lost this ability.

SYNOPSIS

Uncharged molecules are transported across biological membranes either by simple diffusion or by various mediated processes.

Simple diffusion, obeying Fick's laws of diffusion, proceeds dither across the lipid parts of membranes (most organic molecules, drugs) or, rarely, across water-filled channels, possibly only of statistical nature (primarily water).

Mediated processes require the presence of a specific membrane component which recognizes the substrate and either translocates it itself or transmits it to another membrane component, the carrier proper, for transmembrane movement. These processes are characterized by their saturability at high substrate concentrations and usually serve in the transport of sugars, amino acids and other metabolites.

The mediated processes either proceed without participation of metabolic energy (facilitated diffusion) or in coupling with a chemical reaction, such as splitting of ATP (primary active transport) or in coupling with downhill flow of a driving ion, usually H^+ in microorganisms and some plants, or Na^+ in animal tissues and some plants (secondary active transport).

There may be multiple binding sites on the recognition or carrier proteins.

A special type of transport is that by group translocation wherein the chemical nature of the substrate is altered during the transport process (typically the sugar phosphorylation by bacterial phosphotransferases).

5. TRANSPORT OF IONS

"...and perhaps electrical Attraction may
reach to such small distances, even
without being excited by Friction."
Isaac Newton, Opticks

5.1. EQUILIBRIA OF IONS

5.1.1. A simple membrane equilibrium and membrane potentials

We can begin the discussion of ion equilibria by considering a particularly simple practical instance. Let there be two solutions, say 10 mM and 100 mM potassium chloride, separated by a cation-permeable membrane. The membrane may be made, e.g., of a cation-exchange resin and contain pores lined with fixed negative charges. If the pores are sufficiently narrow and the density of fixed charges high, only potassium ions can move through the membrane from one negatively charged site to another, whereas chloride anions are prevented by electrostatic repulsion from entering the pores. In the absence of an external circuit, connecting the two solutions via electrodes and metallic conductors, the system will achieve an equilibrium with respect to ions. Water will not be in equilibrium, being osmotically driven from the more dilute toward the more concentrated solution; its equilibrium, however, can be easily achieved by applying a suitable hydrostatic pressure to the concentrated solution or adjusting the osmolarity of the dilute one with a nonpermeating non-electrolyte.

Equilibria of the two ion species, however, will be of widely different types. The equilibrium of chloride ions, a result of the membrane impermeability, is of a mechanic-electrostatic type and without a drastic change in the membrane structure there is no net transport of chloride ions across the membrane, as well as no exchange observable with tracers. The equilibrium of potassium cations, on the other hand, is a thermodynamical one, being the result of a mutual balance of active tendencies of the system. A very small amount of potassium ions has actually crossed the membrane during the equilibration process charging the capacity of the system to an electrical potential, the gradient of which is exactly equivalent to the concentration gradient of the potassium ions and thus prevents their further diffusional transfer. Net transfer of potassium ions from one side of the membrane to the other is thus not a spontaneous process and no work can be drawn from such a transfer. Hence the partial molal free energy, the so-called electrochemical potential of potassium ions (the maximum amount of work which would be obtained by transferring one mole of potassium ions to some chosen standard state) must be the same in the two solutions. The electrochemical potential may be expressed by

$$\tilde{\mu} = \mu_0 + RT \ln a + zF\varphi \tag{5.1}$$

where μ_0 depends neither on the concentration of the ion nor on the electrical potential, but rather on the nature of the solvent; R is the gas constant, equal to $8.314 \, \text{J mol}^{-1} \, \text{K}^{-1}$ (or voltcoulomb $\text{mol}^{-1} \, \text{K}^{-1}$, joule being equal to voltcoulomb) and T the absolute temperature in degrees of Kelvin. The activity a may be considered as the molar concentration of the ion, corrected for mutual interactions of all ions; $a = fc$, where f is the activity coefficient, in very dilute solutions equal to 1, in ordinary physiological salines having the value of about 0.76. The number of elementary charges per ion or its valency is denoted by z and is a signed quantity ($+1$ for K^+, -1 for Cl^-, $+2$ for Ca^{2+}, etc.), and F is the Faraday number, approximately 96 500 coulomb per mole of univalent ion or per gramequivalent of a polyvalent ion. Finally, φ is the electrical potential; its difference between two points corresponds to the amount of electrical work performed when a unit charge is transferred between the two points. If the charge is 1 coulomb and the potential difference 1 volt, the electrical work done is one voltcoulomb, or joule. Denoting the values corresponding to the concentrated solution by subscript i and those in the dilute solution by o, the equilibrium of an ion of valency z is described by

$$\mu_{0o} + RT \ln a_o + zF\varphi_o = \mu_{0i} + RT \ln a_i + zF\varphi_i. \quad (5.2)$$

Water being the solvent in both solutions, $\mu_{0i} = \mu_{0o}$ and

$$\varphi_i - \varphi_o = \frac{RT}{zF} \ln \frac{a_0}{a_i} = 2.303 \frac{RT}{zF} \log \frac{a_0}{a_i} \quad (5.3)$$

or, if we neglect the difference between the activity coefficients and cancel them out from the ratio of the activities,

$$\varphi_i - \varphi_o = 2.303 \frac{RT}{zF} \log \frac{c_o}{c_i}. \quad (5.4)$$

The values of the coefficient $2.303RT/F$ for various temperatures are given in Table 5.1. In the particular case discussed above, $c_o = 10 \text{ m}M$, $c_i = 100 \text{ m}M$, $\log (c_o/c_i) = -1$ and $z = +1$; hence the electrical potential difference across the membrane, $\varphi_i - \varphi_o$, will be -58.2 mV at 20 °C, -61.6 mV at 37 °C, *etc.*

TABLE 5.1.   Values of $2.3029RT/F$ in mV at different temperatures.
Calculated for $R = 8.3147 . 10^3$ mV . C . mol^{-1} . K^{-1},
$F = 96\,490$ C . mol^{-1} and $T = (t + 273.2)$ °C

t	mV	t	mV
0	54.2	25	59.2
5	55.2	30	60.2
10	56.2	35	61.2
15	57.2	37	61.6
20	58.2		

The equilibrium electrical potential difference (the Nernst – Donnan potential) between two solutions can be measured by connecting a suitable millivoltmeter to the two solutions by electrodes. The electrical potential difference between the electrode and the solution is, in general, a function of the solution composition. In Fig. 5.1**A**, **B**, two limiting cases are shown, the electrodes reacting reversibly either with the ion which is in thermodynamic equilibrium

or with the impermeant ion. To measure directly the electrical potential difference between two solutions one must use a device which does
not contribute by new potential differences to the circuit. This is
usually satisfied by reference electrodes (any electrodes immersed in
a medium of constant composition, calomel electrodes being most
popular), the solutions of which are connected with the measured

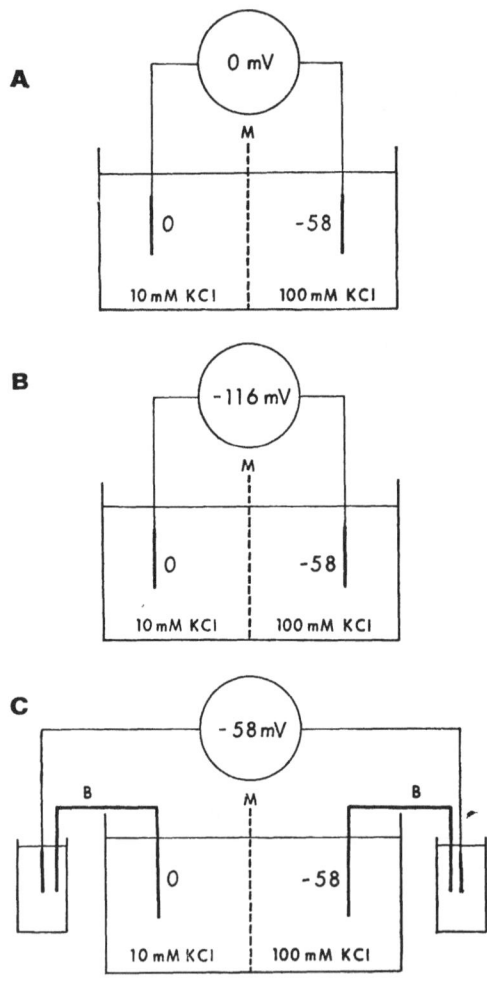

Fig. 5.1. Electrical potential difference measured in a circuit with a cation-
permeable membrane M and (A) cation-sensitive electrodes (potassium amalgam
or potassium-sensitive glass, etc.), (B) anion-sensitive electrodes (silver-silver
chloride), (C) bridges and reference electrodes.

solutions via suitable bridges. The bridges contain a concentrated salt solution with oppositely charged ions of the same mobility (e.g., potassium chloride), the solution being suitably immobilized by, say, an agar gel. The liquid junctions between the bridges and solutions thus develop only small junction potentials and in symmetrical arrangement their difference is likely to be still smaller; hence the possibility to measure the true difference of the electrical potential across a membrane (the so-called membrane potential) as shown in Fig. 5.1C.

The last principle mentioned is of particular importance in biology: the so-called microelectrodes, introduced by Ling and Gerard (1949) are, in fact, microbridges pulled from borosilicate glass and filled usually with $3\,M$ KCl. Using suitable micromanipulators they can be introduced into living cells to measure electrical potential differences across cell membranes.

It is of some interest to calculate the amount of electrical charge (a departure from electroneutrality) responsible for a membrane potential. Cell membrane potentials are of the order of 100 mV or 10^{-1} V; the membrane capacity being about 1 microfarad per cm^2 or 10^{-6} F cm^{-2}, the charge per cm^2 membrane may be calculated from the well-known condenser formula charge $=$ capacity $\cdot$ voltage, as 10^{-7} coulombs per cm^{-2}, which corresponds by the Faraday number (about 96 500 coulomb per mol) to some 10^{-12} mol of an univalent ion. This amount of ions is certainly not detectable by analytical techniques. Presently we will encounter the problem of how such charges are distributed in space (in connection with the distribution of the electrical potential in the so-called diffuse double layers).

The example of the membrane equilibrium discussed above is an especially simple instance of the important Gibbs–Donnan equilibria (with only one impermeant anion and one impermeant cation present); we shall now derive formulae governing Gibbs–Donnan equilibria in a more general case.

5.1.2. Gibbs—Donnan equilibrium

Fig. 5.2 shows a system in which a slightly more complex Gibbs–Donnan equilibrium is established than that discussed in the previous section. The system consists of two compartments; the right one is enclosed by rigid elastic walls (so that a hydrostatic pressure in excess of the atmospheric pressure may be attained in it) and contains nondiffusible anions A^{n-} in the equivalent concentration $[A^-] = n[A^{n-}]$, where $[A^{n-}]$ is the concentration in moles per unit volume. It is immaterial how the nondiffusible anions are restrained from leaving the compartment; they may represent a part of a rigid meshwork or the membrane between the two compartments may be impermeable to them. The solution in the right-hand compartment may be called the Donnan phase. The solution in the left-hand compartment could, of course, also contain nondiffusible anions or cations, without making the derivation of equilibrium relations much more complicated. In our case it contains only diffusible ions, the potassium cation and the chloride anion.

Equilibrium relations of the system can be derived using the conditions of equality of electrochemical potential in the two phases (see eq. 5.2):

$$\mu_{0K^+} + RT \ln a_{K^+_o} + F\varphi_o = \mu_{0K^+} + RT \ln a_{K^+_i} + F\varphi_i,$$
$$\mu_{0Cl^-} + RT \ln a_{Cl^-_o} - F\varphi_o = \mu_{0Cl^-_i} + RT \ln a_{Cl^-_i} - F\varphi_i \qquad (5.5)$$

and the condition of electroneutrality in each of the two phases:

$$a_{K^+_o} - a_{Cl^-_o} = 0,$$
$$a_{K^+_i} - a_{Cl^-_i} - [A^-] = 0. \qquad (5.6)$$

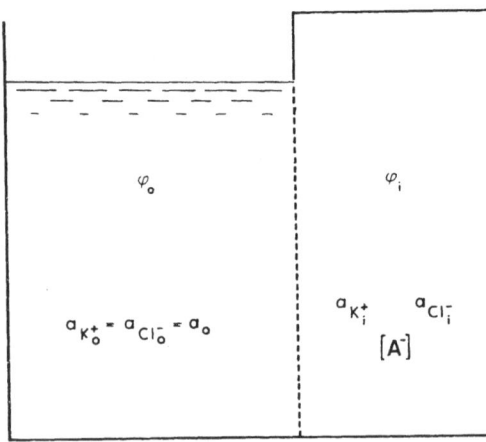

Fig. 5.2. A system in Gibbs—Donnan equilibrium. $[A^-]$, equivalent concentration of a nondiffusible anion; a's, activities of individual ions, φ's, electrical potentials in the two solutions.

From conditions (5.5) it may be immediately derived that

$$\varphi_i - \varphi_o = \frac{RT}{F} \ln \frac{a_{K^+o}}{a_{K^+i}} = \frac{RT}{F} \ln \frac{a_{Cl^-i}}{a_{Cl^-o}} \tag{5.7}$$

or

$$\frac{a_{K^+i}}{a_{K^+o}} = \frac{a_{Cl^-o}}{a_{Cl^-i}} = r \tag{5.8}$$

where r is called the Donnan ratio.

Combining eq. (5.6) and (5.8) and solving the quadratic equations obtained the concentrations of diffusible ions in the Donnan phase may be expressed as functions of the concentration of the nondiffusible anion and of the concentration in the external medium:

$$a_{K^+i} = \frac{1}{2}(\sqrt{[A^-]^2 + 4a_0^2} + [A^-]),$$

$$a_{Cl^-i} = \frac{1}{2}(\sqrt{[A^-]^2 + 4a_0^2} - [A^-]) \tag{5.9}$$

where

$$a_0 = a_{K^+0} = a_{Cl^-0}.$$

The osmotic pressure difference between the two phases can be expressed by using van't Hoff's formula $\pi = RTa$ as

$$\Delta\pi = RT\left(\frac{1}{n}[A^-] + a_{K^+i} + a_{Cl^-i} - 2a_0\right) \tag{5.10}$$

and combination of eq. (5.10) with equations (5.9) yields

$$\Delta\pi = RT\left(\frac{1}{n}[A^-] + \sqrt{[A^-]^2 + 4a_0^2} - 2a_0\right). \tag{5.11}$$

An equally large increment of hydrostatic pressure develops in the Donnan phase to achieve water equilibrium in the system.

Let us now assume that, in addition to potassium and chloride ions, diffusible calcium ions will be present in the system, so that an additional condition of equality of the electrochemical potential will apply:

$$\mu_{0Ca^{2+}} + RT \ln a_{Ca^{2+}o} + 2F\varphi_o =$$
$$= \mu_{0Ca^{2+}} + RT \ln a_{Ca^{2+}i} + 2F\varphi_i \tag{5.12}$$

from which it may be deduced that the Donnan ratio r, defined by eq. (5.8) may be also expressed by

$$r = \frac{\sqrt{a_{Ca^{2+}o}}}{\sqrt{a_{Ca^{2+}i}}} \tag{5.13}$$

or

$$\frac{a_{Ca^{2+}o}}{a_{Ca^{2+}i}} = r^2. \tag{5.14}$$

It may be seen, that the equilibrium ratio of the activities of a bivalent ion is the square of that for a univalent one.

A number of instructive numerical examples of Donnan distributions may be found in the monograph "Electrolytes and Plant Cells" by Briggs and co-workers (1961).

5.1.3. Diffuse electrical double layer

The electrical charge responsible for the electrical potential of a salt solution is represented by a deviation from electroneutrality, by a certain, generally very small, difference between the total amount of cationic and anionic charges in the solution. Unlike the hypothetical electrical charges of electrostatics, ions are subject not only to electrical forces but their electrical attraction to an oppositely charged surface is counteracted by their thermal motion. Interplay of the two factors results in a spatial extension of the charge in solution, the double layer of charges being of a diffuse character, as shown schematically in Fig. 5.3. As a result of this the electrical potential, too, is a function of the space coordinate; its value in the bulk of the solution is not attained by an abrupt step at the membrane surface but rather it decays slowly from the value at the surface which we may call φ_0 to the constant value in the distant bulk, which we may put arbitrarily equal to zero. The aim of the present section is to give an approximate description of this variation of the electrical potential in the diffuse or Gouy–Chapman double layer. Corrections for finite dimensions of ions and specific adsorption of ions at the surface which form the basis of Stern's theory of the double layer are omitted here, the present theory of cell membrane potentials being still too crude for these refinements. The electrical potential difference across the mobile part of the double layer is of importance in description of the electrokinetic phenomena and is commonly called the ζ-potential. A comprehensive and penetrating treatment of the double layer theory may be found in Overbeek (1952).

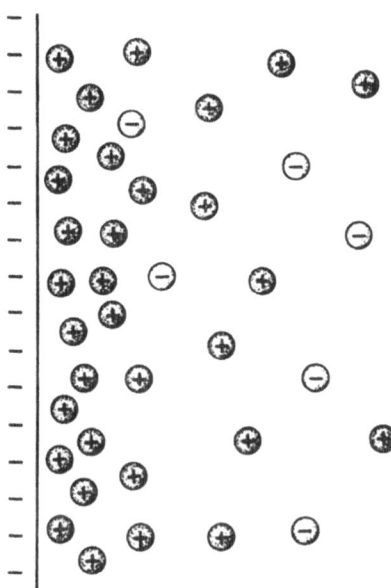

Fig. 5.3. Distribution of ions at a charged surface.

The interplay of electrical forces and random thermal movements of ions is mathematically expressed by a combination of Poisson's electrostatic equation with Boltzmann's statistical law. Poisson's equation is an expression of coulombic interactions and relates the divergence of the electrical potential to charge density. In the unidimensional case we are considering (the surface is presumably flat and all variables change in the direction of the x-axis only) the divergence is expressed by a second derivative and Poisson's equation may be written as

$$\frac{\mathrm{d}^2 \varphi}{\mathrm{d}x^2} = -\frac{\varrho}{\varepsilon} \qquad (5.15)$$

where ϱ is the charge density and ε the absolute dielectric constant. The system of units used is the rationalized SI system (meter — kilogram — second — ampere; rationalized since factors like 4π have been excluded from equations where they were not expected to occur from the geometry). The absolute dielectric constant, ε, is equal to the relative dielectric constant, ε_r (equal to 1 for a vacuum and about 80 for water), multiplied by ε_0, the electric constant, depending only on the choice of units:

$$\varepsilon = \varepsilon_r \varepsilon_0. \qquad (5.16)$$

Boltzmann's statistical law relates the concentration (rigorously activity) of ion j at the point where the electrical potential is φ to its concentration in the distant bulk of the solution, where the electrical potential is set equal to zero:

$$c_j = c_{j\infty}\, e^{-zF\varphi/RT}. \qquad (5.17)$$

For simplicity, we will assume that only potassium chloride is present in the solution so that the concentration of potassium ions in the bulk is equal to the concentration of chloride ion

$$c_{K^+\infty} = c_{Cl^-\infty} = c. \qquad (5.18)$$

Hence, Boltzmann's law written for potassium and chloride ions yields

$$\begin{aligned} c_{K^+} &= c\, e^{-F\varphi/RT}, \\ c_{Cl^-} &= c\, e^{F\varphi/RT}. \end{aligned} \qquad (5.19)$$

The local charge density is equal to the difference between the local concentration of cations and anions multiplied by the Faraday number F

$$\varrho = F(c_{K^+} - c_{Cl^-}) \qquad (5.20)$$

and may be introduced into Poisson's equation (5.15), giving

$$\frac{d^2\varphi}{dx^2} = -\frac{Fc}{\varepsilon}(e^{-F\varphi/RT} - e^{F\varphi/RT}). \qquad (5.21)$$

Equation (5.21) can be solved exactly (as shown by Overbeek, 1952); for the present purpose of a rough estimate of the extent of the diffuse double layer, it is sufficient to solve it for the special case of small potentials φ; hence the approximative character of the solution when higher potential differences are involved. For small exponents the exponential functions may be approximated by the first two terms of its development in series according to Maclaurin's theorem

$$e^y \approx 1 + y.$$

Such an approximation of the two exponential terms in eq. (5.21) results in

$$\frac{d^2\varphi}{dx^2} = \frac{2F^2c}{RT\varepsilon}\varphi = \varkappa^2\varphi. \tag{5.22}$$

As may be easily verified by taking derivatives twice, the function

$$\varphi = \varphi_0 e^{-\varkappa x} \tag{5.23}$$

satisfies eq. (5.22) as well as the boundary conditions, giving $\varphi = \varphi_0$ for $x = 0$ and $\varphi = 0$ for infinite x. The thickness of the double layer can be arrived at by means of δ, the Debye thickness, which is defined as

$$\delta = \frac{1}{\varkappa} \tag{5.24}$$

i.e.

$$\delta = \sqrt{\frac{RT\varepsilon}{2F^2c}} = \sqrt{\frac{RT\varepsilon_r\varepsilon_0}{2F^2c}}. \tag{5.25}$$

As is obvious from eq. (5.23), δ characterizes the exponential decay of the electrical potential with distance x; for $x = \delta$, potential φ_0 decreases to $1/e$, i.e., to 36.8% of its value. Moreover, it can be shown that the capacity of the diffuse double layer is equal to the capacity of a plate condenser with the distance of plates equal to δ. According to the exact solution of eq. (5.21) the decay of the electrical potential in the diffuse layer is only roughly exponential (Overbeek 1952), but it may be still characterized by the quantity δ, commonly called the thickness of the diffuse double layer. The thickness δ may be calculated from the eq. (5.25), using $R = 8.31 \cdot 10^3 \text{ kg m}^2 \text{ s}^{-2} \text{ K}^{-1} \text{ kmol}^{-1}$; $F = 9.65 \cdot 10^7 \text{ A s kmol}^{-1}$; $\varepsilon_0 = 8.85 \cdot 10^{-12} \text{ A}^2 \text{ s}^4 \text{ m}^{-3} \text{ kg}^{-1}$. For 20 °C (= 293 K) this gives

$$\delta = 3.4 \cdot 10^{-11} \sqrt{\frac{\varepsilon_r}{c}}. \tag{5.26}$$

In the SI system the concentration emerges in kmol m^{-3}, which is numerically equal to concentrations in mol l^{-1} or molarity. Using ε_r for water equal to 80, eq. (5.26) gives the values of δ summarized for various concentrations of the uni-univalent electrolyte (KCl in the present example)

Concentration (kmol m^{-3})	Thickness δ (nm)
10^{-5}	96.0
10^{-4}	30.4
10^{-3}	9.6
10^{-2}	3.04
10^{-1}	0.96

When various ions of molarity c_j and valency z_j are present in the solution, a more general form of formula (5.25) can be used to calculate the thickness of the diffuse double layer

$$\delta = \sqrt{\frac{RT\varepsilon}{F^2\, \Sigma\, z_j^2 c_j}} \tag{5.27}$$

whence it follows that the higher the valency of the ions present in the solution, the more compressed the diffuse double layer will be.

Although the above considerations were carried out for a phase boundary between a solid (e.g., a metal) and an ion-containing solution they apply equally well to the dependence of the electrical potential on the space coordinate near the contact of two ionic systems, where double layers are formed. Double diffuse double layers necessarily exist at the boundaries between a membrane and solutions, as expressed by Agin (1967): "The existence of a diffuse layer outside the membrane is at least implicitly accepted by most, but for anything except a metal, a diffuse layer must exist internally as well". Hence the situation encountered at the membrane boundaries is likely to be analogous to that at the interface between two inmiscible liquids, treated theoretically, e.g., by Verwey and Overbeek (1948; see also Overbeek, 1952). If the boundary potential drop due to oriented dipoles is neglected, the course of the electrical potential at the interface may resemble that shown in Fig. 5.4. When the double diffuse double layer is as a whole electrically neutral

$$\int_{-\infty}^{0} \varrho_1(x)\,\mathrm{d}x + \int_{0}^{\infty} \varrho_2(x)\,\mathrm{d}x = 0 \tag{5.28}$$

from which it may be shown, using Poisson's equation (5.15) that

$$\varepsilon_1 \left(\frac{\mathrm{d}\varphi_1}{\mathrm{d}x}\right)_{=0} = \varepsilon_2 \left(\frac{\mathrm{d}\varphi_2}{\mathrm{d}x}\right)_{x=0} \tag{5.29}$$

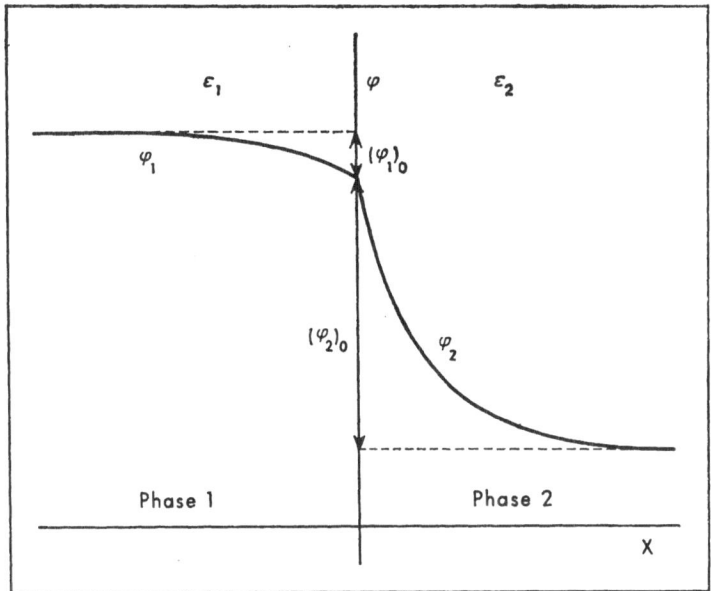

Fig. 5.4. Distribution of the electrical potential at the interface of two liquid phases. (Adapted schematically from Verwey and Overbeek, 1948.)

i.e., the discontinuity in the derivative of the potential at the interface is in this case entirely determined by the different dielectric constants (Verwey and Overbeek, 1948).

To account for the net fixed charge, resulting from the dissociation of ionizable polar head groups of phospholipids, Ciani and co-workers (1973) assume a uniform distribution of the charge at the membrane surface with density σ. Equation (5.29) then becomes

$$\varepsilon_1 \left(\frac{d\varphi_1}{dx} \right)_{x=0} - \varepsilon_2 \left(\frac{d\varphi_2}{dx} \right)_{x=0} = \sigma. \tag{5.30}$$

They also take into account the possible existence of a discontinuous change of electrical potential

$$\varphi_{1(x=0)} - \varphi_{2(x=0)} = a \tag{5.31}$$

resulting from the uniform distribution of a sheet of dipoles on the membrane surface; a is assumed to be a constant, independent of the applied electrical field as well as of the composition of the solutions.

Estimation of the thickness of the internal diffuse double layer within the membrane would involve some uncertainty, the dielectric constant and the concentrations in the regions or channels through which ions permeate not being known.

5.2. ELECTRODIFFUSION AND MEMBRANE POTENTIALS

5.2.1. Introduction

After the equilibrium membrane potentials treated before we shall now deal with potential differences on membranes across which a net ion transport takes place. Unlike the equilibrium potentials, the nonequilibrium ones will be seen to be functions not only of ion concentrations, but also of ion mobilities and partition coefficients between the medium and the membrane. By assuming zero mobilities of all but one ion the formulae describing nonequilibrium potentials must be, of course, reduced to the correct equilibrium relation for the given ion.

For the sake of simplicity it will be assumed in the following that steady-state conditions prevail, i.e., that fluxes of ions are constants independent of time. Those interested more deeply in the kinetic phenomena especially on excitable membranes are referred to the scholarly treatments of this subject by Cole (1965) and Agin (1972). The steady state required by the following derivations will be the result of both the stable properties of the membrane and the stable concentrations at the membrane. It will be immaterial for the derivations how the stability of the concentrations at the membrane will be ensured, three possibilities of achieving steady concentrations of sodium and potassium ions at the two sides of a membrane being shown in Fig. 5.5. In the first case (**a**), the concentrations are steady due to the vastness of the two reservoirs; such a situation is most simply obtained in experiments with artificial membranes. In the second case (**b**), which may be typical of a number of nonpolar cells, the steady concentrations are preserved by the operation of a one-to-one coupled electroneutral sodium-potassium exchange pump. Note that although the net flow of the two ions across the membrane is zero, the passive flows of ions, representing a transport of electrical charge, are not zero. In the third case (**c**), there is a large reservoir at one side of the membrane, whereas the composition of the compartment at the other side is kept constant by a pumping mechanism at the opposite pole of the cell. A situation of this kind may be found in ion-transporting epithelial layers.

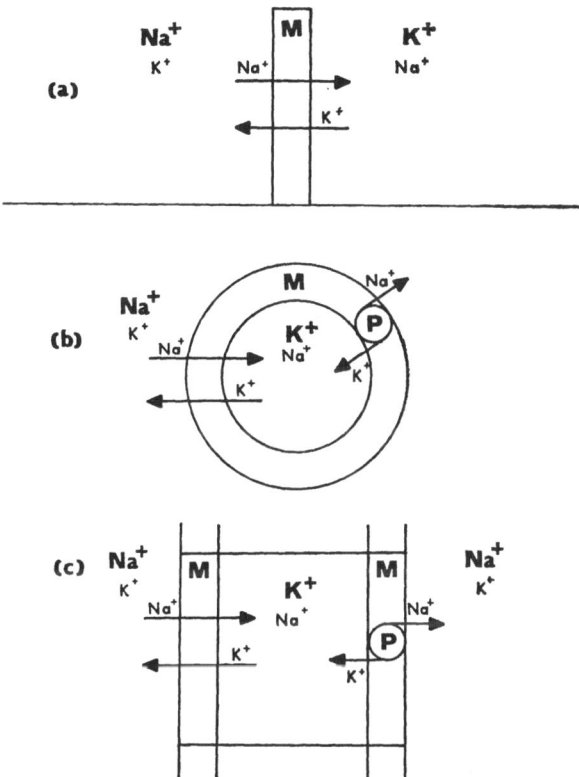

Fig. 5.5. Various mechanisms ensuring steady concentrations of sodium and potassium ions at a membrane. **M** Membrane, **P** pump. The size of the concentration symbols corresponds roughly to the concentration levels. Further explanations in the text.

To derive formulae describing the differences of the electrical potential between the solutions at the two sides of the membrane we will consider first the electrodiffusion in the membrane proper and neglect the steps of electrical potential at its two boundaries. The membrane (or rather the regions of the membrane through which the ions permeate) will be treated as a continuous medium. Such an approach, when applied to the thin cell membrane, is certainly not free of theoretical objections; if the penetration of an ion across a layer corresponds to only a small number of transitions across energy barriers, a discontinuous approach based on the absolute rate theory rather than a continuous one appears to be warranted. Although there are already some promising attempts in this direction (e.g., Ciani, 1965), the practical applicability of the theory may be still rather remote. Following Planck, Henderson, and Goldman we shall calculate the difference of the electrical potential between two planes inside a continuous layer as a function of ion concentrations at the two planes and of the ion mobilities in the layer. In this way we shall arrive at an approximate value of the potential difference across the membrane proper, provided that various simplifying assumptions be introduced and the concentrations of ions at the surface, just inside the membrane, be known (see Fig. 5.6).

Later we shall consider the difference of electrical potential across the two boundaries, between the membrane and the adjacent solu-

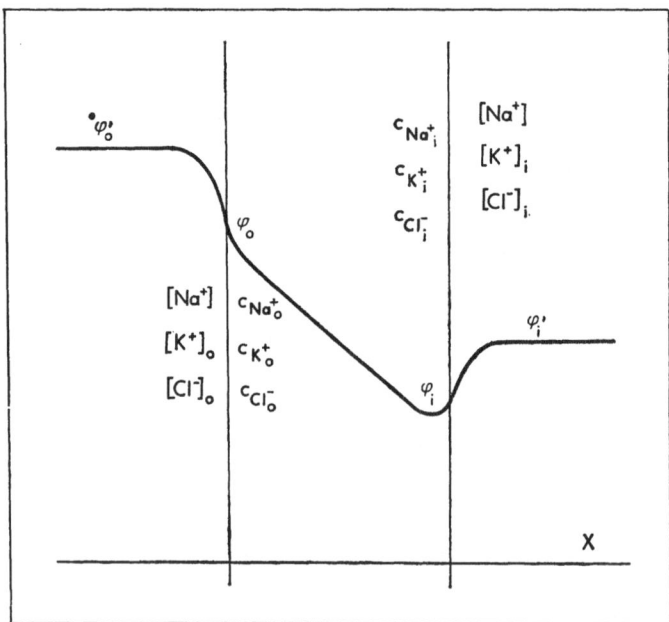

Fig. 5.6. Electrical potential differences across a membrane and its boundaries, the concentration symbols shown being as used in the appropriate equations in the text. The potential profile is shown highly schematically.

tions, derive formulae for the potential difference across the whole membrane and inquire how this picture is changed by the presence of an electrogenic pump, i.e., of an active mechanism translocating a net electrical charge across the membrane.

5.2.2. The electrodiffusion equation—general considerations

When discussing the equilibrium membrane potentials we observed that the condition of thermodynamic equilibrium of an ion is that its electrochemical potential be the same at any point of the space available for its movement. From this observation one is led to the conclusion that the driving force of an ion flux is the gradient of its electrochemical potential in space. Even this assumption may represent only an approximation for the cell membrane, for it apparently requires both the condition of local electroneutrality (for the electrochemical potential to be defined at a point) and the condition of thermal equilibrium (for Einstein's interpretation of the diffusion coefficient $D = RTU$) to be satisfied if it is to be valid strictly (see Agin, 1967, for discussion and references).

When the electrochemical potential changes in space in one direction only, say, in the direction of the x-axis, it will be a function solely of time t and of x, and its gradient will be equal to the partial derivative with respect to x:

$$\text{grad } \tilde{\mu}(t, x) = \frac{\partial \tilde{\mu}}{\partial x}. \tag{5.32}$$

Thus the general transport equation due to Teorell (1953)

FLUX = MOBILITY . CONCENTRATION . DRIVING FORCE

will have for the transport of an ion j the form*

$$J_j = -c_j U_j \frac{\partial \tilde{\mu}_j}{\partial x}. \tag{5.33}$$

The electrochemical potential may be expressed by (see p. 244)

$$\tilde{\mu}_j = \mu_{j0} + RT \ln a_j + z_j F \varphi \tag{5.34}$$

---

* The minus sign stems from the fact that when the derivative is a positive quantity, $\tilde{\mu}$ increases in the direction of the positive x-axis and the flux proceeds in the opposite direction.

where $a_j = fc_j$ is the activity of the ion, its molar concentration multiplied by the activity coefficient. Introducing this expression into eq. (5.33) we obtain

$$J_j = -c_j U_j \left(RT \frac{\partial \ln a_j}{\partial x} + z_j F \frac{\partial \varphi}{\partial x} \right). \tag{5.35}$$

Now, the well-known identity from differential calculus $d \ln y = dy/y$ can be used, so that

$$d \ln a = \frac{da}{a} = \frac{d(fc)}{fc} = \frac{f}{f} \frac{dc}{c} = \frac{dc}{c}$$

and eq. (5.35) is changed into

$$J_j = -RTU_j \frac{\partial c_j}{\partial x} - z_j F U_j c_j \frac{\partial \varphi}{\partial x} \tag{5.36}$$

which is the familiar Nernst–Planck equation. The first term on the right-hand side describes pure diffusion, the second term is the expression for migration in an electrical field or electric drift. The Nernst–Planck differential equation of electrodiffusion may thus be considered as a superposition of Fick's and Ohm's laws. It may be noted that not the activity coefficient but only its dependence on x was neglected in the above derivation.

A rigorous derivation of a more general differential equation describing the ionic flux was carried out by methods of steady-state thermodynamics by Schlögl (1964). Schlögl's derivation is described in detail in section 5.2.3. The Schlögl equation may be written

$$J_j = c_j v - RTU_j \frac{\partial c_j}{\partial x} - z_j F U_j c_j \frac{\partial \varphi}{\partial x} - RTU_j \frac{\partial \ln f_j}{\partial x} -$$

$$- U_j M_j c_j \left(\frac{\bar{V}_j}{M_j} - \frac{\bar{V}_n}{M_n} \right) \frac{\partial p}{\partial x} \tag{5.37}$$

where $\bar{V}_j$ and $\bar{V}_n$ are the partial molal volumes and M_j and M_n the molecular weights of the ion and of the solvent, respectively, and v is the volume flow. Comparing the Nernst–Planck equation (5.36) with Schlögl's equation (5.37) it can be seen that the former neglects (apart from the mutual interactions between ions, omitted explicitly even in Schlögl's derivation) (1) flow of ion due to volume flow, (2) dependence of activity coefficient on the x-coordinate and (3) pressure diffusion described by the last term of the Schlögl equation. It may be seen that in pressure diffusion, ions with a specific volume greater than that of solvent molecules move in the direction of the negative pressure gradient, whereas those less voluminous move in the opposite direction.

Expressing the electrochemical potential or each of its components in $J\,mol^{-1}$, flux in $mol\,cm^{-2}\,s^{-1}$ and concentration in $mol\,cm^{-3}$, mobility U is seen to have the dimension of $cm^2\,s^{-1}\,J^{-1}$ mol. In electrochemistry, mobility u is often used with the dimension of $cm^2\,s^{-1}\,V^{-1}$. The relation between the two mobilities is

$$u_j = |z_j|\,FU_j.$$

The equation of electrodiffusion (5.36) relates the value of the flux to the values of differential quotients, which are not amenable to a direct experimental measurement; only finite differences of concentrations or of the electrical potential over finite distances can be measured. In order to establish the relation between the flux and the finite differences, electrodiffusion equation has to be integrated over a finite distance. There are, however, two unknown functions of x in the equation and an additional condition is required to carry out the integration. Two assumptions concerning the nature of these functions have been widely used. According to the first, originally applied by Henderson (1907, 1908), the concentration of each individual ion is a linear function of the distance; according to the other, due to Goldman (1943), such a linear function is represented by the electrical potential. (The first condition is satisfied in a situation for which it was formerly intended, i.e. in mixed boundaries, and hence it is not likely to be applicable to unmixed systems like membranes. The second condition will be seen to be satisfied in locally electroneutral systems when the ionic strength is a constant independent of the space coordinate and will be used as a simplifying assumption in derivations of Goldman's formulae, which can be used as approximations to describe cell membrane potentials in steady state.)

When, however, the aim of the calculation is to supply only a formula for the electrical potential difference across a continuous layer as a function of ionic concentrations at its boundaries, rather than expressions for fluxes of individual ions, there is a different, although less simple, approach. The mathematical procedure is due to Planck (1890) and involves only the assumption of local, or microscopic, electroneutrality. We will see later that the condition of local electroneutrality may not be satisfied in cell membranes (for a discussion see Agin, 1967). Still, it will be attempted in the following to describe Planck's derivation in a great detail both for its intrinsic merit (Cole, 1965, speaks about its beauty and power mostly lost in the numerous rederivations, which can be fully appreciated in the original) and to show under which conditions Planck's transcendental equation for the electrical potential difference (Planck, 1890) reduces to the explicit formula of Goldman (1943).

Thus the following three sections will be concerned, respectively, with the derivation of Schlögl's differential equation of electrodiffusion, of Henderson's formula and of Planck's formula for the difference of electrical potential in continuous ionic systems. Those interested more in derivations of formulae directly applicable to cell membranes may skip the following three sections.

5.2.3. Schlögl's (1964) derivation of the general differential equation of electrodiffusion

The derivation, based on the principles of steady-state thermodynamics of continuous systems, aims at calculating the entropy production in the liquid phase inside membrane pores. The liquid phase contains n components, the relative abundance of each component being expressed by its partial density $\varrho_j = m_j/V$, the "mean velocity" of each component being denoted by $\mathbf{v}_j$. The product $\varrho_j\mathbf{v}_j$ then corresponds to the mass flow of component j. "Barycentric velocity" (the velocity of the centre of mass) is defined as

$$\mathbf{v} \equiv \sum_{j=1}^{n} \varrho_j \mathbf{v}_j / \varrho \tag{5.38}$$

where $\varrho = \Sigma \varrho_j$ and the relative movement of component j with respect to the movement of the common centre of mass is characterized by vector $\mathbf{M}_j$

$$\mathbf{M}_j = \varrho_j(\mathbf{v}_j - \mathbf{v}). \tag{5.39}$$

From (5.38) and (5.39) it follows that

$$\sum_{j=1}^{n} \mathbf{M}_j = \sum_{j=1}^{n} \varrho_j \mathbf{v}_j - \sum_{j=1}^{n} \varrho_j \cdot \sum_{j=1}^{n} \varrho_j \mathbf{v}_j / \varrho = \sum_{j=1}^{n} \varrho_j \mathbf{v}_j - \varrho \sum_{j=1}^{n} \varrho_j \mathbf{v}_j / \varrho = 0. \tag{5.40}$$

Assuming that no chemical reactions take place, the following equations of continuity hold

$$\frac{\partial \varrho_j}{\partial t} = -\operatorname{div}(\varrho_j \mathbf{v}_j) \tag{5.41}$$

$$\frac{\partial \varrho}{\partial t} = -\operatorname{div}(\varrho \mathbf{v}) \tag{5.42}$$

expressing the fact that no substance is lost – spatial change of flow is equal to temporal change of density. Furthermore, since ϱ (unlike ϱ_j) in condensed systems depends only little on composition, equation

$$\operatorname{div} \mathbf{v} \approx 0 \tag{5.43}$$

is assumed to be approximately valid.

Assuming the system to be in thermal equilibrium, entropy production per unit volume of the system, σ, multiplied by absolute temperature (the so-called dissipation function) is given by

$$T\sigma = T\sigma_{\text{visc}} - \sum_{j=1}^{n} \mathbf{M}_j \operatorname{grad} \frac{\tilde{\mu}_j}{M_j} \tag{5.44}$$

where σ_{visc} is the contribution to entropy production due to the dissipation of convective energy by friction, $\tilde{\mu}_j$ the electrochemical potential of component j, M_j its molecular weight and hence $\tilde{\mu}_j/M_j$ the electrochemical potential per unit of mass of component j. Using eq. (5.39), the second right-hand term of eq. (5.44) may be rewritten as

$$-\sum_{j=1}^{n} \mathbf{M}_j \operatorname{grad} \frac{\tilde{\mu}_j}{M_j} = -\sum_{j=1}^{n} \mathbf{v}_j \varrho_j \operatorname{grad} \frac{\tilde{\mu}_j}{M_j} + \mathbf{v} \sum_{j=1}^{n} \varrho_j \operatorname{grad} \frac{\tilde{\mu}_j}{M_j}. \tag{5.45}$$

The sum in the second right-hand term of eq. (5.45) may be transformed using the following thermodynamical considerations:

The first law of thermodynamics for an open system which performs only volume work can be written as

$$dU = T dS - p dV + \sum_{j=1}^{n} \mu_j dn_j \tag{5.46}$$

where μ_j is the chemical potential of component j. From the definition of Gibbs free energy

$$G \equiv U - TS + pV$$

a mathematical identity follows by differentiation

$$dG \equiv dU - S\,dT - T\,dS + V\,dp + p\,dV. \qquad (5.47)$$

Adding identity (5.47) to the first law (5.46) we obtain an expression for the first law in the form

$$dG = -S\,dT + V\,dp + \sum_{j=1}^{n} \mu_j\,dn_j. \qquad (5.48)$$

The Gibbs free energy is an extensive variable of state; if, at constant temperature and pressure, the number of moles of each component is increased by a certain factor, the Gibbs free energy of the system is increased by the same factor. Hence the Gibbs free energy is a homogeneous function of the first degree of the number of moles n_j and, according to Euler's theorem, may be written as

$$G = \sum_{j=1}^{n} n_j \left(\frac{\partial G}{\partial n_j} \right)_{p,\,T} = \sum_{j=1}^{n} n_j \mu_j. \qquad (5.49)$$

By differentiating eq. (5.49) we obtain

$$dG = \sum_{j=1}^{n} \mu_j\,dn_j + \sum_{j=1}^{n} n_j\,d\mu_j \qquad (5.50)$$

and, on subtracting eq. (5.48) from this, we have

$$\sum_{j=1}^{n} n_j\,d\mu_j = -S\,dT + V\,dp \qquad (5.51)$$

which is the well-known Gibbs – Duhem equation. Dividing this equation by V and observing that

$$\sum_{j=1}^{n} \frac{n_j}{V}\,d\mu_j = \sum_{j=1}^{n} \frac{n_j M_j}{V} \frac{d\mu_j}{M_j} = \sum_{j=1}^{n} \frac{m_j}{V} \frac{d\mu_j}{M_j} = \sum_{j=1}^{n} \varrho_j \frac{d\mu_j}{M_j}$$

we obtain

$$\sum_{j=1}^{n} \varrho_j \frac{d\mu_j}{M_j} = -\frac{S}{V}\,dT + dp. \qquad (5.52)$$

If the particles of component j are charged the gradient of their chemical potential balances in equilibrium the force exerted on them by the electrical potential gradient; in other words, their electrochemical potential $\tilde{\mu}_j = \mu_j + z_j F \varphi$ has the same value everywhere in the system and (5.52) may be written as

$$\sum_{j=1}^{n} \varrho_j \frac{\mathrm{d}\tilde{\mu}_j}{M_j} = -\frac{S}{V} \mathrm{d}T + \mathrm{d}p + \sum_{j=1}^{n} \frac{\varrho_j z_j F}{M_j} \mathrm{d}\varphi. \qquad (5.53)$$

Defining $\varrho_{el} = \Sigma \varrho_j z_j F / M_j$ (the electrical space charge per unit volume) and proceeding from differentials to gradients, we obtain

$$\sum_{j=1}^{n} \varrho_j \,\mathrm{grad}\, \frac{\tilde{\mu}_j}{M_j} = -\frac{S}{V} \,\mathrm{grad}\, T + \mathrm{grad}\, p + \varrho_{el} \,\mathrm{grad}\, \varphi \qquad (5.54)$$

where the first term on the right-hand side vanishes in thermal equilibrium. Under this condition, eq. (5.45) may be written as

$$-\sum_{j=1}^{n} \mathbf{M}_j \,\mathrm{grad}\, \frac{\tilde{\mu}_j}{M_j} = -\sum_{j=1}^{n} \mathbf{v}_j \varrho_j \,\mathrm{grad}\, \frac{\tilde{\mu}_j}{M_j} + \mathbf{v} \,\mathrm{grad}\, p + \mathbf{v} \varrho_{el} \,\mathrm{grad}\, \varphi.$$
$$(5.55)$$

This equation is equivalent with

$$-\sum_{j=1}^{n} \mathbf{M}_j \,\mathrm{grad}\, \frac{\tilde{\mu}_j}{M_j} = \mathrm{div}\,(\mathbf{v}p) + \varrho_{el} \,\mathrm{div}\,(\mathbf{v}\varphi) -$$
$$-\sum_{j=1}^{n} \mathrm{div}\left(\varrho_j \mathbf{v}_j \frac{\tilde{\mu}_j}{M_j}\right) - \sum_{j=1}^{n} \frac{\tilde{\mu}_j}{M_j} \frac{\partial \varrho_j}{\partial t} \qquad (5.56)$$

as shown by the following consideration: $\mathrm{div}\,(\mathbf{v}p) = p \,\mathrm{div}\, \mathbf{v} + \mathbf{v} \,\mathrm{grad}\, p$ where $p \,\mathrm{div}\, \mathbf{v} = 0$ (from eq. 5.43) and, similarly, $\varrho_{el} \,\mathrm{div}\,(\mathbf{v}\varphi) = \varrho_{el}\varphi \,\mathrm{div}\, \mathbf{v} + \varrho_{el}\mathbf{v} \,\mathrm{grad}\, \varphi$ where again $\varrho_{el}\varphi \,\mathrm{div}\, \mathbf{v} = 0$, and, finally,

$$-\sum_{j=1}^{n} \mathrm{div}\left(\varrho_j \mathbf{v}_j \frac{\tilde{\mu}_j}{M_j}\right) = -\sum_{j=1}^{n} \varrho_j \mathbf{v}_j \,\mathrm{grad}\, \frac{\tilde{\mu}_j}{M_j} - \sum_{j=1}^{n} \frac{\tilde{\mu}_j}{M_j} \,\mathrm{div}\,(\varrho_j \mathbf{v}_j)$$

where the first term on the right-hand side appears in eq. (5.55), whereas the second is (by eq. 5.41) equal to $+ \Sigma \dfrac{\tilde{\mu}_j}{M_j} \dfrac{\partial \varrho_j}{\partial t}$. (Hence it cancels out with the last term of eq. 5.56.)

Eq. (5.56) is now introduced into the dissipation function (5.44); integration over the volume between two cross-sections of the membrane at x and $x + \Delta x$ is carried out. The integration may be carried out over the whole membrane since outside the pores, in the matrix of the membrane, the integrands vanish. When the divergence theorem

$$\int_V \mathrm{div}\, \mathbf{v} \,\mathrm{d}V = \int_S \mathbf{v} \,\mathrm{d}\mathbf{S}$$

is used, integration yields

$$T\int_V \sigma \, \mathrm{d}V = T\int_V \sigma_{\mathrm{visc}} \, \mathrm{d}V - \sum_{j=1}^n \int_V \frac{\tilde{\mu}_j}{M_j} \frac{\partial \varrho_j}{\partial t} \, \mathrm{d}V +$$

$$+ \int_S p\mathbf{v} \, \mathrm{d}\mathbf{S} + \varrho_{\mathrm{el}} \int_S \varphi \mathbf{v} \, \mathrm{d}\mathbf{S} - \sum_{j=1}^n \int_S \varrho_j \mathbf{v}_j \frac{\tilde{\mu}_j}{M_j} \, \mathrm{d}\mathbf{S}. \qquad (5.57)$$

In a steady state, the second term on the right-hand side is considered to be negligibly small (representing only rather small temporal fluctuations). The surface integrals are replaced by integrals over the two cross-sections of the membrane (the edge area of thickness Δx being small) which, in their turn, are assumed to be equivalent to integrals over the corresponding temporally and spatially fluctuating equipotential surfaces. Hence the quantities p, φ and $\tilde{\mu}$, which correspond to the medium in the pores of the membrane and are functions of x only, can be factored out before the integration symbols. The remaining surface integrals have the values

$$\int_S \mathbf{v} \, \mathrm{d}\mathbf{S} = \pm A\mathbf{v}$$

$$\int_S \varrho_j \mathbf{v}_j \, \mathrm{d}\mathbf{S} = A J_j M_j$$

where A is the membrane surface, $\mathbf{v}$ the barycentric velocity of the fluid in pores calculated over the whole membrane cross-section, J_j the molar flow of component j again expressed across the whole cross-section and M_j its molecular weight. The surface vectors $\mathrm{d}S$ being directed outwards from the volume of integration, their direction is opposite at x and at $x + \Delta x$. Hence the minus sign corresponds to distance x and the plus sign to $x + \Delta x$. Introducing the values of the surface integrals into eq. (5.57) and replacing the resulting differences $p_{x+\Delta x} - p_x$ with $\dfrac{\mathrm{d}p}{\mathrm{d}x} \Delta x$, etc., we obtain after division by the volume of the considered membrane layer, $A \Delta x$, the dissipation function per unit volume of the membrane

$$T\sigma = T\sigma_{\mathrm{visc}} + \mathbf{v}\left(\frac{\mathrm{d}p}{\mathrm{d}x} + \varrho_{\mathrm{el}} \frac{\mathrm{d}\varphi}{\mathrm{d}x}\right) - \sum_{j=1}^n J_j \frac{\mathrm{d}\tilde{\mu}_j}{\mathrm{d}x}. \qquad (5.58)$$

However, it may be shown, by "Gedanken" experiments using devices consisting of a closed (hence $\Sigma \mathrm{d}U = 0$) isothermal ($T = $ const.) system with one isobaric ($p = $ const.) and one isochoric ($V = $ const.) subsystem, that the whole rate of entropy production when ions are transported across membrane is accounted for by the dissipation function

$$T\sigma = -\sum_{j=1}^{n} J_j \frac{d\tilde{\mu}_j}{dx} \qquad (5.59)$$

(see Schlögl, 1964).Comparing eq. (5.58) and (5.59) we have

$$T\sigma_{\text{visc}} = v\left(-\frac{dp}{dx} - \varrho_{\text{el}} \frac{d\varphi}{dx}\right). \qquad (5.60)$$

At this stage, Schlögl defines the mean mass flow of component j as

$$M_j = \frac{1}{A} \int_S \varrho_j(v_j - v) \, dS = J_j M_j - \varrho_j v \qquad (5.61)$$

or,

$$J_j = \frac{M_j}{M_j} + \frac{\varrho_j v}{M_j}. \qquad (5.62)$$

Introducing (5.60) and (5.62) into (5.58) we get

$$T\sigma = v\left(-\frac{dp}{dx} - \varrho_{\text{el}} \frac{d\varphi}{dx}\right) + v\left(\frac{dp}{dx} + \varrho_{\text{el}} \frac{d\varphi}{dx}\right) - $$
$$- \sum_{j=1}^{n} \frac{M_j}{M_j} \frac{d\tilde{\mu}_j}{dx} - \sum_{j=1}^{n} \frac{\varrho_j v}{M_j} \frac{d\tilde{\mu}_j}{dx} \qquad (5.63)$$

and using eq. (5.53) for an isothermal system

$$\sum_{j=1}^{n} \varrho_j \frac{d\tilde{\mu}_j}{M_j} = dp + \varrho_{\text{el}} \, d\varphi \qquad (5.64)$$

the second and fourth term on the right-hand side of eq. (5.63) are seen to cancel out and the dissipation function takes the form

$$T\sigma = v\left(-\frac{dp}{dx} - \varrho_{\text{el}} \frac{d\varphi}{dx}\right) - \sum_{j=1}^{n} \frac{M_j}{M_j} \frac{d\tilde{\mu}_j}{dx}. \qquad (5.65)$$

The expression may be seen to represent the sum of the products of fluxes with their conjugate local (x-dependent) driving forces. We can now write the linear relations of steady-state thermodynamics, according to which each flow is linearly dependent not only on its conjugate force, but also on forces conjugate to flows with which the former flow interacts. Attention is called here by Schlögl to the Curie principle, according to which processes of unlike tensorial order do not interact in isotropic systems. The production of entropy by inner frinction σ_{visc} may be represented as a sum of two second-order tensors (even if their tensorial character is obscured by an averaging integration procedure in the present derivation) whereas gradients of the electrochemical potentials are obviously vectors, *i.e.* first-order tensors. Hence the barycentric velocity expressed across the whole cross-section of the membrane (the volume flow) will be given by the simple hydrodynamic relation

$$v = k\left(-\frac{dp}{dx} - \varrho_{\text{el}}\frac{d\varphi}{dx}\right) \tag{5.66}$$

whereas the flows of individual substances j will be related only to gradients of the electrochemical potentials per unit mass

$$M_j = -\sum_{j=1}^{n} L_{jk} \frac{1}{M_k} \frac{d\tilde{\mu}_k}{dx} \qquad (j = 1, ..., n). \tag{5.67}$$

According to eq. (5.40)

$$\sum_{j=1}^{n} M_j = 0 \tag{5.68}$$

and hence, as follows on introducing (5.67) into (5.68)

$$\sum_{j=1}^{n} L_{jk} = 0 \qquad (k = 1, ..., n). \tag{5.69}$$

In view of Onsager's reciprocity relations, $L_{jk} = L_{kj}$, also the condition,

$$\sum_{k=1}^{n} L_{jk} = 0 \qquad (j = 1, ..., n) \tag{5.70}$$

is valid, or

$$\sum_{k=1}^{n-1} L_{jk} = -L_{jn} \qquad (j = 1, ..., n) \tag{5.71}$$

where $j = 1, ..., n - 1$ denote solutes and $j = n$ the solvent. Hence eq. (5.67) may be written as

$$M_j = -\sum_{k=1}^{n} L_{jk}\left(\frac{1}{M_k}\frac{d\tilde{\mu}_k}{dx} - \frac{1}{M_n}\frac{d\tilde{\mu}_n}{dx}\right) \qquad (j = 1, ..., n - 1). \tag{5.72}$$

Neglecting now any possible interactions between individual solutes,

Schlögl sets $L_{jk} = 0$ with the exception of L_{jj} (and L_{jn} which, however, does not appear in eq. (5.72) explicitly). Hence

$$M_j = -L_{jj}\left(\frac{1}{M_j}\frac{d\tilde{\mu}_j}{dx} - \frac{1}{M_n}\frac{d\tilde{\mu}_n}{dx}\right) \qquad (5.73)$$

and, using eq. (5.62), the molar flow J_j of the ionic solute may be expressed as

$$J_j = c_j v - \frac{L_{jj}}{M_j}\left(\frac{1}{M_j}\frac{d\tilde{\mu}_j}{dx} - \frac{1}{M_n}\frac{d\tilde{\mu}_n}{dx}\right) \qquad (5.74)$$

where $c_j = \varrho_j/M_j$ is the molar concentration.

Writing now the electrochemical potential as

$$\tilde{\mu}_j = \mu_{0j} + \bar{V}_j p + z_j F\varphi + RT\ln a_j$$

where $\bar{V}_j$ is the partial molal volume and a_j, the activity of an ion, may be expressed as $f_j c_j$ (activity coefficient multiplied by molar concentration), and considering a_n (activity of abundant solvent) as approximately constant, we get (for $z_n = 0$)

$$d\tilde{\mu}_j - d\tilde{\mu}_n = (\bar{V}_j - \bar{V}_n)\,dp + z_j F\,d\varphi +$$
$$+ RT\,d\ln c_j + RT\,d\ln f_j.$$

Finally, on introducing the last expression into eq. (5.74), we obtain the Schlögl equation

$$J_j = c_j v - RTU_j\frac{dc_j}{dx} - z_j FU_j c_j\frac{d\varphi}{dx} - \qquad (5.75)$$
$$-RTU_j c_j\frac{d\ln f_j}{dx} - U_j M_j c_j\left(\frac{V_j}{M_j} - \frac{V_n}{M_n}\right)\frac{dp}{dx} \qquad (j = 1, \ldots, n).$$

5.2.4. Henderson's equation—potential difference across a continuous layer with constant concentration gradients of individual ions

Henderson's condition of linear dependence of concentration on the distance for each ionic species, valid more or less strictly in mixed boundaries, may be written as

$$c_j = c_{jo} + \frac{c_{ji} - c_{jo}}{d}x \qquad (5.76)$$

so that the concentration gradient for each ionic species is

$$\frac{dc}{dx} = \frac{c_{ji} - c_{jo}}{d}.$$ (5.77)

Introducing these two expressions into the Nernst–Planck equation (5.36), written for the steady-state conditions in ordinary derivatives, we have

$$J_j = -RTU_j \frac{dc}{dx} - z_j F U_j c_j \frac{d\varphi}{dx} =$$

$$= -RTU_j \frac{c_{ji} - c_{jo}}{d} - z_j F U_j \left(c_{jo} + \frac{c_{ji} - c_{jo}}{d} x \right) \frac{d\varphi}{dx}.$$ (5.78)

Now, in the absence of an external circuit, no current flows across the membrane in the steady state

$$\sum_{j=1}^{n} z_j J_j = 0$$ (5.79)

i.e.

$$RT \frac{1}{d} \left(\sum_{j=1}^{n} z_j U_j c_{ji} - \sum_{j=1}^{n} z_j U_j c_{jo} \right) +$$

$$+ \frac{d\varphi}{dx} F \left[\sum_{j=1}^{n} z_j^2 U_j c_{jo} + \frac{x}{d} \left(\sum_{j=1}^{n} z_j^2 U_j c_{ji} - \sum_{j=1}^{n} z_j^2 U_j c_{jo} \right) \right] = 0.$$ (5.80)

After separation of variables, integration can be carried out easily

$$\int_{\varphi_o}^{\varphi_i} d\varphi = -\frac{RT}{F} \frac{1}{d} \int_0^d \left(\sum_{j=1}^{n} z_j U_j c_{ji} - \sum_{j=1}^{n} z_j U_j c_{jo} \right) \times$$

$$\times \frac{dx}{\left[\sum_{j=1}^{n} z_j^2 U_j c_{jo} + \frac{x}{d} \left(\sum_{j=1}^{n} z_j^2 U_j c_{ji} - \sum_{j=1}^{n} z_j^2 U_j c_{jo} \right) \right]}$$ (5.81)

yielding

$$\varphi_i - \varphi_o = -\frac{RT \sum_{j=1}^{n} z_j U_j c_{ji} - \sum_{j=1}^{n} z_j U_j c_{jo}}{F \sum_{j=1}^{n} z_j^2 U_j c_{ji} - \sum_{j=1}^{n} z_j^2 U_j c_{jo}} \ln \frac{\sum_{j=1}^{n} z_j^2 U_j c_{ji}}{\sum_{j=1}^{n} z_j^2 U_j c_{jo}}$$ (5.82)

which is Henderson's equation for the difference of electrical potential across a layer in which the concentration gradients of individual ionic species may be considered to be constant.

5.2.5. Planck's procedure—potential difference across a microscopically electroneutral continuous layer

Apart from the validity of the equation of electrodiffusion (5.36) Planck's procedure assumes the validity of the condition of local electroneutrality and further that all ions present are of the same valency (say, all univalent). The condition of local or microscopic electroneutrality is expressed by

$$\sum_{j=1}^{n} c_{j+} = \sum_{j=1}^{n} c_{j-} = c \qquad (5.83)$$

where c_{j+} is the local concentration of a cation, c_{j-} the local concentration of an anion and c the total concentration of cations or anions at any point. The total concentration of cations being equal to the total concentration of anions, the same must be true about their derivatives with respect to time

$$\sum_{j=1}^{n} \frac{\partial c_{j+}}{\partial t} = \sum_{j=1}^{n} \frac{\partial c_{j-}}{\partial t}. \qquad (5.84)$$

Planck (1890) relates the time derivatives of concentrations at first to fluxes and then to gradients of concentrations and of the electrical potential. The flux of an ion across a plane of unit area, perpendicular to the x-axis is J at the coordinate x and $J + (\partial J/\partial x) \Delta x$ at the coordinate $(x + \Delta x)$. Accumulation of the ion between the two planes during the time Δt is thus

$$-\frac{\partial J}{\partial x} \Delta x \, \Delta t$$

per unit area which, in the volume of Δx cm^3 (between the two unit areas), amounts to a change of concentration

$$\Delta c = -\frac{\partial J}{\partial x} \Delta t$$

and in the limit

$$\frac{\partial c}{\partial t} = -\frac{\partial J}{\partial x}. \qquad (5.85)$$

The flux J, however, is given by the electrodiffusion equation (5.36) and hence we can write for cations $(z = +1)$

$$\frac{\partial c_{j+}}{\partial t} = RTU_{j+} \frac{\partial^2 c_{j+}}{\partial x^2} + FU_{j+} \frac{\partial}{\partial x_t}\left(c_{j+} \frac{\partial \varphi}{\partial x}\right) \qquad (5.86)$$

and for anions

$$\frac{\partial c_{j-}}{\partial t} = RTU_{j-}\frac{\partial^2 c_{j-}}{\partial x^2} - FU_{j-}\frac{\partial}{\partial x}\left(c_{j-}\frac{\partial \varphi}{\partial x}\right). \tag{5.87}$$

Introducing equations (5.86) and (5.87) into (5.84) we have

$$RT\sum_{j=1}^{n} U_{j+}\frac{\partial^2 c_{j+}}{\partial x^2} + F\sum_{j=1}^{n} U_{j+}\frac{\partial}{\partial x}\left(c_{j+}\frac{\partial \varphi}{\partial x}\right) =$$

$$= RT\sum_{j=1}^{n} U_{j-}\frac{\partial^2 c_{j-}}{\partial x^2} - F\sum_{j=1}^{n} U_{j-}\frac{\partial}{\partial x}\left(c_{j-}\frac{\partial \varphi}{\partial x}\right) \tag{5.88}$$

which can be integrated term by term, giving

$$RT\sum_{j=1}^{n} U_{j+}\frac{\partial c_{j+}}{\partial x} + F\sum_{j=1}^{n} U_{j+}c_{j+}\frac{\partial \varphi}{\partial x} =$$

$$= RT\sum_{j=1}^{n} U_{j-}\frac{\partial c_{j-}}{\partial x} - F\sum_{j=1}^{n} U_{j-}c_{j-}\frac{\partial \varphi}{\partial x} \tag{5.89}$$

or

$$RT\frac{\partial U}{\partial x} + FU\frac{\partial \varphi}{\partial x} = RT\frac{\partial V}{\partial x} - FV\frac{\partial \varphi}{\partial x} \tag{5.90}$$

where

$$U = \sum_{j=1}^{n} U_{j+}c_{j+} \quad \text{and} \quad V = \sum_{j=1}^{n} U_{j-}c_{j-}. \tag{5.91}$$

Eq. (5.90) may be solved for the local gradient of electrical potential, giving

$$\frac{\partial \varphi}{\partial x} = -\frac{RT}{F}\frac{\dfrac{\partial(U - V)}{\partial x}}{U + V} \tag{5.92}$$

This equation cannot be integrated, since the dependence of ionic concentrations (and hence also of U and V) on x is not known. It is for this reason that further derivation is restricted to steady-state conditions, eq. (5.92) being preserved to be used later. In a steady state, the time derivatives of concentrations vanish, so that (5.86) and (5.87) may be written as

$$RT\frac{d^2 c_{j+}}{dx^2} + F\frac{d}{dx}\left(c_{j+}\frac{d\varphi}{dx}\right) = 0, \tag{5.93}$$

$$RT\frac{d^2 c_{j-}}{dx^2} - F\frac{d}{dx}\left(c_{j-}\frac{d\varphi}{dx}\right) = 0. \tag{5.94}$$

The concentrations and the electrical potential in a steady state being functions of x only, ordinary derivatives instead of partial derivatives are used. Equations (5.93) and (5.94) may now be integrated, giving

$$RT \frac{dc_{j+}}{dx} + Fc_{j+} \frac{d\varphi}{dx} = A_j, \qquad (5.95)$$

$$RT \frac{dc_{j-}}{dx} - Fc_{j-} \frac{d\varphi}{dx} = B_j. \qquad (5.96)$$

Summation over all ionic species yields

$$RT \frac{dc}{dx} + Fc \frac{d\varphi}{dx} = A, \qquad (5.97)$$

$$RT \frac{dc}{dx} - Fc \frac{d\varphi}{dx} = B \qquad (5.98)$$

where

$$A = \sum_{n=1}^{n} A_j, \qquad B = \sum_{n=1}^{n} B_j \qquad (5.99)$$

and c is the total concentrations of cations or anions, defined by eq. (5.83). Adding the eq. (5.97) and (5.98) we obtain

$$2RT \frac{dc}{dx} = A + B \qquad (5.100)$$

which can be integrated readily, yielding

$$2RTc = (A + B) x + \text{const.} \qquad (5.101)$$

It is thus seen that from the condition of local electroneutrality and from the condition of steady state it follows that the total concentration of ions is a linear function of the distance. If d is the thickness of the continuous layer, c_0 the total concentration of cations or anions at $x = 0$ and c_i the concentration at $x = d$, the total local concentration at any x between zero and d may be expressed by

$$c = c_o + \frac{c_i - c_o}{d} x. \qquad (5.102)$$

Subtracting, on the other hand, eq. (5.98) from eq. (5.97) we have

$$2Fc \frac{d\varphi}{dx} = A - B. \qquad (5.103)$$

Combining equations (5.102) and (5.103) we obtain

$$\frac{d\varphi}{dx} = \frac{(A - B)d}{2F[(c_i - c_o)x + c_o d]} \tag{5.104}$$

which integrates from $x = 0$ to $x = d$ to

$$\varphi_i - \varphi_o = \frac{(A - B)d}{2F(c_i - c_o)} \ln \frac{c_i}{c_o} \tag{5.105}$$

which may be written

$$\varphi_i - \varphi_o = \frac{RT}{F} \ln \xi \tag{5.106}$$

where

$$\xi = \left(\frac{c_i}{c_o}\right)^{\frac{(A-B)d}{2(c_i-c_o)RT}} \tag{5.107}$$

It is, however, necessary to exclude the constants **A** and **B** from this expression so that the dependence of ξ (and hence, also of $\varphi_i - \varphi_o$) only on concentrations and mobilities of ions is obtained. To this end, Planck multiplies each of the equations (5.95) and (5.96) with the appropriate mobility, summing afterwards over all cation and anion species and using symbols defined by (5.91):

$$\sum_{j=1}^{n} U_{j+}A_j = RT \frac{dU}{dx} + FU \frac{d\varphi}{dx} \tag{5.108}$$

$$\sum_{j=1}^{n} U_{j-}B_j = RT \frac{dV}{dx} - FV \frac{d\varphi}{dx}. \tag{5.109}$$

Using eq. (5.105), it is easy to prove that the right-hand sides of eq. (5.108) and (5.109) are equal. Hence the same must be true about their left-hand sides:

$$\sum_{j=1}^{n} U_{j+}A_j = \sum_{j=1}^{n} U_{j-}B_j = C. \tag{5.110}$$

Now, the eq. (5.104) and (5.110) are introduced into each of eq. (5.108) and (5.109) yielding two linear first-order equations:

$$\frac{dU}{dx} + U \frac{(A - B)d}{2RT[(c_i - c_o)x + c_o d]} = \frac{C}{RT}, \tag{5.111}$$

$$\frac{dV}{dx} - V \frac{(A - B)d}{2RT[(c_i - c_o)x + c_o d]} = \frac{C}{RT}. \tag{5.112}$$

The form of these equations is

$$\frac{dy}{dx} + f(x) y = g(x)$$

and hence the form of their solution is

$$y = e^{-\int f(x)\,dx} \left[\int e^{\int f(x)\,dx} g(x)\,dx + \text{const.} \right]$$

i.e.

$$U = \frac{2C[(c_i - c_o)x + c_o d]}{2(c_i - c_o)RT - (A - B)d} +$$

$$+ [(c_i - c_o)x + c_o d]^{-\frac{(A-B)d}{2(c_i-c_o)RT}} . \text{const.} \qquad (5.113)$$

$$V = \frac{2C[(c_i - c_o)x + c_o d]}{2(c_i - c_o)RT + (A - B)d} +$$

$$+ [(c_i - c_o)x + c_o d]^{\frac{(A-B)d}{2(c_i-c_o)RT}} . \text{const.} \qquad (5.114)$$

At $x = 0$, $U = U_0$ and $V = V_0$; at $x = d$, $U = U_i$ and $V = V_i$; equations (5.113) and (5.114) are written in pairs (for each of the two boundary conditions), the constants of the last integration are eliminated and ξ defined by eq. (5.107) is introduced

$$\xi U_i - U_o = \frac{2Cd(\xi c_i - c_o)}{2(c_i - c_o)RT + (A - B)d}, \qquad (5.115)$$

$$V_i - \xi V_o = \frac{2Cd(c_i - \xi c_o)}{2(c_i - c_o)RT - (A - B)d}. \qquad (5.116)$$

C is eliminated by division of the two equations

$$\frac{\xi U_i - U_o}{V_i - \xi V_o} = \frac{2(c_i - c_o)RT - (A - B)d}{2(c_i - c_o)RT + (A - B)d} \frac{\xi c_i - c_o}{c_i - \xi c_o} \qquad (5.117)$$

and $(A - B)$ is eliminated by substitution from eq. (5.107)

$$\frac{\xi U_i - U_o}{V_i - \xi V_o} = \frac{\ln\frac{c_i}{c_o} - \ln \xi}{\ln\frac{c_i}{c_0} + \ln \xi} \frac{\xi c_i - c_o}{c_i - \xi c_o}. \qquad (5.118)$$

This is the final equation of Planck; introducing in to it the total concentrations defined by eq. (5.83) and the quantities U and V defined by equations (5.91) the quantity ξ is determined from which the potential difference across a continuous layer may be calculated using the simple eq. (5.106). It is not easy to calculate ξ from eq. (5.118); due to its transcendent form, ξ occurs both inside and outside the argument of the logarithmic function. The easiest solution is perhaps a graphical one (MacInnes, 1961): both the left-hand side and the right-hand side of eq. (5.118) is plotted in the same graph as a function of various assumed values of ξ. The point of the intersection of the two curves then corresponds to the value of which solves the equation (5.118) for given values of U_o, V_o, U_i, V_i, c_o and c_i.

In order to learn when the condition of local electroneutrality is strictly satisfied in steady-state ionic systems, the Poisson equation may be used, a general expression for electrostatic interactions which, for a unidimensional case, may be written as

$$\frac{d^2\varphi}{dx^2} = -\frac{\varrho}{\varepsilon} \tag{5.119}$$

where φ is the electrical potential, ϱ the charge density and ε the absolute dielectric constant. When local electroneutrality prevails, the charge density is zero:

$$F \sum_{j=1}^{n} z_j c_j = \varrho = 0 \tag{5.120}$$

(where F is again the Faraday number, z_j the valency of the j-th ionic species and c_j its concentration) so that the Poisson equation turns into the Laplace equation

$$\frac{d^2\varphi}{dx^2} = 0 \tag{5.121}$$

which by integration gives

$$\frac{d\varphi}{dx} = \text{const.} \tag{5.122}$$

As stressed by Agin (1967) any attempt to calculate potential profiles in a regime using condition (5.120) but not the condition (5.122) is physically inadmissible. Now, following Polissar (1954), we multiply the Nernst−Planck differential equation of electrodiffusion

(5.81) which can be written for steady-state conditions (c and φ being functions of x only) as

$$J_j = -RTU_j \frac{dc_j}{dx} - z_j F U_j c_j \frac{d\varphi}{dx} \qquad (5.123)$$

by z_j/FU_j and carry out sumation over all ionic species present. After rearrangement we obtain

$$\frac{d\varphi}{dx} \sum_{j=1}^{n} z_j^2 c_j = -\frac{RT}{F} \frac{d}{dx} \sum_{j=1}^{n} z_j c_j - \frac{1}{F} \sum_{j=1}^{n} \frac{z_j J_j}{U_j}. \qquad (5.124)$$

The first term on the right-hand side vanishes when local electroneutrality prevails, the second does so only either when there is no electrodiffusional flow J_j (equilibrium), or when all ions have the same mobility (mobility then can be taken out before the summation sign and the sum of electric currents $z_j J_j$ is zero uder open-circuit conditions). Now, if local electroneutrality prevails, at least some of the ionic species present have different mobilities and the system is in a steady state rather than in equilibrium (so that $J_j = $ const. $\neq 0$) and also mobilities are constants independent of x, eq. (5.124) than shows

$$\frac{d\varphi}{dx} = \frac{k}{\displaystyle\sum_{j=1}^{n} z_j^2 c_j}. \qquad (5.125)$$

This equation is compatible with the Poisson equation for zero charge density (5.122) only when $\sum_{j=2}^{n} z_j^2 c_j$, the local ionic strength, is a constant independent of x, and hence only under this condition is the condition of local electroneutrality strictly satisfied. When only univalent ions are present in significant concentrations, eq. (5.125) becomes equivalent to eq. (5.103) of Planck and the requirement for a strict total electroneutrality is that the total concentration of ions is a constant independent of x. In a locally electroneutral steady-state system, however, the total concentration is always a linear function of x, as shown by eq. (5.101) and hence using eq. (5.102) we can see that the total concentration of ions is a constant independent of x when it is the same at two different points, i.e., when $c_i = c_o$.

The condition of equality of total concentrations ($c_i = c_o$) not only makes the condition of local electroneutrality valid and hence also the condition of constant field exactly satisfied but, at the same time, it greatly simplifies the transcendental equation of Planck

(5.118) by converting it into the equation of Goldman. On introducing the condition $c_i \approx c_o$ into eq. (5.118) its right-hand side becomes equal to one, so that

$$\zeta = \frac{U_o - V_i}{U_i + V_o} \qquad (5.126)$$

and, using eq. (5.106) and equations (5.91) we have

$$\varphi_i - \varphi_o = \frac{RT}{F} \ln \frac{U_o + V_i}{U_i + V_o} = \frac{RT}{F} \ln \frac{\sum\limits_{j=1}^{n} U_{j^+} c_{j^+o} + \sum\limits_{j=1}^{n} U_{j^-} c_{j^-i}}{\sum\limits_{j=1}^{n} U_{j^+} c_{j^+i} + \sum\limits_{j=1}^{n} U_{j^-} c_{j^-o}}$$

$$(5.127)$$

which indeed is the equation for electrical potential difference by Goldman (1943).

However, the above derivations which were carried out for a layer of a continuous and uniform medium, can probably be applied to the interior of a membrane or of ion-permeable regions of a membrane only when the membrane is so thick that the diffuse double layers extending inside it from the phase boundaries may be neglected — in the double layers the condition of local electroneutrality is violated. As we have seen already (p. 253) the effective thickness of an electrical diffuse double layer (the Debye length) is directly proportional to the square root of the relative dielectric constant and inversely proportional to the square of the concentration beyond the extent of the double layer. The relative dielectric constant in the ion-permeable regions of cell membranes may be quite low (between 1 for a vacuum and about 80 for water) but if the concentration of ions in these regions is low, too, the effective thickness of the double layer may be greater than the thickness of the cell membrane. Under this condition, the assumption of constant field again represents a useful approximation (Cole, 1965), which is the better the smaller the ratio of membrane thickness to the effective thickness of the double layer. Thus we can recognize two cases where the constant-field condition for the valitidy of the Goldman equation (5.127) is approximately satisfied: (1) in thick membranes when the local ionic strength is approximately constant; (2) in thin membranes when the concentration of ions in the membrane or in the ion-permeable membrane regions is low.

The Goldman equation (5.127) can be derived in a much simpler way if the constant-field assumption is introduced at the very beginning. This derivation is shown in the next section.

5.2.6. Goldman's procedure—potential difference across a continuous layer with constant field

The constant-field condition may be expressed by

$$\frac{\partial \varphi}{\partial x} = \frac{\varphi_i - \varphi_o}{d} \qquad (5.128)$$

where $\varphi_i - \varphi_o$ is the electrical potential difference across a continuous layer (say, across the membrane proper) and d is the thickness of the layer. We can introduce this condition into the differential equation of electrodiffusion (5.36), assuming, moreover, steady-state conditions, so that J_j will be a constant and c_j a function of x only (hence ordinary derivatives):

$$J_j = -RTU_j \frac{dc_j}{dx} - z_j FU_j c_j \frac{\varphi_i - \varphi_o}{d} \qquad (5.129)$$

we separate the variables and integrate across the layer (from $x = 0$, $c_j = c_{j0}$, to $x = d$, $c_j = c_j$):

$$\int_0^d dx = -RTU_j \int_{c_{jo}}^{c_{ji}} \frac{dc_j}{J_j + z_j FU_j \dfrac{\varphi_i - \varphi_o}{d} c_j}$$

which gives

$$d = \frac{RTd}{z_j F(\varphi_i - \varphi_o)} \ln \frac{J_j + z_j FU_j \dfrac{\varphi_i - \varphi_o}{d} c_{ji}}{J_j + z_j FU_j \dfrac{\varphi_i - \varphi_o}{d} c_{jo}}.$$

The last equation may be solved for J_j, yielding

$$J_j = z_j FU_j \frac{\varphi_i - \varphi_o}{d} \cdot \frac{c_{ji} - c_{jo} e^{-z_j F(\varphi_o - \varphi_i)/RT}}{e^{-z_j F(\varphi_o - \varphi_i)/RT} - 1}. \qquad (5.130)$$

This is the familiar constant-field expression for the steady-state flux of an individual ionic species, applicable in this form to the inside of a continuous layer. It is of interest to see how this equation is reduced, when the absence of either the concentration gradient or of the electrical potential gradient is assumed.

In the first case, when $c_{ji} = c_{jo} = c_j$, it yields

$$J_j = -z_j FU_j c_j \frac{\varphi_i - \varphi_o}{d} \qquad (5.131)$$

which describes the steady-state migration of an ion in the electrical field.

In the other case, however, when $\varphi_i - \varphi_o$ is zero, eq. (5.130) is an indeterminate quantity 0/0. L'Hôpital's rule has to be used, according to which

$$\lim \frac{f(x)}{g(x)} = \lim \frac{f'(x)}{g'(x)},$$

the limit of the ratio of two functions is equal to the limit of the ratio of their derivatives. We consider the right-hand side of eq. (5.130) as a quotient of two functions of the variable $\varphi_i - \varphi_o$, we calculate the derivatives of the numerator and of the denominator and in the quotient of the two derivatives we let the variable $\varphi_i - \varphi_o$ vanish. In this way we obtain the correct steady-state formula

$$J_j = RTU_j \frac{c_{jo} - c_{ji}}{d} \tag{5.132}$$

describing the diffusion of an ion in the absence of the electrical field.

We may observe a certain asymmetry of the formula (5.130); the external concentration of the ion c_{jo} is multiplied by an exponential factor, whereas its internal concentration, c_{ji}, is not. This is a natural result of our choice of the external potential φ_o, as a reference; the exponential factor multiplying the internal concentration thus contains $\varphi_o - \varphi_o = 0$ in the exponent and hence reduces to 1.

To calculate the value of the electrical potential difference across a constant-field layer, use can be made of the condition that when there is no external electric circuit, the sum of the individual ionic currents across the layer must be zero,

$$\sum_{j=1}^{n} z_j F J_j = 0. \tag{5.133}$$

Simple formulae for the electrical potential difference are obtained when all ions present in significant concentrations are of the same valency (say, univalent), for then the denominators in the flux formulae of the type of (5.130) can be brought to the same form and therefore may be cancelled out from equations of the type of (5.133). Thus the current carried across the layer by an univalent cation will be given by

$$FJ_{j^+} = F^2 U_{j^+} \frac{\varphi_i - \varphi_o}{d} \frac{c_{j^+i} - c_{j^+o} e^{-F(\varphi_i - \varphi_o)/RT}}{e^{-F(\varphi_i - \varphi_o)/RT} - 1} \tag{5.134}$$

and that carried by an univalent anion by

$$FJ_{j-} = -F^2 U_{j-} \frac{\varphi_i - \varphi_o}{d} \frac{c_{j-i} - c_{j-o} e^{F(\varphi_i - \varphi_o)/RT}}{e^{F(\varphi_i - \varphi_o)/RT} - 1} =$$

$$= F^2 U_{j-} \frac{\varphi_i - \varphi_o}{d} \frac{c_{j-i} e^{-F(\varphi_i - \varphi_o)/RT} - c_{j-o}}{e^{-F(\varphi_i - \varphi_o)/RT} - 1} \qquad (5.135)$$

From the condition (5.133) that under open-circuit conditions the sum of individual currents is zero we have (when only univalent ions are present)

$$\sum_{j=1}^{n} FJ_{j+} + \sum_{j=1}^{n} FJ_{j-} = 0. \qquad (5.136)$$

It follows that

$$\sum_{j=1}^{n} U_{j+} c_{j+i} - \sum_{j=1}^{n} U_{j+} c_{j+o} e^{-F(\varphi_i - \varphi_o)/RT} =$$

$$= \sum_{j=1}^{n} U_{j-} c_{j-i} e^{-F(\varphi_i - \varphi_o)/RT} - \sum_{j=1}^{n} U_{j-} c_{j-o} \qquad (5.137)$$

or

$$e^{-F(\varphi_i - \varphi_o)/RT} = \frac{\sum\limits_{j=1}^{n} U_{j+} c_{j+i} + \sum\limits_{j=1}^{n} U_{j-} c_{j-o}}{\sum\limits_{j=1}^{n} U_{j+} c_{j+o} + \sum\limits_{j=1}^{n} U_{j-} c_{j-i}} \qquad (5.138)$$

i.e.

$$\varphi_i - \varphi_o = \frac{RT}{F} \ln \frac{\sum\limits_{j=1}^{n} U_{j+} c_{j+o} + \sum\limits_{j=1}^{n} U_{j-} c_{j-i}}{\sum\limits_{j=1}^{n} U_{j+} c_{j+i} + \sum\limits_{j=1}^{n} U_{j-} c_{j-o}}.$$

For example, if only potassium, sodium and chloride ions are present in significant concentrations, the electrical potential difference across a constant-field layer will then be given by

$$\varphi_i - \varphi_o = \frac{RT}{F} \ln \frac{U_{K^+} c_{K^+o} + U_{Na^+} c_{Na^+o} + U_{Cl^-} c_{Cl^-i}}{U_{K^+} c_{K^+i} + U_{Na^+} c_{Na^+i} + U_{Cl^-} c_{Cl^-o}}. \qquad (5.139)$$

An important (although rather obvious) observation is now in order: ions which are in thermodynamic equilibrium do not enter into eq. (5.138); their fluxes are identical in both directions and the current carried by them is therefore zero and need not be taken into account (refer to the condition of zero net total electric current across an open-circuited membrane, eq. 5.136). Thus, e.g., when in the above system with sodium, potassium and chloride ions the chloride ions are in thermodynamic equilibrium everywhere inside the continuous layer,

$$\varphi_i - \varphi_o = \frac{RT}{F} \ln \frac{U_{K^+}c_{K^+o} + U_{Na^+}c_{Na^+o}}{U_{K^+}c_{K^+i} - U_{Na^+}c_{Na^+i}} = \frac{RT}{F} \ln \frac{c_{Cl^-i}}{c_{Cl^-o}} . \quad (5.140)$$

5.2.7. Constant-field equation for potential difference across the whole membrane

When accepting formula (5.139) or (5.140) as an expression for the electrical potential difference across the membrane proper we are faced with the task of relating the concentrations of ions at the boundaries inside the membrane to those in the adjacent media as well as of accounting for the steps of the electrical potential between the media and the membrane (see once more Fig. 5.6).

The assumption of equilibrium between the membrane and the adjacent media simplifies the derivation, but even under such an assumption it would be incorrect to relate the concentration of an ion just inside the membrane to its concentration in the adjacent medium by a constant partition coefficient. Ions are charged particles and their equilibrium distribution between two phases is thus a function not only of the chemical nature of the two media but also of the electrical potential difference between them. Such electrical potential difference between phases of different chemical composition can never be measured experimentally for the very reason that other forces beside the potential gradient are operative in transitions of really existing electrical charges (ions) between the two phases. The difference between the pair of such phase potentials, however, does contribute to the overall potential difference across the membrane and this is why these individual steps must be considered (Polissar, 1954).

The equilibrium of, say, the potassium ion at the outer surface of the membrane is expressed by equality of its electrochemical potentials and may be written (referring again to Fig. 5.6 and neglecting activity coefficients) as

$$\mu'_{0K^+} + RT \ln [K^+]_o + F\varphi'_o = \mu_{0K^+} + RT \ln c_{K^+o} + F\varphi_o \quad (5.141)$$

where μ'_{0K^+} and μ_{0K^+} are standard potentials (the part of partial concentration and of the electrical potential) for the medium and the membrane, respectively. From eq. (5.141) it follows that

$$\ln c_{K^+{}_o} = \frac{\mu'_{0K^+} - \mu_{0K^+}}{RT} + \ln [K^+]_o + \frac{F(\varphi'_o - \varphi_o)}{RT} \quad (5.142)$$

or

$$c_{K^+{}_o} = e^{\frac{\mu'_{0K}{}^+ - \mu_{0K}{}^+}{RT}} [K^+]_o \, e^{\frac{F(\varphi'_o - \varphi_o)}{RT}} = k_{K^+}[K^+]_o \, e^{-FE_o/RT} \quad (5.143)$$

where $k_{K^+} = e^{\frac{u_0'{}_K{}^+ - \mu_{0K}{}^+}{RT}}$ is the partition (or distribution) coefficient of potassium ions and $E_o = \varphi_o - \varphi'_o$. Analogously, for concentrations of all ions considered in the membrane it may be written

$$c_{Na^+{}_o} = k_{Na^+}[Na^+]_o \, e^{-FE_o/RT}$$
$$c_{Na^+{}_i} = k_{Na^+}[Na^+]_i \, e^{FE_i/RT}$$
$$c_{K^+{}_o} = k_{K^+}[K^+]_o \, e^{-FE_o/RT}$$
$$c_{K^+{}_i} = k_{K^+}[K^+]_i \, e^{FE_i/RT}$$
$$c_{Cl^-{}_o} = k_{Cl^-}[Cl^-]_o \, e^{FE_o/RT}$$
$$c_{Cl^-{}_i} = k_{Cl^-}[Cl^-]_i \, e^{-FE_i/RT} \quad (5.144)$$

Introducing relations (5.144) into eq. (5.139) for the potential difference across the membrane proper, $\varphi_i - \varphi_o$, and calculating the overall membrane potential $E = \varphi'_i - \varphi'_o = E_o + \varphi_i - \varphi_o + E_i$, we make use of the identity

$$E_o + E_i \equiv \frac{RT}{F} \ln \left(e^{FE_o/RT} \cdot e^{FE_i/RT} \right) \quad (5.145)$$

to obtain

$$E = \frac{RT}{F} \ln \frac{U_{K^+}k_{K^+}[K^+]_o + U_{Na^+}k_{Na^+}[Na^+]_o + U_{Cl^-}k_{Cl^-}[Cl^-]_i \, e^{F(E_o - E_i)/RT}}{U_{K^+}k_{K^+}[K^+]_i + U_{Na^+}k_{Na^+}[Na^+]_i + U_{Cl^-}k_{Cl^-}[Cl^-]_o \, e^{F(E_o - E_i)/RT}} \quad (5.146)$$

Finally, we may introduce permeability coefficients

$$P_{K^+} = RTU_{K^+}k_{K^+}/d$$
$$P_{Na^+} = RTU_{Na^+}k_{Na^+}/d \quad (5.147)$$
$$P_{Cl^-} = RTU_{Cl^-}k_{Cl^-}/d$$

with which eq. (5.146) takes the form

$$E = \frac{RT}{F} \ln \frac{P_{K^+}[K^+]_o + P_{Na^+}[Na^+]_o + P_{Cl^-}[Cl^-]_i \, e^{F(E_o - E_i)/RT}}{P_{K^+}[K^+]_i + P_{Na^+}[Na^+]_i + P_{Cl^-}[Cl^-]_o \, e^{F(E_o - E_i)/RT}}$$

(5.148)

or, when chloride ions are in a passive-flux equilibrium or when in the membrane their permeability is negligibly low

$$E = \frac{RT}{F} \ln \frac{P_{K^+}[K^+]_o + P_{Na^+}[Na^+]_o}{P_{K^+}[K^+]_i + P_{Na^+}[Na^+]_i}$$

(5.149)

If P_{K^+} is much greater than the other permeability coefficients, formulae (5.148) or (5.149) reduce to that for the equilibrium potential of potassium ions, for other terms in the nominator and the denominator may be neglected and the potassium permeability coefficients cancel out. This state of affairs apparently corresponds to many instances of resting potentials of muscle and nerve membranes. If, on the other hand, the sodium permeability of such membranes increases, the membrane potential decreases and even changes its sign, the distribution of sodium ions across the membranes being opposite to that to that of potassium ions and becoming more and more important in the generation of the membrane potential. Such a change of membrane permeability forms the basis of the membrane potential change, called the action potential, according to the well-known sodium theory of Hodgkin and Huxley (see, e.g., Katz, 1966).

5.2.8. Constant-field equation for steady-state membrane potential in the presence of an electrogenic sodium pump

Formula (5.149) has been derived under the assumption that active sodium transport, if present, is electroneutral, i.e., that the sodium pump performs obligatory coupled exchange of one sodium ion for one potassium ion. However, it has been shown by Thomas (1972) that, in a steady state, eq. (5.149) includes even the case of an electrogenic sodium pump. The following assumptions must be made. (1) The net passive flux of sodium and potassium ion is described by eq. (5.130), or, more generally, by any equation of the form

$$J = U(c_i - c_o \, e^{-F(\varphi_i - \varphi_o)/RT}) f(\varphi_i - \varphi_o)$$

(5.150)

where $f(\varphi_i - \varphi_o)$ is an arbitrary function of the potential difference across the membrane proper, $\varphi_i - \varphi_o$.

(2) Due to the steady-state condition, the ion concentrations are steady and hence the sum of net passive flux J and net active flux J_{act} is zero for each of the two ions:

$$J_{K^+} + J_{K^+act} = 0, \qquad (5.151)$$

$$J_{Na^+} + J_{Na^+act} = 0. \qquad (5.152)$$

(3) Finally, there is a given coupling ratio r giving the number of sodium ions pumped from the cell for each of the potassium ions taken actively in:

$$rJ_{K^+act} + J_{Na^+act} = 0. \qquad (5.153)$$

From (5.151), (5.152) and (5.153) it follows that also

$$rJ_{K^+} + J_{Na^+} = 0. \qquad (5.154)$$

Now, introducing the explicit expressions for net passive fluxes (5.150) we obtain

$$rU_{K^+}(c_{K^+i} - c_{K^+o}\, e^{-F(\varphi_i - \varphi_o)/RT})f(\varphi_i - \varphi_o) +$$
$$+\, U_{Na^+}(c_{Na^+i} - c_{Na^+o}\, e^{-F(\varphi_i - \varphi_o)/RT})f(\varphi_i - \varphi_o) = 0 \qquad (5.155)$$

where $f(\varphi_i - \varphi_n)$ may be cancelled and the electrical potential difference across the membrane proper, $\varphi_i - \varphi_o$, can be expressed as follows:

$$\varphi_i - \varphi_o = \frac{RT}{F}\ln\frac{rU_{K^+}c_{K^+o} + U_{Na^+}c_{Na^+o}}{rU_{K^+}c_{K^+i} + U_{Na^+}c_{Na^+i}}. \qquad (5.156)$$

Finally, the same procedure of accounting for the electrical potential steps at the membrane boundaries and relating the ion concentrations inside the membrane to those in the adjacent media as used in the ast section yields

$$E = \frac{RT}{F}\ln\frac{rP_{K^+}[K^+]_o + P_{Na^+}[Na^+]_o}{rP_{K^+}[K^+]_i + P_{Na^+}[Na^+]_i} \qquad (5.157)$$

where r is the number of sodium ions actively extruded per one potassium ion actively transported in the opposite direction. When the pump is electroneutral (coupled one-to-one) $r = 1$ and (5.157) reduces to (5.149). When, on the other hand, the pump functions in a purely electrogenic manner, r tends to infinity, the sodium terms become negligible and the expression reduces to that for equilibrium potassium

potential. An important conclusion may be drawn from the last
derivation: in a steady state, the electrogenic or electroneutral character
of the pump cannot be inferred from the knowledge of the membrane
potential and of equilibrium potentials of individual ions alone. As is
obvious from eq. (5.157) the steady-state value of the membrane
within the limits given by the equilibrium potentials. Variations of the
membrane potential within these limits may result equally well from
variations of passive permeabilities and of the coupling ratio. Hence
the character of the pump must be established by other means, by
a direct search for coupling interdependence of individual active
fluxes or for nonsteady-state potentials outside the range of equilibrium
potentials.

5.2.9. The Hodgkin-Horowicz equation

There is an alternative approach to the description of membrane
potentials, based on equivalent circuits or electrical analogues of
the membrane and discussed, e.g., by Dainty (1960) and Finkelstein and Mauro (1963). Although known for years, this approach
has not been fully appreciated until a recent penetrating study by
Jaffe (1974). He showed the relations of the type of eq. (5.159) to be

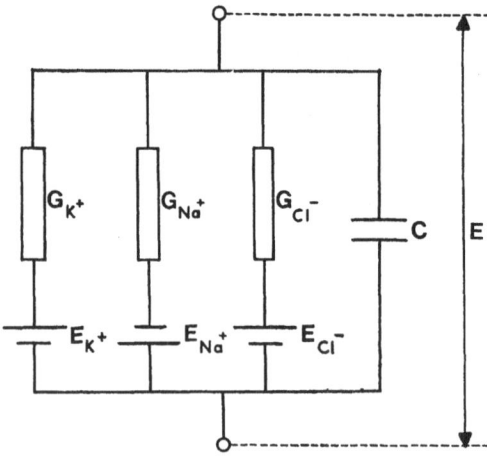

Fig. 5.7. Equivalent circuit for a cell membrane. C Membrane capacitance, G's
conductances for individual ion species, E_j's electromotive forces, equal to equilibrium potentials of the individual species (e.g., $E_{K^+} = (RT/F) \ln ([K^+]_o/[K^+]_i)$).

better suited for describing potential differences across membranes of many, perhaps all, cells than are Goldman-type equations; in many voltage-concentration relations found experimentally and described in the literature the conductances in eq. (5.159) behave as constants whereas the permeabilities in Goldman-type equations do not. Thus, different ion species appear to use relatively separate pathways across cell membranes. Jaffe calls such relations the Hodgkin – Horowicz equations, since Hodgkin and Horowicz (1959) used them explicitly.

The equations may be derived as follows. Let us consider an equivalent circuit for the membrane system with potassium, sodium and chloride ions, the steady state of which was described above using the Goldman equation (5.148) or (5.149). The circuit is shown in Fig. 5.7. The membrane potential E is seen to be generated by electromotive forces (equal to equilibrium potentials of individual ionic species) which are connected in series with corresponding integral conductances. The current through each of the circuit branches is given by

$$I_j = G_j(E_j - E) \qquad (5.158)$$

and from the condition that the sum of the currents be zero under open-circuit conditions when the capacity charge is constant (steady state) the membrane potential may be expressed as

$$E = \frac{G_{K^+}E_{K^+} + G_{Na^+}E_{Na^+} + G_{Cl^-}E_{Cl^-}}{G_{K^+} + G_{Na^+} + G_{Cl^-}}. \qquad (5.159)$$

It may be seen that if the conductance of a certain ionic species exceeds considerably the other conductances, the membrane potential will be close to the equilibrium potential of that species. When an electrogenic pumping mechanism is present another branch with electromotive force of the pump and the pump conductivity has to be included in the circuit.

Theoretical description of cellular ion transport is likely to profit considerably in the near future from the comprehensive physical theory of ion transport across modified artificial phospholipid membranes, developed by Markin and Chizmadjev (1974). A very interesting approach to the theoretical description of ion transport across the cell membrane is represented by the stochastic method of Györgyi and Sugar (1974), used by the authors to describe the alkali cation transport in erythrocyte membranes.

5.3. CHEMICAL NATURE
OF ION-TRANSLOCATING SYSTEMS

Most translocations of the major cations are accomplished at the expense of ATP in a reaction catalyzed by one or another membrane-bound adenosinetriphosphatase.

5.3.1. Na, K-Adenosinetriphosphatase

The enzyme that has aroused greatest interest and about which perhaps more is known than about all the other transport agencies taken together is the sodium-potassium stimulated, magnesium-activated adenosinetriphosphatase; for the most recent cross section of views see the Annals of the New York Academy of Sciences, **242**, 1−741 (1974).

The Na, K-ATPase is a widely distributed plasma membrane enzyme, occurring in all animal organs so far tested, as well as in protozoans, higher plants and some algae (Bonting, 1970), with the notable exception of some erythrocytes (dog) and of yeasts. In bacteria it plays a minor role (if any) in Na^+ and K^+ transport. All the Na, K-ATPase systems and preparations have a number of properties in common. Thus, they all require Mg^{2+} for activity; the apparent half-saturation constant for Na^+ is of the order of 10 mM, that for K^+ of the order of 1 mM, there being probably two sets of sites, the i-sites preferring sodium, the o-sites preferring potassium (*cf.* Skou, 1972); the pH optimum is at about 7.5; they are all inhibited by various glycosides, typically ouabain (it causes 50 % inhibition at concentrations ranging from 10^{-7} to 10^{-4} M); they are all slightly stimulated by 10^{-9} to 10^{-8} M ouabain*; the ratio of cations transported under optimal conditions per ATP molecule split is 2.6 ±0.19, reflecting the frequently observed exchange of 3 Na^+ for 2 K^+ per 1 ATP.

ATP can be partly replaced with other nucleotides, the relative efficiencies being as follows. ATP : dATP : CTP : ITP : GTP : UTP = = 100 : 49 : 2.3 : 24 : 0.6 : 0.6.

The requirement for both Na^+ and K^+ is shown in Fig. 5.8, the different sodium : potassium affinity ratios of the two sets of sites being

---

* There appear to be two binding sites for ouabain in the eel electric organ but only one in cat brain.

reflected in the asymmetry of the curve. The coupling ratio is, however, rather flexible and under suitable conditions one can find an electro-neutral exchange of Na^+ for K^+. Moreover, there exists a $Na^+ : Na^+$ exchange as well as a $K^+ : K^+$ exchange, the various possibilities depending in a complicated way on the ratios of Na^+ to K^+ outside and inside cells but most distinctly also on intracellular concentrations of ATP, ADP and inorganic phosphate. (e.g., Garrahan and Glynn, 1967). Actually, an activity series for the "potassium" site has been established as $Tl^+ > K^+ > Rb^+ > NH_4^+ > Cs^+ > Li^+$.

Magnesium can be replaced with less efficiency by Mn^{2+} or Co^{2+} but inhibition ensues in the presence of Fe^{2+}, Ca^{2+}, Sr^{2+}, Ba^{2+}, Be^{2+}, Zn^{2+} or Cu^{2+}.

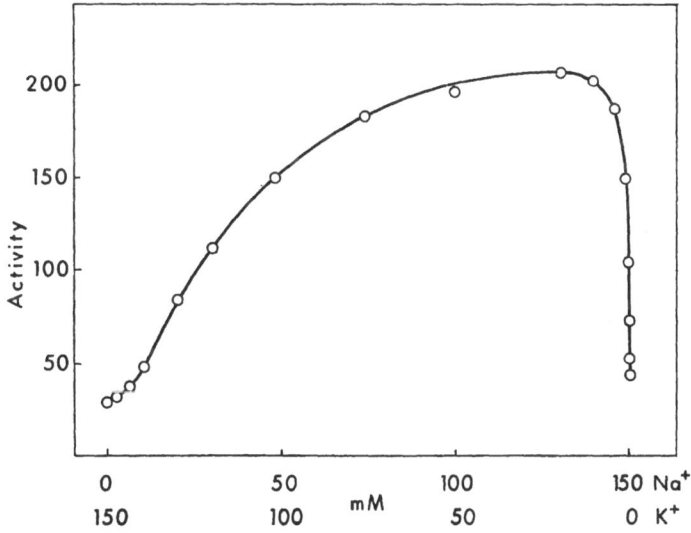

Fig. 5.8. The effect of sodium and potassium ions on the Na, K-adenosinetri-phosphatase from ox brain microsomes. The experiment was carried out with 6 mM Mg^{2+}, 3 mM ATP, at pH 7.4 and 37 °C. The ATPase activity is expressed in μmol liberated P per hour per mg protein. (Redrawn from Skou, 1972.)

As new observations are reported, the existing models of ATPase action are being amended and enlarged so that up to the present about a dozen serious suggestions have been advanced in this direction. The general consensus is that there are several partial reactions involved:

$$E_1 + ATP \xrightleftharpoons[]{Mg_2^{+}, Na^{+}} E_1{\sim}P + ADP$$

$$E_1{\sim}P \xrightleftharpoons[]{Mg_2^{+}} E_2{-}P$$

$$E_2{-}P + K^{+} \xrightleftharpoons{} K{-}E_2{-}P$$

$$K{-}E_2{-}P + H_2O \xrightleftharpoons{} K{-}E_2 + phosphate$$

$$K{-}E_2 \xrightleftharpoons{} E_2 + K^{+}$$

$$E_2 \xrightleftharpoons[]{Na^{+}} E_1$$

The Na^{+}-dependent phosphorylase activity thus includes the formation of a phosphorylated enzyme intermediate, formerly assumed to involve the phosphate bond at the γ-carboxyl of glutamic acid, now known to be at the β-carboxyl group of L-aspartic acid (Post and Orcutt, 1973). All the steps are seen to be reversible and, indeed, under suitable conditions, ATP can be generated by the Na^{+} pump running in reverse. There are apparently at least two binding sites for Mg^{2+}, one with a high affinity for the cation which is required for the ATP-ADP exchange reaction, another with a much lower affinity, which is required for the actual hydrolysis giving rise to inorganic phosphate.

The K^{+}-dependent phosphatase activity has been viewed by some to represent an independent enzyme entity.

The property that is germane to the transport function is the vectorial orientation of the enzyme complex in the membrane. In the above set of reactions, E_1 is thought to be exposed to the inside of the cell where it binds Na^{+}, then moves over to the other side (this being apparently an exergonic step) where it binds K^{+} to transport it inward. However, there is evidence that sites reactive with sodium as well as potassium are present at both membrane faces (Hoffman and Tosteson, 1971).

It is rather attractive to view the ATPase molecule as an oligomeric enzyme, this being supported by the purification attempts as well as by cooperative kinetics of interactions between sodium (low) and potassium (high) (e.g., Robinson, 1970). A membrane-spanning tetramer with an internal cavity (involving half-site reactivity in analogy with enzyme kinetics) was suggested in this connection by Stein and co-workers (1973) and a flip-flop mechanism of action of an oligomer (in analogy to the "tight" and "relaxed" state of monomers of allosteric enzyme kinetics) by Repke and Schön (1973). The last-named model has been developed to a high level of sophistication

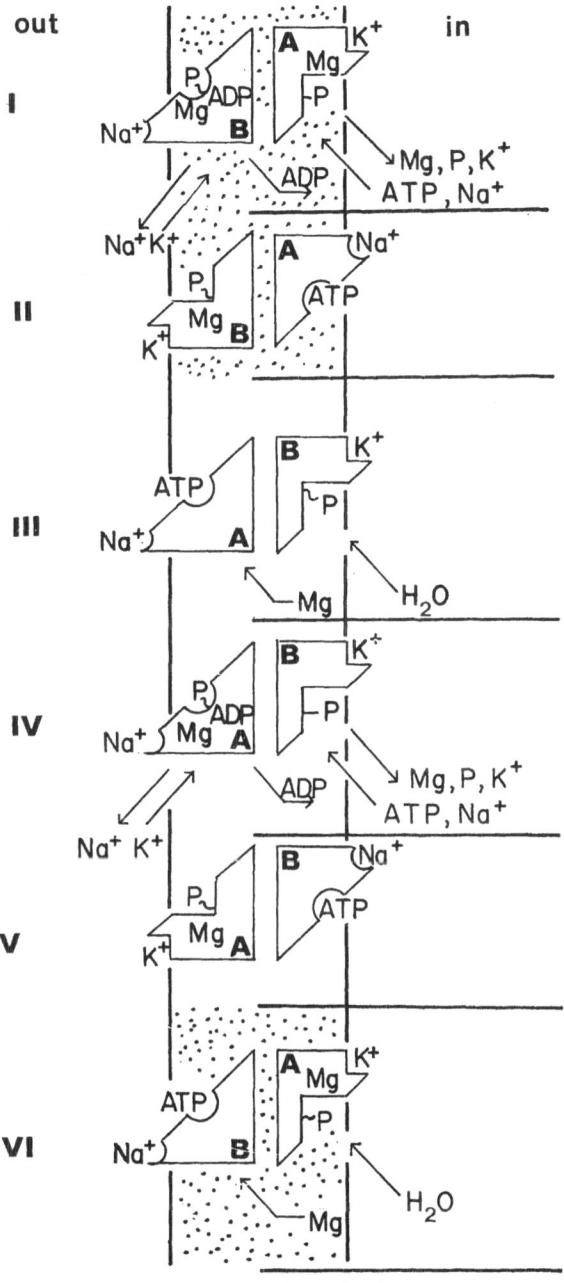

Fig. 5.9. The sequence of steps in the flip-flop model of Na, K-ATPase as suggested by Repke and Schön (1973), based on two subunits **A** and **B**.

and shown to be thermodynamically least demanding. Due to the tight coupling between the two states which flip from one membrane face to the other in unison with a symmetrical set of monomers, the changes in free energy involved in the individual steps are cut down to values roughly one-fourth those without the flip-flop coupling. The model is shown schematically in Fig. 5.9.

Ouabain is believed primarily to bind to the E_2 form but possibly allotopically with respect to Na^+ or K^+. There are peculiar aspects of this inhibition, such as its being time-dependent; this is apparently due to ouabain stabilizing (and inactivating) one of the enzyme conformations which is thus gradually removed from the membrane transport cycle.

A number of Na, K-ATPases have been isolated from membranes and purified to a considerable degree, the extent of optimal purification being rather difficult to determine as enzyme activity is frequently lost as lipids are removed or the quaternary structure of the molecule is disrupted. The apparent molecular weight of the whole preparation ranges from 190 000 to 560 000, depending on the source and method of determination (e.g., Kepner and Macey, 1968; Atkinson et al., 1971; Uesugi et al., 1971; Kyte, 1971). However, sodium dodecyl sulfate electrophoresis of the preparations shows it to be composed of probably two types of polypeptide chains, the "α" one having a molecular weight of over 100 000, the "β" one about 50 000. The whole molecule may thus be an $\alpha_2\beta_2$ tetramer or a $\alpha_2\beta_4$ hexamer or possibly have a more complicated quaternary structure. The latest work of Hokin (1974) proceeding from the shark rectal gland and the eel electric organ, shows the light chain to be a glycoprotein and the oligomer to be perhaps of the $\alpha_4\beta_2$ type. It is probably the heavy chain which is phosphorylated during ATPase operation (Avruch and Fairbanks, 1972; Nakao et al., 1974).

The isolated preparations show an affinity for Mg^{2+}, Na^+, K^+ and ouabain in varying degrees and always one for ATP ($K_D =$ $= 0.2$ mM), ADP ($K_D = 0.5-2$ mM) but practically none for CTP, GTP, ITP and UTP ($K_D = 0.1-1$ M). The binding of ATP is affected by univalent as well as bivalent cations.

Preparations of Na, K-ATPase have been combined with artificial phospholipid membranes and marked effects on conductance for Na^+ and K^+ have been observed (e.g., Jain et al., 1972). Shamoo (1974) shows that two α and one β polypeptide chains must be combined in an artificial system to produce the ionophoretic effect.

5.3.2. Ca-Adenosinetriphosphatase

The transport of calcium ions has been studied with special emphasis on the intestine and on sarcoplasmic reticulum, the implication of an ATPase in the transport being clear in the case of muscle reticulum. The mechanism resembles that of Na,K-ATPase:

$$E_1 + ATP \rightleftharpoons E_1\text{---}ATP$$
$$E_1\text{---}ATP + 2Ca^{2+} \rightleftharpoons E_1\text{---}ATP\text{---}Ca_2$$
$$E_1\text{---}ATP\text{---}Ca_2 \rightleftharpoons E_1\sim P\text{---}Ca_2 + ADP$$
$$E_1\sim P\text{---}Ca_2 \rightarrow E_2\text{---}P\text{---}Ca_2$$
$$E_2\sim P\text{---}Ca_2 \xrightleftharpoons{Mg^{2+}} E_2 + \text{phosphate} + 2\,Ca^{2+}$$
$$E_2 \rightleftharpoons E_1$$

The E_1 form is assumed to occur on the surface of the microsomes while E_2 exists at the inside face of the microsomal membrane.

Like the Na,K-ATPase, it has a requirement for phospholipids and for Mg^{2+} (although here it aids primarily in the phosphatase-type reaction).

The most acceptable explanation of the various kinetic data is one of a half-site reactivity enzyme capable of existing in two conformations.

A calcium pump has now been reconstituted in vesicles of sarcoplasmic reticulum by addition of isolated ATPase and dioleyllecithin (Warren et al., 1974).

Calcium-binding proteins have been isolated from the intestine (Fullmer and Wasserman, 1973) and from the sarcoplasmic reticulum (Ostwald and MacLennan, 1974). The Ca^{2+}-binding intestinal protein has been obtained from various animals and has a molecular weight ranging from 12 000 in the cow and the rat to 28 000 in the chick. It is produced in response to vitamin D_3, has a dissociation constant with Ca^{2+} of $10^{-6}\,M$ and binds probably four calcium ions per molecule. It is not altogether clear whether it might be a part of a Ca^{2+}-dependent ATPase as suggested by Melancon and DeLuca (1970).

The six Ca^{2+}-binding proteins from sarcoplasmic reticulum (including the two components of calsequestrin) are distinct from the Ca^{2+}-ATPase, range in molecular weight from 20 000 to 55 000 and in the K_D with Ca^{2+} from $5 . 10^{-3}\,M$ to $2.5 . 10^{-6}\,M$. However, their implication in calcium transport does not seem likely.

5.3.3. Other adenosinetriphosphatases

There is some evidence that even Mg^{2+}, Mn^{2+} and other bivalent cations may use the energy of ATP for their transport but the chemical nature of the process is not yet fully understood.

It is rather likely that Mg^{2+}-activated ATPases are active particularly in microorganisms where they translocate H^+ ions outward (probably $2 H^+/ATP$). A number of Mg^{2+}-activated ATPases have been isolated and purified and it is surmised that some may be the active H^+ translocators, the best example being provided by the work on *Streptococcus faecalis* (Schnebli and Abrams, 1970; Harold *et al.*, 1969; Abrams and Baron, 1967) where the Mg^{2+}-ATPase was shown to have a molecular weight of 385 000 (apparently an $\alpha_6\beta_6$ oligomer) and to resemble in its inhibitor sensitivity the cation (particularly H^+) transport in the intact cells.

Various ATPases were found to be implicated in the transport of anions, the best understood being the translocation of bicarbonate in gastric oxyntic cells (e.g., Wiebelhaus *et al.*, 1971). The enzyme is also activated by Mg^{2+}.

Transport of chloride anions, in some animal and plant cells, often correlates with an ATPase activity although little is known about the mechanism.

5.3.4. Ion-binding proteins

Besides the Ca^{2+} binding, vitamin D-dependent protein isolated from the intestine of various animals, the Ca^{2+}-binding proteins and the Mg^{2+}-activated proteins from sarcoplasmic reticulum (Meissner, 1973), there are two membrane proteins known from bacteria to be involved in ion transport. One is the sulfate-binding protein from *Salmonella typhimurium* (Pardee, 1967) of molecular weight of 32 000, with a K_D *in vitro* of $3 . 10^{-5}$ M. The other is a protein from *Escherichia coli*, probably a component of the high-affinity system for phosphate transport (Medweczky and Rosenberg, 1970). It has a molecular weight of 42 000 and a K_D *in vitro* of $7 . 10^{-7}$ M.

5.3.5. Transport of ferric ions

Several systems of membrane transport of iron are known in some detail.

The one in *Mycobacterium smegmatis* (Ratledge and Marshall, 1972) involves the iron-chelating compound mycobactin

which transports ferric ions inward where they are reduced to the ferrous form by NADH. Another system is of relatively wide distribution in bacteria and includes one or another of the chelating agents sideramines which act as membrane carriers (Franklin, 1974)

In *Escherichia coli*, Fe^{3+} ions are transported in a chelate with entero-bactin (enterochelin) which is a cyclic form of 2,3-dihydroxy-

N-benzoylserine, or in a chelate with citrate. The former system also occurs in *Salmonella typhimurium* (O'Brien and Gibson, 1970; Young *et al.*, 1967; Pollack and Neilands, 1970).

In *Bacillus megaterium*, Fe^{3+} ions enter cells in a coordination complex with schizokinen (Mullis *et al.*, 1971):

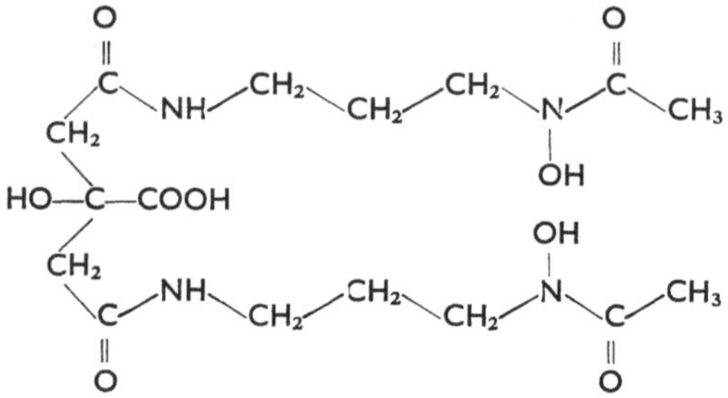

In animal organs, particularly in the gut, the transmembrane movement of iron (mainly in the ferrous form) is apparently a function of the intracellular concentration of apoferritin (molecular weight 450 000) which contains as much as 23% Fe by weight, and of the plasma concentration of transferrin (or siderophilin), a Fe^{2+}-binding globulin of molecular weight of 90 000, with 5.5% content of carbohydrate. Whether the proteins are involved in the transmembrane movement itself remains unclear.

5.3.6.   Ionophores

Although their role in natural ion transport is disputable, there exist a number of compounds, mostly of bacterial or fungal origin, which permit the passage across membranes of various ions, either very selectively or with a broad spectrum. They are generally designated as ionophores.

The membranes sensitive to their attack include mitochondria, chloroplasts, Gram-positive and some Gram-negative bacteria, erythrocytes, as well as artificial phospholipid membranes. However, it cannot be excluded that other membranes may be susceptible to the ionophoretic action of some of these agents.

The chemical nature of ionophores is rather heterogeneous. They may be divided into peptides, depsipeptides, macrotetrolides and cyclic polyethers.

The peptides can be either linear (gramicidin A) or cyclic (gramicidin S, tyrocidin A, PV peptide, antamanide, alamethicin).

Gramicidin A

$$HCO-\text{L-val}-\text{gly}-\text{L-ala}-\text{D-leu}-\text{L-ala}-\text{D-val}-\text{D-val}-$$
$$-\text{D-val}-\text{L-tyr}-\text{D-leu}-\text{L-tyr}-\text{D-leu}-\text{L-tyr}-\text{D-leu}-$$
$$-\text{L-tyr}-NHCH_2CH_2OH$$

Gramicidin S

L-val — L-orn — L-leu — D-phe — L-pro
| |
L-pro — L-glu — L-asp — D-phe — L-phe

Tyrocidin A

L-val — L-orn — L-leu — D-phe — L-pro
| |
L-tyr — L-glu — L-asp — D-phe — L-phe

PV peptide (artificially synthesized)

L-val — D-pro — D-val — L-pro — L-val — D-pro
| |
L-pro — D-val — D-pro — L-val — L-pro — D-val

Antamanide

L-ala — L-phe — L-phe — L-pro — L-pro
| |
L-pro — L-pro — L-val — L-phe — L-phe

Alamethicin

AIB — Glu — Pro — AIB — Ala — AIB — Ala — Gln — AIB*
| |
| Gln
|
AIB — Val — Pro — AIB — Leu — Gly — AIB — Val

The depsipeptides are probably the most important group of ionophores and include valinomycin, various enniatins and beauvericin.

Valinomycin

$$-[\text{L-lactate}-\text{L-val}-HIV-\text{D-val}]_{\overline{3}}$$

Enniatin A

$$-[HIV-\text{L-Me-ile}]_{\overline{3}}$$

* AIB, α-aminoisobutyric acid; HIV, D-hydroxyisovaleric acid; Me, methyl.

Enniatin B

$$[HIV–L\text{-}Me\text{-}val]_3$$

Enniatin C

$$[HIV–L\text{-}Me\text{-}leu]_3$$

Beauvericin

$$[HIV–L\text{-}Me\text{-}phe]_3$$

The macrotetrolides or actins are derivatives of nonactinic acid with four lactonated moieties in the molecule:

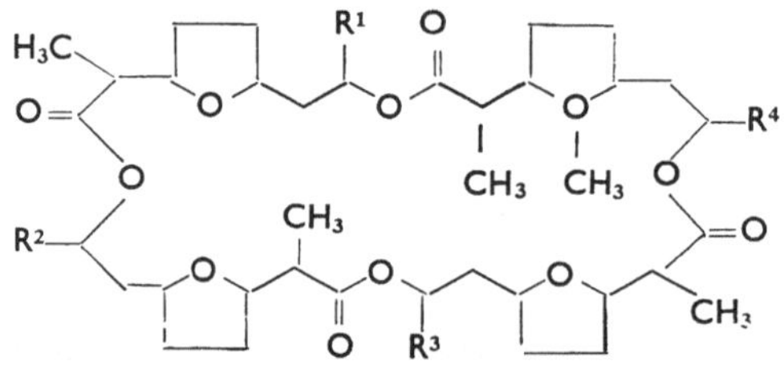

When $R^1 = R^2 = R^3 = R^4 = CH_3$, it is nonactin; when $R^1 = R^2 = R^3 = CH$ and $R^4 = C_2H_5$, it is monactin; when $R^1 = R^3 = CH_3$ and $R^2 = R^4 = C_2H_5$, it is dinactin; when $R^1 = CH_3$ and $R^2 = R^3 = R^4 = C_2H_5$, it is trinactin.

The cyclic polyethers are all artificial compounds of the basic formula

$$[CH_2CH_2O]_n$$

with various substituents at the methylene groupings. They are often called crowns, thus dibenzo-18-crown-6 is

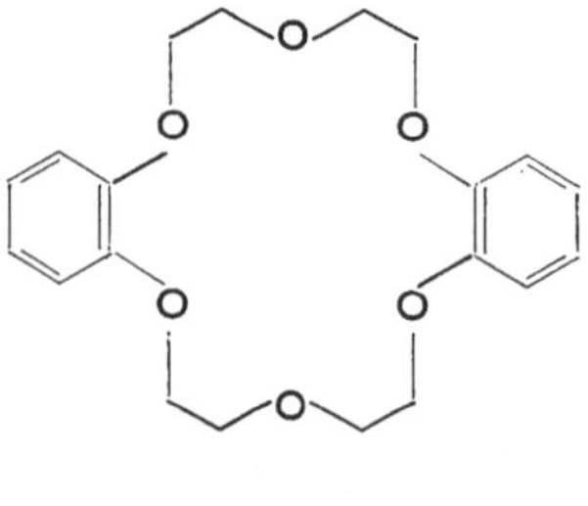

There are crown-type ionophores which are highly selective toward a particular ionic species. Thus

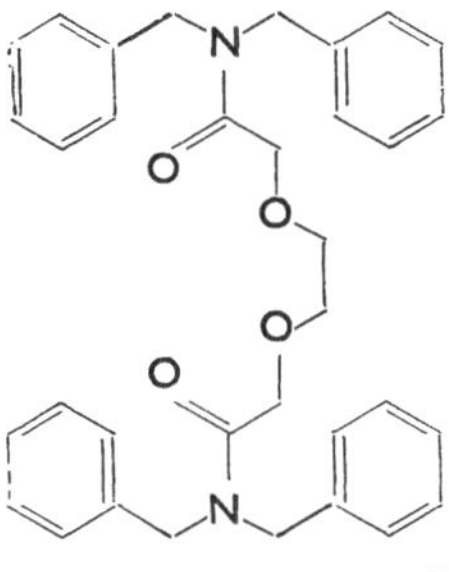

for Na$^+$

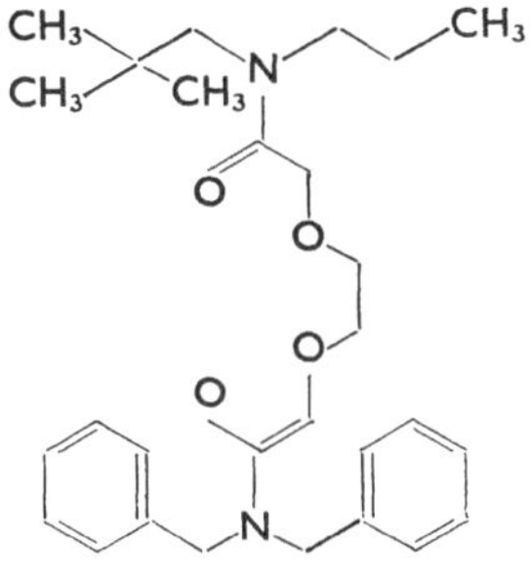

for Ca^{2+},

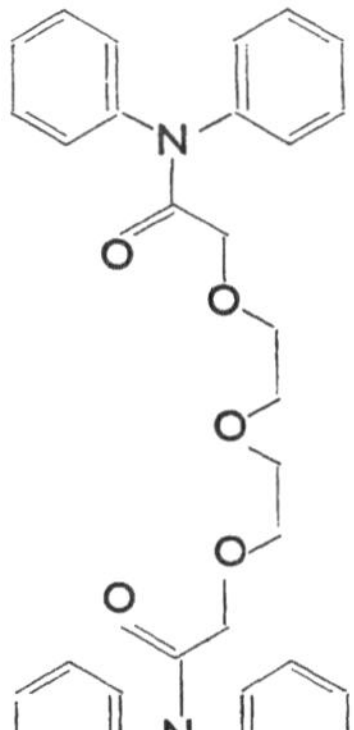

for Ba^{2+}

Natural analogues to the cyclic polyethers become cyclic only on formation of a metal-binding complex (monensin or nigericin).

Monensin

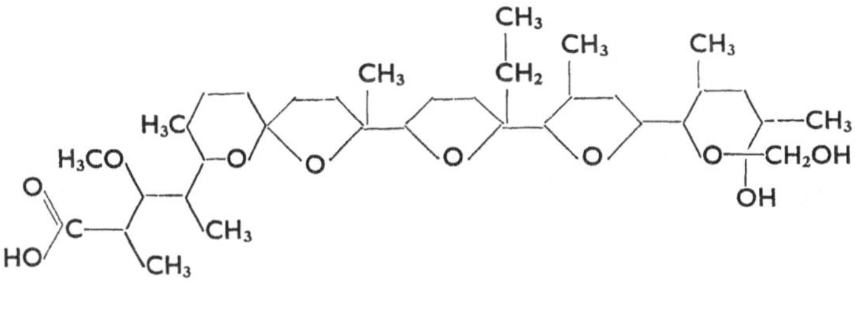

Nigericin

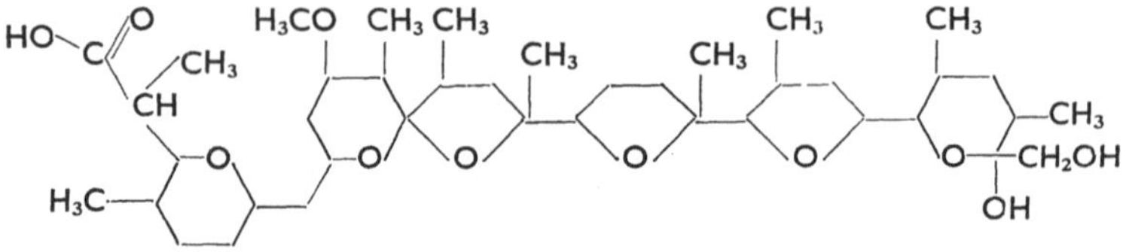

A special ionophore has been prepared that binds preferentially bivalent cations and was designated as X-537A (*cf.* Pressman and de Guzman, 1974). The relative transport capacity mediated by it for various ions is as follows: $Ca^{2+} : Cs^+ : Ba^{2+} : Sr^{2+} : Rb^+ : Na^+ =$ $= 100 : 99 : 75 : 45 : 38 : 33$.

X-537A

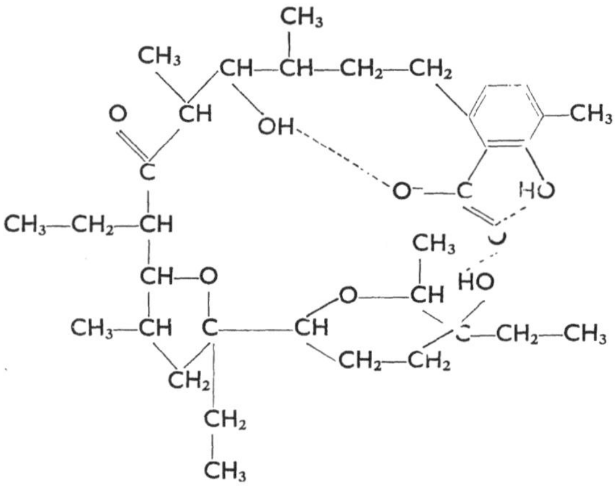

It is a characteristic of all ionophores that they are attached to the plasma or organelle membrane without any particular requirement for the lipid composition.*

Most ionophores, in particular valinomycin and the enniatins, actins and crowns, are known to form complexes with alkaline metal ions, some with remarkable selectivity especially for K^+. Thus the stability constant of the valinomycin-K^+ complex (in ethanol) is $1.9 . 10^6 M^{-1}$, while that of the valinomycin-Na^+ complex (in methanol) only $12 M^{-1}$. Others, such as the enniatins and actins, display a difference of one order of magnitude in favor of K^+ while

---

* It is perhaps no coincidence that cyclic peptides of a completely different field of interest are similarly attached to plasma membranes—the phallotoxins and amatoxins found in the deadly toadstools of the genus *Amanita* which are now known to disrupt membranes causing leakage of Ca^{2+} and K^+ from cells.

monensin actually prefers Na^+ over K^+. The binding of ion in a co-ordination complex is accompanied by a conformational change of the molecule (*cf.* Ivanov *et al.*, 1969; Dobler *et al.*, 1969; Kilbourn *et al.*, 1967).

The resulting effect of the above ionophores is a facilitation of transport of one or more cations which, under some conditions, may result in an electrogenic building up of a membrane potential. Several examples of action of valinomycin on bacteria or bacterial vesicles, resulting in a "short-circuiting" for K^+ ions, K^+ efflux and generation of a transient membrane potential are cited in support of the membrane electrical potential difference acting as a driving force in the translocation of various nonelectrolytes.

Nigericin and monensin, having the same lipid solubility in the protonated and nonprotonated forms, facilitate the exchange (antiport) of K^+ for H^+ and of Na^+ for H^+, respectively. Unlike with valinomycin which in complex with a cation bears a positive charge, both monensin and nigericin are electrically neutral when carrying a cation.

In contrast with all the above-mentioned agents, gramicidin A apparently forms channels across the lipid membranes but these are not as large as to permit hydrodynamic flow (in distinction to amphotericin B, filipin and others; *cf.* p. 205). In fact, the channels can be somewhat cation-specific, the specificity sequence being:

$$H^+ > NH_4^+ \geq Cs^+ > Rb^+ \geq K^+ > Na^+ > Li^+$$

(*cf.* Hladky and Haydon, 1970). Unlike the previous ionophores, whose action ceases as membranes freeze (assume the crystalline state), the action of gramicidin A is not markedly affected even at the transition temperature.

For a most scholarly review of the subject of ionophores the reader is referred to Koryta (1973).

5.4. DISTRIBUTION AND ROLE OF ION TRANSPORT

As suggested in the introduction, transport of ions by cellular membranes is intimately connected with other life processes and serves them in many ways. On the one hand, the transport regulates the electrolyte composition of the cell interior and the free energy stored in the form of ion gradients across the cell membrane can be used in several ways; on the other hand, in more highly developed organisms the transcellular transport across specialized epithelial cells serves the homeostasis of the organism as a whole.

In the animal kingdom a prominent role is played by the active transport of sodium and potassium ions, mediated by membrane Na,K-ATPase (see p. 288), commonly called the sodium pump. Two important properties of the sodium pump, which are not often stressed, may be mentioned here: the pump is a reversible mechanism and the sodium-potassium coupling ratio seems to be a function of the opposing gradients. The reversibility of the pump is obvious from the existence of a critical energy barrier across which no further net extrusion of sodium from the cell can proceed; as found by Conway (Conway and Mullaney, 1961), pumping of sodium from muscle fibers is prevented by a sufficiently high gradient of sodium electrochemical potential. Moreover, in erythrocytes, ATP synthesis was shown to result from a backflow of sodium ions through the pump (Glynn and Lew, 1970). Thus, even if the passive permeability of the cell membrane to sodium ions (the sodium leak) were nil, the sodium pump would create only a definite sodium gradient in accordance with the Le Chatelier principle – a reversible mechanism is opposed and finally balanced by the gradient which it creates. An interesting positive feedback, a change in the sodium-to-potassium coupling ratio, seems to oppose the Le Chatelier effect. In snail neurones, the pump becomes more coupled and hence less electrogenic when it operates against a greater gradient (Kostyuk et al., 1972). Similarly, in epithelial cells of the anuran urinary bladder, the sodium pump appears to be electrogenic when it extrudes sodium ions downhill, into a choline or potassium solution (Essig and Leaf, 1963; Frazier and Leaf, 1963) but obligatorily tightly coupled when it extrudes the same ions against a gradient into a sodium solution (Janáček et al., 1972). Coupling diminishes the opposing effect of electrical forces and thus, at the same energy consumption rate, reduces the effort (see p. 178).

Perhaps the most spectacular exploitation of the sodium electrochemical potential gradient across the cell membrane is displayed in the generation of the action potential in nerve and muscle fibres. As explained by the well-known sodium theory (see p. 284) a sudden pronounced increase in membrane permeability to sodium which takes place in the membrane depolarized by local currents, belonging to a propagated electrical impulse, brings about a repolarization of the membrane. The polarity change, called the action potential, amplifies the signal and promotes its further propagation along cable-like structures.

The sodium leak, governed by the sodium gradient and membrane sodium permeability, apparently plays in many cases a more important role than that of a primitive overflow feedback dissipating the metabolic energy, being coupled to the inflow of nutrients and thus permitting their accumulation in cells by a secondary active transport (see p. 299). Thus the flow of sugars (Crane, 1965) and of amino acids (Schultz and Curran, 1970) across the mucosal membranes of intestinal cells as well as the accumulation of amino acids by ascites tumor cells (Heinz, 1972) are, according to considerable experimental evidence, governed by the sodium electrochemical potential gradient across the membrane.

The great importance of transcellular ion transport is best displayed in specialized organs like kidney, the remarkable function of which is assisted and the study of which is rendered difficult by the great morphological complexity of the organ. Simpler tissues, notably skins and urinary bladders of anurans, frogs and toads, have been extensively used in studies of transcellular ion transport. These organs not only survive for many hours in simple salt solutions at room temperature; they may be used as partitions in split chambers with identical solutions at the two sides and the electrical potential difference across the tissue reduced to zero by passing a definite electrical current across it from an external battery. This procedure forms the basis of the ingenious short-circuit current technique introduced by Ussing and Zerahn (1951). The current required to bring the spontaneous potential across the tissue to zero is an exact and convenient measure of active ion flows across the preparation, for there is no gradient of any electrochemical potential across the tissue under the above conditions and hence net passive ion flows come to a stop. With this technique it was demonstrated that in most anuran skins and in anuran bladders, sodium ions are almost exclusively subject to an *in vitro* active transcellular transport, since potassium ions pumped in the opposite direction return passively into their original compartment. An exception is represented by the skin of the South American frog, *Leptodactyllus ocellatus*, performing *in vitro* a considerable transcellular active chloride transport (Zadunaisky et al. 1963). Pharmacological actions of drugs and stimulatory effects of hormones (neurohypophysial hormones, aldosterone, insulin, etc.) were conveniently studied in this manner and metabolic relations of the active sodium transport were established. As shown by Vieira and co-workers (1972) each frog skin has its own stoichiometric ratio of sodium ions actively transported per molecule of oxygen consumed for the transport, which is changed neither by transport stimulation with antidiuretic hormone, nor by transport inhibition with ouabain (in individual skins ratios from 7.1 to 30.9 were found). On the other hand, it was demonstrated by Sarkadi and Schubert (1972) that the ratio varies with temperature, medium pH and sodium osmolarity, being lowest under physiological conditions, where the efficiency of energy conversion exhibited the highest value.

An interesting feature of ion transport in plant cells, as compared with the above mechanisms found in animals, is the relative importance of ATP-independent anionic pumps. Thus in the alga *Nitella translucens*, together with an active accumulation of potassium ions dependent on ATP synthesized by photophosphorylation an active accumulation of chloride anions directly coupled to the second photosynthetic system was demonstrated by MacRobbie (1965).

In microorganisms, including bacteria and yeasts, alkaline cations are translocated by systems not coupled to the Na, K-ATPase, the general observation being that of an $H^+ - K^+$ exchange. It appears that a proton-generating ATP-ase may be instrumental in the process. The transport of bivalent cations utilizes specific carrier systems, energized mostly by ATP. Anions, such as chloride, bromide, iodide, nitrate, are distributed probably passively across the membrane according to the existing membrane potential; on the other hand, phosphate and sulfate appear to require specific binding proteins as well as metabolic energy for transport.

SYNOPSIS

Thermodynamic characterization of membrane equilibria of ions in terms of measurable quantities, concentrations and electrical potential differences, permits to recognize stationary distributions of ions resulting from active transport. Combination of electrostatic and statistical laws gives an estimate of double-layer thickness and, thereby, a vivid picture of electrical relations and ion distributions in the proximity of membrane boundaries.

Steady-state membrane potentials across cellular membranes may be described by approximate solutions of the differential equation of electrodiffusion, accounting both for the steps of the electrical potential at the membrane boundaries and for electrogenic active ion transport. An alternative description of membrane potentials can be approached using electrical analogues.

The mechanism of active ion transport is linked to the biochemistry and comparative physiology of membrane ATPases. Mediated transport of ions, both natural and induced, is the result of chemical properties of a heterogeneous group of chelating agents, often called ionophores.

Active ion transport, distributed throughout the living world, creates ionic gradients serving as free energy reservoirs available for various physiological functions and contributes to the homeostasis of living systems. In animal cells the ATPase-mediated sodium and potassium transport appears to be of prevailing importance, in plant cells anion transport directly coupled to photosynthesis may be equally important, in microorganisms an exchange of K^+ for H^+ driven by various energy sources is almost ubiquitous.

6. TRANSPORT OF WATER

*"All Birds, Beasts and Fishes, Insects,
Trees, and other Vegetables, with their
several Parts, grow out of Water and
watry Tinctures and Salts, and by
Putrefaction return again into watry
Substances."*

Isaac Newton, Opticks

6.1. STEADY-STATE THERMODYNAMICS
OF WATER PERMEATION

The whole field of water transport in cells and tissues was recently reviewed in an extensive and scholarly manner by House (1974). The content of the present chapter is much more modest being mostly limited to the phenomenological description of water transport across membranes by the methods of steady-state thermodynamics (see p. 172), as developed by Kedem and Katchalsky (1958).

The two most common and principal causes of water movement across a membrane from one compartment into another are the following: (1) An excess of hydrostatic pressure in the first compartment; (2) an excess of concentration of a solute to which the membrane is less permeable than to water, in the second compartment. The second type of water movement, known as osmosis, involves a number of interesting theoretical problems. The theory of osmotic water flow is relatively simple when the membrane behaves as ideally semipermeable, i.e., when it is permeable to water and completely impermeable to the solute. In general, however, the membrane is permeable, although not in the same degree, to both water and the solute. As may be easily envisaged, when water and the solute follow inside such a leaky membrane a common pathway (say, water-filled pores) the flows of the two interact frictionally and the magnitude of each depends on the degree of interaction.

As discussed previously (p. 173), interactions of this type between steady flows are satisfactorily treated by the thermodynamics of the steady state, which assumes each generalized flow in the system to be proportional to all generalized forces in the system. This approach is not free of theoretical criticism when applied to thin cell membranes which, according to Schlögl (1969), cannot be considered as a continuous phase. Still, for the time being, the treatment by Kedem and Katchalsky (1958) appears to provide the best, even if approximative, quantitative description of water flows across biological cell membranes. The simple case of ideally semipermeable membranes will be seen to be a limiting case of this description.

For simplicity, let there be only one solute present in the system, so that there are two flows across the membrane: flux of water, J_w, and flux of solute, J_s. Both of them are proportional to the gradient of chemical potential of water and the gradient of chemical potential of the solute:

$$J_w = L_{ww}\left(-\frac{\mathrm{d}\mu_w}{\mathrm{d}x}\right) + L_{ws}\left(-\frac{\mathrm{d}\mu_s}{\mathrm{d}x}\right), \qquad (6.1)$$

$$J_s = L_{ss}\left(-\frac{\mathrm{d}\mu_s}{\mathrm{d}x}\right) + L_{sw}\left(-\frac{\mathrm{d}\mu_w}{\mathrm{d}x}\right). \qquad (6.2)$$

As shown by Katchalsky (1961) the equations may be integrated, by the method of Kirkwood (1954), for steady-state conditions, when the fluxes are independent of the x-coordinate, giving

$$J_w = L'_{ww}\,\Delta\mu_w + L'_{ws}\,\Delta\mu_s, \qquad (6.3)$$

$$J_s = L'_{ss}\,\Delta\mu_s + L'_{sw}\,\Delta\mu_w \qquad (6.4)$$

where $\Delta\mu$'s are the differences of chemical potential of water and solute across the membrane, Onsager's law (p. 174) is satisfied by the new phenomenological cross coefficients

$$L'_{ws} = L'_{sw}. \qquad (6.5)$$

It is now important to develop equations (6.3) and (6.4) in terms of directly measurable quantities, differences of solute concentration, Δc_s, and of the hydrostatic pressure, Δp, across the membrane. Assuming that equilibrium prevails at the membrane surfaces, the differences of chemical potentials between the two surfaces will be the same as those between the two solutions. As long as the solutions are considered ideal, the chemical potentials may be expressed as proportional to the logarithm of the mole fraction of the substance and also, by its partial molal volume, to hydrostatic pressure. The differences of chemical potential across the membrane thus can be written as

$$\mu_w = \bar{V}_w \,\Delta p + RT\,\Delta \ln x_w, \tag{6.6}$$

$$\mu_s = \bar{V}_s \,\Delta p + RT\,\Delta \ln x_s. \tag{6.7}$$

Since

$$x_w = \frac{c_w}{c_w + c_s} \quad \text{and} \quad x_s = \frac{c_s}{c_w + c_s} \tag{6.8}$$

we may write

$$\Delta \ln x_w = \Delta \ln \frac{c_w}{c_w + c_s} = -\Delta \ln \frac{c_w + c_s}{c_w} =$$

$$= -\Delta \ln\left(1 + \frac{c_s}{c_w}\right) \approx -\Delta \frac{c_s}{c_w} = -\frac{\Delta c_s}{\tilde{c}_w} \tag{6.9}$$

where $\tilde{c}_w$ is the mean concentration of water in the two solutions. The approximation $\ln(1 + x) \approx x$, valid for small x, was used here. Similarly,

$$\ln x_s = \Delta \ln \frac{c_s}{c_w + c_s} \approx \ln \frac{c_s}{c_w} =$$

$$= \Delta(\ln c_s - \ln c_w) \approx \Delta \ln c_s = \frac{\Delta c_s}{\tilde{c}_s} \tag{6.10}$$

where $\tilde{c}_s$ is the mean concentration of the solute in the two solutions, approximations $c_w + c_s \approx c_w$ (since $c_w \gg c_s$) and $\Delta \ln c_w \approx 0$ having been used. Hence equations (6.6) and (6.7) may be written as

$$\Delta\mu_w = \bar{V}_w \,\Delta p - \frac{RT\,\Delta c_s}{\tilde{c}_w} \tag{6.11}$$

$$\Delta\mu_s = \bar{V}_s \,\Delta p + \frac{RT\,\Delta c_s}{\tilde{c}_s}. \tag{6.12}$$

The dissipation function (see p. 174) which represents the sum of products of individual fluxes and their conjugate forces may then be written as

$$\frac{d_i S}{dt} T = J_w \, \Delta\mu_w + J_s \, \Delta\mu_s = J_w \left(\overline{V}_w \Delta p - \frac{RT \Delta c_s}{\widetilde{c}_w} \right) +$$

$$+ J_s \left(V_s \, \Delta p + \frac{RT \Delta c_s}{\widetilde{c}_s} \right) = (J_w \overline{V}_w + J_s \overline{V}_s) \, \Delta p +$$

$$+ \left(\frac{J_s}{\widetilde{c}_s} - \frac{J_w}{\widetilde{c}_w} \right) RT \, \Delta c_s = J_V \, \Delta p + J_D RT \, \Delta c_s. \qquad (6.13)$$

Two new fluxes were thus defined; volume flow J_V

$$J_V = J_w \overline{V}_w + J_s \overline{V}_s \qquad (6.14)$$

conjugate to the force Δp, and exchange flow, J_D, which corresponds to the velocity of solute relative to water in the membrane

$$J_D = \frac{J_s}{\widetilde{c}_s} - \frac{J_w}{\widetilde{c}_w} \qquad (6.15)$$

conjugate to the force $RT \, \Delta c_s$, which is called osmotic pressure. The phenomenological equations may now be written in terms of new flows and forces as

$$J_V = L_p \, \Delta p + L_{pD} RT \, \Delta c_s, \qquad (6.16)$$

$$J_D = L_D RT \, \Delta c_s + L_{pD} \, \Delta p \qquad (6.17)$$

since Onsager's relation $L_{pD} = L_{Dp}$ is again valid.

The cross coefficient L_{pD} shows how large a volume flow is generated by osmotic pressure $RT \, \Delta c_s$ on a given membrane and how large an ultrafiltration is brought about on the same membrane by the difference in hydrostatic pressure Δp. With very coarse membranes, such as those of sintered glass, both osmosis and ultrafiltration are negligible, and hence $L_{pD} = 0$ for such membranes. On the other hand, with ideally semipermeable membranes, volume flow J_V is given solely by the volume flow of water

$$J_V = J_w \overline{V}_w \qquad (6.18)$$

and

$$J_D = -\frac{J_w}{c_w} \qquad (6.19)$$

since J_s in eq. (6.15) is zero for ideally semipermeable membranes. Assuming that the solutions are dilute, $1/\bar{c}_w \approx \bar{V}_w$ (≈ 18 cm^3 mol^{-1}), and from equations (6.18) and (6.19) it follows that

$$J_V + J_D = 0 \tag{6.20}$$

for ideally semipermeable membranes. Combining eq. (6.20) with equations (6.16) and (6.17) we obtain

$$(L_p + L_{pD})\, \Delta p + (L_D + L_{pD})\, RT\, \Delta c_s = 0. \tag{6.21}$$

This equation is satisfied for an ideally semipermeable membrane for all values of hydrostatic and osmotic pressure differences, so that, for such membranes, obviously

$$L_p = -L_{pD} = L_D. \tag{6.22}$$

Thus, eq. (6.16) for an ideally semipermeable membrane becomes

$$J_V = L_p(\Delta p - RT\, \Delta c_s). \tag{6.23}$$

This is an interesting equation since it shows that in an ideally semipermeable membrane the volume flow is proportional to the difference in hydrostatic pressure by the same coefficient as to the difference in osmotic pressure. This fact implies that hydrostatic and osmotic pressures generate volume flow by the same mechanism. Thus in a porous membrane, where hydrostatic pressure produces a mass flow through pores, osmotic flow must proceed by the same mechanism, rather than each water molecule diffusing separately in the concentration gradient of water and encountering on its way considerably higher friction. The origin of the osmotic mass flow was explained by Dainty and Meares (Dainty, 1963). No solute can penetrate into the pores of an ideally semipermeable membrane and hence there is a juxtaposition of a layer of solution and of a layer of pure water at the pore opening. The concentration of water being higher in the latter layer, the jumps of water molecules from it into the solution are more frequent than in the opposite direction and leave behind more vacancies. Hence the density and, as a result of this, the hydrostatic pressure, is less in this part of the pore than in the surroundings. The difference in the hydrostatic pressure is relieved by mass flow from the other end of the pore, rather than from the juxtaposed solution, where the movement of water is opposed by its concentration gradient. The osmotic pressure is thus seen to produce the mass flow *via* the hydrostatic pressure. The hydrostatic pressure is, however, localized only in the proximity of pore openings, does not distort the membrane and makes much higher pressures practicable in osmotic experiments than in hydrostatic experiments.

Coefficient L_p is called the hydraulic conductivity of the membrane and characterizes completely the permeability properties of an ideally semipermeable membrane. The general membrane, permeable also to the solute, requires for its description the knowledge of two other coefficients, L_D and L_{pD}, or better still, of two related coefficients, ω and σ, which are more convenient in the interpretation of experimental data. Their meaning and relation to the former coefficients may be derived as follows:

The flow of solute, J_s, is usually measured experimentally rather than exchange flow, J_D. In dilute solutions the flow of solute can be expressed using eq. (6.14), in which $J_s \bar{V}_s$ is very small as compared with $J_w \bar{V}_w$ and may be neglected, and eq. (6.15) as

$$J_s = (J_V + J_D)\, \tilde{c}_s. \tag{6.24}$$

The flow of solute, J_s, is often measured with tracers in the absence of volume flow, J_V. For $J_V = 0$ it follows from eq. (6.16) that

$$\Delta p = -\frac{L_{pD}}{L_p}\, RT\, \Delta c_s. \tag{6.25}$$

Substituting from eq. (6.25) and (6.17) into (6.24) for $J_V = 0$ we obtain

$$J_s = \frac{L_p L_D - L_{pD}^2}{L_p}\, \tilde{c}_s RT\, \Delta c_s. \tag{6.26}$$

Coefficient ω was introduced by Kedem and Katchalsky (1958) as

$$\omega = \frac{L_p L_D - L_{pD}^2}{L_p}\, \tilde{c}_s \tag{6.27}$$

so that eq. (6.26) becomes

$$J_s = RT\omega\, \Delta c_s \tag{6.28}$$

where $RT\omega$ is the permeability coefficient of a nonelectrolyte. Staverman's reflection coefficient σ was defined by Kedem and Katchalsky as

$$\sigma = -\frac{L_{pD}}{L_p}. \tag{6.29}$$

With this definition eq. (6.25) becomes

$$\Delta p = \sigma RT \, \Delta c_s. \qquad (6.30)$$

According to eq. (6.30) the volume flow across a membrane is zero when Δp, the observed osmotic pressure, is equal to the theoretical osmotic pressure $RT \, \Delta c_s$ multiplied by the reflection coefficient. For an ideally semipermeable membrane $\sigma = 1$, since then $L_p = L_{pD}$ (eq. 6.22). In leaky membranes $0 < \sigma < 1$ and the observed osmotic pressure, measured as hydrostatic pressure preventing volume flow, is less than the theoretical osmotic pressure. Phenomenological equations may now be rewritten using the new coefficients. From equations (6.16) and (6.29) the volume flow is

$$J_V = L_p(\Delta p - \sigma RT \, \Delta c_s) \qquad (6.31)$$

and the flow of solute can be expressed by introducing eq. (6.29) into (6.27), so that

$$\omega = (L_D - L_p\sigma^2) \, \tilde{c}_s \qquad (6.32)$$

and by combining equations (6.24), (6.31), (6.17) with (6.29) and (6.32), yielding

$$J_s = \tilde{c}_s L_p(1 - \sigma) \, \Delta p + \left[\omega - \tilde{c}_s L_p(1 - \sigma) \, \sigma \right] RT \, \Delta c_s. \quad (6.33)$$

Finally, equations (6.33) and (6.31) combined give

$$J_s = RT\omega \, \Delta c_s + \tilde{c}_s(1 - \sigma) \, J_V. \qquad (6.34)$$

Equations (5.31) and (6.34) describe completely the simple transport of solute and water across a membrane.

Eq. (6.31) may be generalized to account for the presence of the difference of an impermeant solute concentration across the membrane, Δc_i:

$$J_V = L_p(\Delta p - RT \, \Delta c_i - \sigma RT \, \Delta c_s). \qquad (6.35)$$

In the absence of a hydrostatic pressure difference the volume flow vanishes when

$$RT \, \Delta c_i = \sigma RT \, \Delta c_s \qquad (6.36)$$

so that the reflection coefficient can be estimated as

$$\sigma = -\frac{\Delta c_i}{\Delta c_s}. \qquad (6.37)$$

Two more points deserve our attention. Especially with highly permeable solutes the effect of unstirred layers (see p. 198) becomes very important and the difference of solute concentration, Δc_s, between the solutions at the membrane surfaces can be considerably less than that between the bulks of the two solutions, so that appropriate corrections have to be applied (see p. 199).

Further it is of importance to realize, that although frictional interactions in the common pathway of water and the solute is the principal cause reducing the value of the reflection coefficient, the reflection coefficient of a permeating solute is always slightly less than one, even when the solute pathway and water pathway are distinct. Let us assume that the pathways in some membrane are indeed distinct. Then the volume flow across the solute-impermeable pathway is given solely by the volume flow of water and, moreover, the reflection coefficient is equal to unity for this pathway. Eq. (6.31) for this pathway becomes

$$J_w \bar{V}_w = L_p(\Delta p - RT\,\Delta c_s). \tag{6.38}$$

The volume flow of solute across the water-impermeable pathway under the condition of zero total volume flow (under which the reflection coefficient is measured) is, according to eq. (6.34), equal to

$$J_s \bar{V}_s = \bar{V}_s RT\omega\,\Delta c_s. \tag{6.39}$$

Since the sum of the two flows is zero (being equal to J_V), it follows that

$$\frac{p}{RT\,\Delta c_s} = 1 - \frac{\bar{V}_s\omega}{L_p} \tag{6.40}$$

which according to eq. (6.30), is equal to the measured value of the reflection coefficient. Hence the conclusion about a common pathway (pores) for solute and water is justified only when

$$\sigma < 1 - \frac{\bar{V}_s\omega}{L_p}. \tag{6.41}$$

6.2. THE STATE OF WATER IN CELLS

The accessibility of intracellular water to solutes penetrating across the cell membrane is of importance in considerations of equilibrium distribution between the external medium and the cell interior. The history of experimental approaches to the problem together with much useful information on the properties of water was recently reviewed by House (1974). The present chapter restricts itself to the description of the most recent development in two principal experimental approaches to the elucidation of the state of water in cells, *viz.*, comparison of properties of intracellular and extracellular water by nuclear magnetic resonance, and direct measurement of the distribution of various solutes between the medium and the cell water.

Considerable progress in NMR studies of the state of intracellular water was achieved by Civan and Shporer (1972) who used $H_2^{17}O$ as the NMR probe to examine the state of water in the frog striated muscle. $H_2^{17}O$ not only resembles the properties of ordinary water better than ^2H_2O used previously but also, because of the large quadrupolar interaction of the ^{17}O nucleus in $H_2^{17}O$, it is particularly sensitive to changes in molecular movement. The authors found the total intensity of the ^{17}O signal of $H_2^{17}O$ in frog striated muscle to represent only about three-fourths of the maximum anticipated intensity. As shown by the authors, however, an immobilization of approximately one-fourth of the intracellular water by adsorption or binding to a solid phase represents only one (and apparently not the most probable) of the possible explanations of the reduction of the NMR signal in the muscle. As shown by the same authors (Shporer and Civan, 1972) already for NMR spectra of ^{23}Na, alternative interpretations are possible, *viz.*, tumbling of a part of intracellular water molecules in anisotropic regions or a rapid exchange of free water molecules with a bound water fraction which may be very small.

These possibilities appear to be in better agreement with new data on nonelectrolyte distribution in muscle. As observed by Miller (1974), the distribution ratio between intracellular water (the content of which was determined by drying overnight at 98 °C) and external medium in mouse diaphragm is frequently equal to unity (for methanol, ethylene glycol, glycerol, 2-deoxyerythritol, 2-deoxy-D-ribose and, at 34 °C in the presence of insulin, also for D-xylose). The finding indicates that there is no appreciable non-solvent water in this tissue. Distribution ratios for other substances which are less than one appear to be adequately explained in terms of intracellular compartmentation, with membranes surrounding the individual compartments which are either impermeable for the substance in question or which extrude it actively.

SYNOPSIS

Whereas the water flow across an ideally semipermeable membrane can be adequately described as being proportional to the difference of hydrostatic and osmotic pressure across the membrane, in membranes which are leaky for solutes this simple description fails mainly because it cannot account for interactions between water flow and solute flow in the membrane. The best contemporary approach to the phenomenological description of water flow across membranes permeable to solutes is offered by the formalism of steady-state thermodynamics. The reflection coefficient introduced by this theory is a quantitative expression of the fact that the osmotic pressure of a permeating substance is less than that of a nonpermeating one at the same concentration, especially if there is a common path for water and solute in the membrane and the two flows can interact.

Concerning the state of water in cells, the modern interpretation of nuclear magnetic resonance studies does not appear to contradict the new data on the distribution of various nonelectrolytes between the cell and its surroundings: practically all cell water can act as a solvent and its reduced accessibility for solutes may be explained by intracellular compartmentation and active transport.

7. TRANSPORT BY SPECIAL MECHANISMS

*"At least, I see nothing of Contradiction
in all this."*
        Isaac Newton, Opticks

There is considerable evidence showing that, along with carrier-mediated processes of transport of organic substrates or ions there are rather special mechanisms which permit the uptake or release of substances both in the low and in the high-molecular weight range.

7.1. OLIGOPEPTIDE PERMEASES

The first to be dealt with is actually a carrier-type mechanism which, however, can handle molecules of up to a molecular weight of nearly one thousand. Two such systems, oligopeptide permeases, are known from *Escherichia coli* (for a review see Payne and Gilvarg, 1971) and from *Salmonella typhimurium* (Ames *et al.*, 1973).

The operability of the system is somewhat surprising in view of findings with other compounds (in particular sugars) which cannot ride on a carrier once they exceed a certain size although they are bound to it (e.g., maltose binding but no transport in human erythrocytes; Beneš and Kotyk, 1976). On the other hand, the ability to

transport oligopeptides may have some bearing on the rather un-
expected finding of a freely reversible uptake of histone by *Escherichia
coli* (Pavlasová and Stejskalová, 1972).

7.2. PINOCYTOSIS

There is another way of carrying large molecules across the cell
membranes which is by nature indiscriminate but which, nonetheless,
may function only after a certain stimulus is received. The mechanism
is called pinocytosis and may be described morphologically as the
transport of vesicles enclosing liquid drops, either from the cytoplasm
outward (exocytosis; particularly in cells excreting protein products,

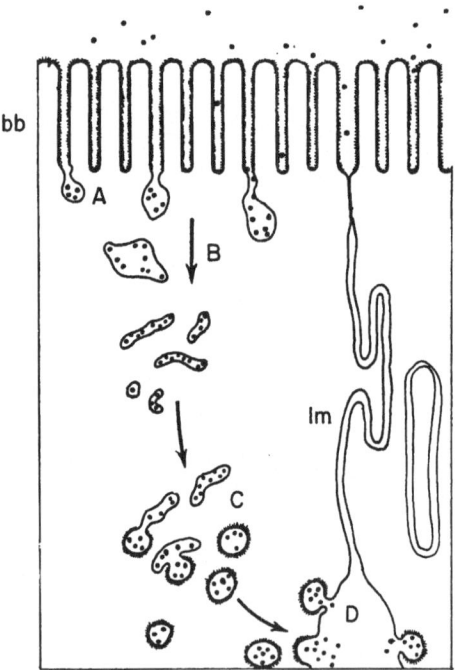

Fig. 7.1.   Diagram of antibody transport across the brush-border membranes (bb)
of the small intestine of a newborn rat. At A, antibodies are bound selectively to
specialized pits in the membrane (*cf.* Fig. 2.35D); at B, the antibodies are absorbed
by pinocytosis; at C, the pinocytotic vesicles turn into spherical coated vesicles;
at D, the vesicles discharge the antibodies at the lateral membrane (Im) by reverse
pinocytosis. (Adapted from Rodewald, R. (1973). *J. Cell Biol.* **58,** 189.)

such as the pancreatic acinar cells) or from the medium inward (endocytosis; most pronouncedly in protozoans but apparently also in various animal cells, notably in some tissue culture cells exhibiting an avidity for certain components of the surrounding medium, and

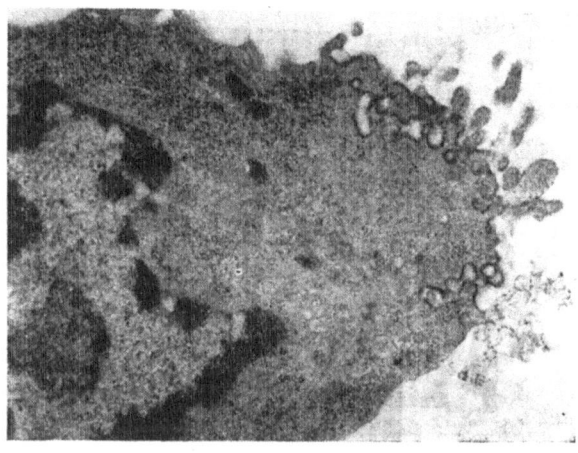

━━━━━━━━ 1000 nm

Fig. 7.2. An activated spleen lymphocyte from newborn pig showing abundant endocytosis of extraneous material and formation of pinocytotic vesicles. (Courtesy of Dr. I. Trebichavský, Institute of Microbiology, ČSAV, Prague.)

in the cells lining the blood capillaries). The sequence of events of pinocytosis, both inward and outward, is shown schematically in Fig. 7.1.

Inward pinocytosis obviously serves mainly the nutrition of cells and absorption by epithelial lining and sets in only after a certain food stimulus is received (apparently the outer envelope receptors are involved). Although this stimulus may be quite specific the cell engulfs with the drop of extracellular fluid inadvertently all that is contained in it (Fig. 7.2). This fact can be made use of for bringing into unwanted (e.g., cancer) cells displaying an increased endocytosis such compounds as will destroy them.

Outward pinocytosis (Fig. 7.3) is a process typical of all hormone-producing cells (cf. Fig. 2.54) and has been observed as the sole transport process for both acetylcholine and epinephrine from the nerve end plate to the muscle receptors across the synaptic

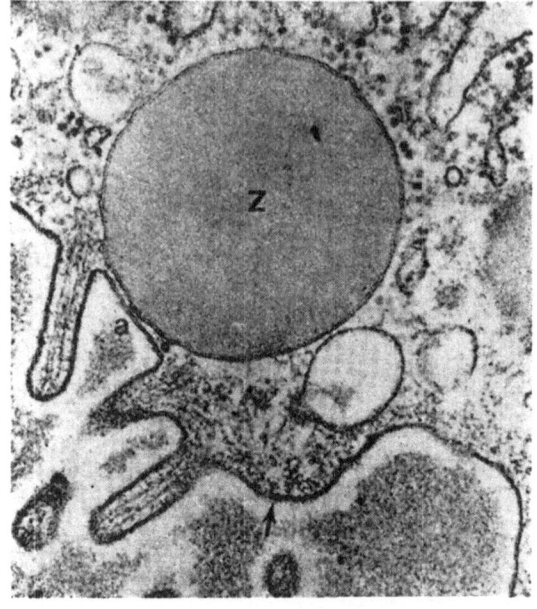

——— 200 nm

Fig. 7.3. Discharge of a zymogen granule (Z) from a pancreatic exocrine cell into the lumen of the duct. The point of confluence of the granule with the plasma membrane is shown at a. (Taken with permission from Jamieson, 1972, as in Fig. 2.54**B**.)

cleft. This fact brings out an important, previously unsuspected, ability of pinocytosis, *viz.* to proceed at a rapid rate, comparable with that of carrier transport so that a nerve impulse may be transmitted (Fig. 7.4).

7.3. UPTAKE OF NUCLEIC ACIDS AND SPECIAL PROTEINS

There is now little doubt that in a number of instances, both nucleic acids and proteins enter into cells after having traversed the membrane as such. A case in point is the transfer of genetic information (encoded in deoxyribonucleic acid) into a recipient competent bacterial cell.*

* Analogous processes probably take place in some animal cells.

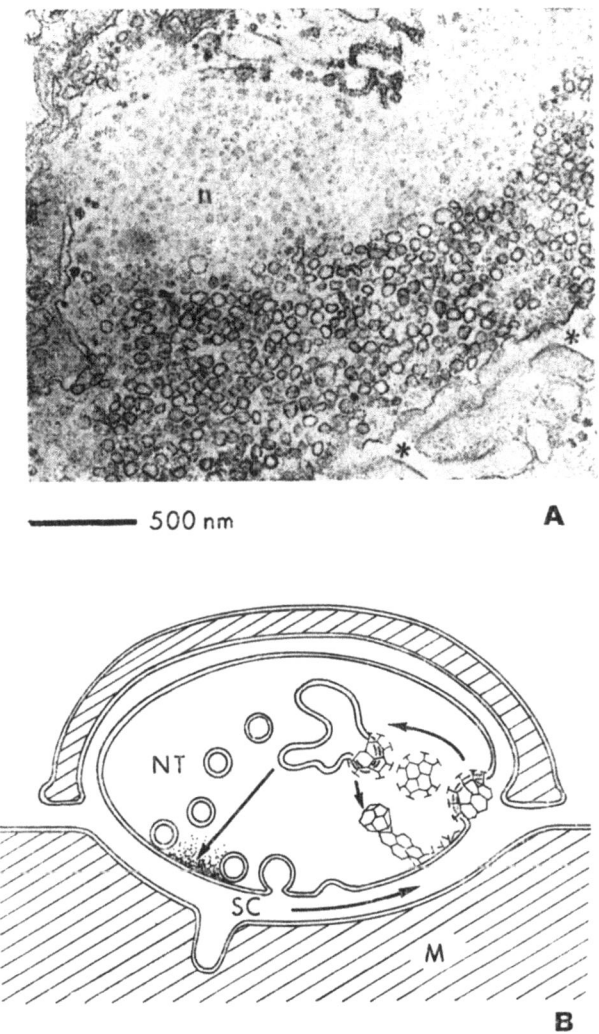

━━━━━━ 500 nm   **A**

B

Fig. 7.4. **A**. A portion of the neuromuscular junction and muscle fiber of the pectoral muscle of a frog. Numerous synaptic vesicles are present in the nerve terminal (n) about to discharge into the synaptic cleft (marked with asterisks). (Reproduced from Ceccarelli, B., Hurlblut, W. P., and Mauro, A. (1973). *J. Cell Biol.* **57** 499.) **B**. Schematic drawing of the path of a sinaptic vesicle from the nerve terminal (NT) through the sympatic cleft (SC) and back. The content of the vesicle (acetylcholine or noradrenaline) is discharged in the cleft and trapped by the muscle (M) surface receptors. (Adapted from Heuser, J. E. and Reese, T. S. (1973). *J. Cell Biol* **57**, 315.)

If we disregard processes such as conjugation (which involves direct contact of heterosexual cells), transduction (which requires the penetration of the cell by a phage particle) there are transformation and transfection wherein a free DNA molecule of another bacterium or a phage, respectively, penetrates the competent cell.

The state of competence itself requires the transmission of an endogenously produced and plasma membrane-bound protein called the activator or competence factor, to a noncompetent cell. The competence factor has a molecular weight of about 10 000 and is rather species-specific. At the bacterial surface it is bound to a receptor of about 50 000 mol. wt. and in an interaction with it "sets the stage" for the admission of the transformation DNA. Protein and RNA synthesis are required for the preparatory process. An agglutinin is formed at this stage which modifies the surface features of the cytoplasmic membrane.

After arriving at the cytoplasmic membrane (apparently in regions of cell envelope growth; cf. the mesosome) the DNA is adsorbed to the surface similarly to a virus particle — an irreversible energy-independent process, whereupon it is drawn in by an energy-requiring process and, if homologous, is incorporated into the bacterial chromosome. The process is completed in a matter of seconds or at most minutes. A molecular description of the penetration of DNA through the membrane is not possible at present (for more information see, e.g., Tomasz, 1973).

Likewise, the uptake of proteins, be it colicins or interferon, obviously requires specific receptor sites at the membrane surface which then (possibly by a cooperative process) alter the local permeability in such a way that even the large molecules can be drawn in. It is of interest that some of the receptors involved in nucleic acid uptake are actually nonresponsive to binding of polynucleotides of molecular weights lower than 10^5.

Analogies between the state of membrane proteins (e.g., Na,K-adenosinetriphosphatase) and the uptake activity toward large molecules or even viruses stress the necessity of regarding the membrane as an interacting continuum (cf. Koch and Fehér, 1973).

One is tempted to visualize a part of a mosaic fluid membrane to be dislodged upon binding the large molecule in such a way as to permit its penetration which may be accomplished either by a coordinated binding-and-releasing process or simply by the local physical forces, such as surface tension or tendency toward maximum hydrophobic-hydrophobic interactions in the membrane. Once the protein is bound to the receptor which is mobile in the membrane, one should not be surprised by very fast reorientation or rotation of the whole complex—relaxation times of $10^1 - 10^2 \, \mu s$ have been calculated for molecules of mol. wt. about 50 000 in lipid membranes (Sackmann *et al.*, 1973).

Although this is obviously in the realm of speculations one should keep in mind the dynamic character of the membrane and the possibility of fast-spreading (cooperatively transmitted) signals along its surface which can, within fractions of a second, alter the permeability properties either generally or at a site responsible for a particular process. But there is much to be learned yet and we should strive with zest and zeal to learn it.

SYNOPSIS

Large molecules, such as proteins and nucleic acids, generally do not use membrane carriers but cross the cell membrane either by pinocytosis or by specific, enzyme-driven processes. Pinocytosis, however, may serve the transport even of small molecules, such as neuro-transmitters at the synaptic cleft.

REFERENCES

Abercrombie, M. and Heaysman, J. E. M. (1953). *Exptl. Cell Res.* **5**, 111.

Abragam, A. (1961). *The Principles of Nuclear Magnetism.* Oxford University Press, London.

Abrahamsson, S., Pascher, I., Larsson, K., and Karlsson, K. A. (1972). *Chem. Phys. Lipids* **8**, 152.

Abrams, A. and Baron, C. (1967). *Biochemistry* **6**, 225.

Agin, D. (1967). *Proc. Natl. Acad. Sci. USA* **57**, 1232.

Agin, D. (1972). In: *Foundations of Mathematical Biology* (ed. by Rosen, R.), vol. I, p. 253. Academic Press, New York—London.

Aksamit, R. and Koshland, D. E. Jr. (1972). *Biochem. Biophys. Res. Commun.* **48**, 1348.

Ames, G. F. and Lever, J. (1970). *Proc. Nat. Acad. Sci.* **66**, 1096.

Ames, G. F. and Lever, J. (1972). *J. Biol. Chem.* **247**, 4309.

Ames, B. N., Ames, G. F.—L., Young, J. D., Tsuchiya, D., and Lecocq, J. (1973). *Proc. Natl. Acad. Sci. USA* **70**, 456.

Anraku, Y. (1968*a*). *J. Biol. Chem.* **243**, 3116.

Anraku, Y. (1968*b*). *J. Biol. Chem.* **243**, 3123.

Anraku, Y. (1968*c*). *J. Biol. Chem.* **243**, 3128.

Atkinson, A., Gatenby, A. D., and Lowe, A. C. (1971). *Nature* **233**, 145.

Avruch, J. and Fairbanks, G. (1972). *Proc. Natl. Acad. Sci. USA* **69**, 1216.

Baddiley, J., Blumson, N. L., and Douglas, L. J. (1968). *Biochem. J.* **110**, 565.

Bangham, A. D. (1968). In: *Progress in Biophysics and Molecular Biology* (ed. by Butler, J. A. V. and Noble, D.). Pergamon Press, New York.

Bangham, A. D., Hill, M. W. and Miller, N. G. A. (1974). In: *Methods in Membrane Biology* (ed. by Korn, E. D.), vol. **1**, p. 1. Plenum Press, New York—London.

Barash, H. and Halpern, Y. S. (1971). *Biochem. Biophys. Res. Commun.* **45**, 681.

Baraud, J., Mawrice, A., and Napias, C. (1970). *Bull. Soc. Chim. Biol.* **52**, 421.

Beneš, I. and Kotyk, A. (1976). *Can. J. Biochem.*, **54**, 99.

Benson, A. A. (1968). In: *Membrane Models and the Formation of Biological Membranes* (ed. by Bolis, L. and Pethica, D. A.). North Holland, Amsterdam.

Berger, E. A. and Heppel, L. A. (1972). *J. Biol. Chem.* **247,** 7684.

Berman, M., Shahn, E., and Weiss, M. F. (1962). *Biophys. J.* **2,** 275.

Bervaes, J. C. A. M. and Kuiper, P. J. C. (1975). Personal communication.

Betz, A. L., Gilboe, D. D. and Drewes, L. R. (1975). *Biochim. Biophys. Acta* **401,** 416.

Beychok, S. (1967). In: *Poly-α-amino Acids—Protein Models for Conformation Studies* (ed. by Fasman, G. D.). Marcel Dekker, New York.

Blasie, J. K. and Worthington, C. R. (1969). *J. Mol. Biol.* **39,** 417.

Bloj, B., Morero, R. O., and Farias, R. W. (1973). *FEBS Lett.* **38,** 101.

Bodnaryk, R. P. (1972). *Canad. J. Biochem.* **50,** 524.

Bonting, S. L. (1970). In: *Membranes and Ion Transport* (ed. by Bittar, E. E.), vol. I, p. 257. Willey Interscience, London.

Boos, W., Gordon, A. S., Hall, R. E., and Price, H. D. (1972). *J. Biol. Chem.* **247,** 917.

Bretscher, M. S. (1973). *Science* **181,** 622.

Briggs, G. E., Hope, A. B., and Robertson, R. N. (1961). *Electrolytes and Plant Cells.* Blackwell Scientific Publications, Oxford.

Britton, H. G. (1966). *J. Theoret. Biol.* **10,** 28.

Bruckdorfer, K. R., Demel, R. A., de Gier, J., and van Deenen, L. L. M. (1969). *Biochim. Biophys. Acta* **181,** 334.

Caccam, J. F., Jackson, J. J., and Eylar, E. N. (1969). *Biochem. Biophys. Res. Commun.* **35,** 505.

Cantrell, A. C. (1973). *Biochim. Biophys. Acta* **311,** 381.

Caspar, P. L. D. and Kirschner, D. A. (1971). *Nature New Biol.* **231,** 46.

Chapman, D. (1969). *Lipids* **4,** 251.

Chapman, D., Kamat, V. B., de Gier, J., and Penkett, S. A. (1968). *J. Mol. Biol.* **31,** 101.

Chapman, D. and Wallach, D. F. H. (1968). In: *Biological Membranes* (ed. by Chapman, D.) p. 125. Academic Press, London—New York.

Chapman, D. and Dodd, G. H. (1971). In: *Structure and Function of Biological Membranes* (ed. by Rothfield, L. I.), p. 13. Academic Press, New York—London.

Chapman, D. and Urbina, J. (1971). *FEBS Lett.* **12,** 169.

Cho, K. Y. and Salton, M. R. J. (1966). *Biochim. Biophys. Acta* **116,** 73.

Ciani, S. (1965). *Biophysik* **2,** 368.

Ciani, S. M., Eisenman, G., Laprade, R., Szabo, G. (1973). In: *Membranes* (ed. by Eisenman, G.), Vol. **2,** p. 61. Marcel Dekker, New York.

Civan, M. M. and Shporer, M. (1972). *Biophys. J.* **12,** 404.

Clarke, M. (1971). *Biochem. Biophys. Res. Commun.* **45,** 1063.

Cleland, W. W. (1963). *Biochim. Biophys. Acta* **67,** 104.

Colacicco, G. (1972). *Ann. N. Y. Acad. Sci.* **195,** 224.

Cole, K. S. (1965). *Physiol. Rev.* **45,** 340.

Coleman, R. (1973). *Biochim. Biophys. Acta* **300,** 1.

Collander, R. (1949). *Physiol. Plant.* **2,** 300.

Conway, E. J. and Mullaney, M. (1961). In: *Membrane Transport and Metabolism* (ed. by Kleinzeller, A. and Kotyk, A.), p. 117. Academic Press, New York.

Cotman, C., Brown, D. H., Harrell, B. W., and Anderson, N. G. (1970). *Arch. Biochem. Biophys.* **136,** 436.

Crane, R. K. (1965). *Fed. Proc.* **24**, 1000.

Cronan, J. E. Jr., Birge, C. H., and Vagelos, P. R. (1969). *J. Bacter.* **100**, 601.

Cronan, J. E. Jr., Roy, T. K., and Vagelos, R. P. (1970). *Proc. Natl. Acad. Sci. USA* **65**, 737.

Cuppoletti, J. and Segel, I. H. (1975). *J. Theoret. Biol.* **53**, 125.

Curran, P., Hajjar, J. J., and Glynn, I. M. (1970). *J. Gen. Physiol.* **55**, 297.

Curtis, A. S. G. (1967). *The Cell Surface.* Logos Press and Academic Press, London.

Dainty, J. (1960). *Symp. Soc. Exptl. Biol.* **14**, 140.

Dainty, J. (1963). *Adv. Bot. Res.* **1**, 279.

Davson, H. and Danielli, J. F. (1952). *The Permeability of Natural Membranes.* Cambridge Univ. Press, Cambridge.

Dawson, R. M. C., Herrington, N., and Lindsay, D. B. (1960). *Biochem. J.* **77**, 226.

Deák, T. and Kotyk, A. (1968). *Folia Microbiol.* **13**, 212.

Dehlinger, P. J. and Schimke, R. T. (1971). *J. Biol. Chem.* **246**, 2574.

de Kruyff, B., van Dijck, P. W. M., Goldbach, R. W., Demel, R. A., and van Deenen, L. L. M. (1973). *Biochim. Biophys. Acta,* **330**, 269.

Denbigh, K. G. (1951). *The Thermodynamics of the Steady State.* Methuen & Co., Ltd. London.

Dennis, V. W., Stead, N. W., and Andreoli, T. E. (1970). *J. Gen. Physiol.* **55**, 375.

De Robertis, E. (1971). *Science,* **171**, 963.

Desai, J. D. and Modi, V. V. (1975). *Experientia* **31**, 160.

Dewey, M. M. and Barr, L. (1970). In: *Current Topics in Membranes and Transport* (ed. by Bronner, F. and Kleinzeller, A.), vol. I, p. 1. Academic Press, New York—London.

Diesterhaft, M. and Freese, E. (1974). *Fed. Proc.* **33**, Abstr. 964.

Djerassi, C. (1960). *Optical Rotatory Dispersion.* Mc Graw—Hill, New York.

Dobler, M., Dunitz, J. D., and Krajewski, J. (1969). *J. Mol. Biol.* **42**, 603.

Dreyer, W. J., Papermaster, D. S., and Kühn, H. (1972). *Ann. N. Y. Acad. Sci.* **195**, 61.

Edsall, J. T. and Wyman, J. (1958). *Biophysical Chemistry.* Academic Press, New York.

Einstein, A. (1905). *Ann. Physik* **17**, 549.

Engelman, D. M. and Morowitz, H. J. (1968). *Biochim. Biophys. Acta* **150**, 385.

Ernster, L. and Kuylenstierna, B. (1970). In: *Membranes of Mitochondria and Chloroplasts* (ed. by Racker, E.), p. 172. Van Nostrand—Reinhold, New York.

Essig, A. and Leaf, A. (1963). *J. Gen. Physiol.* **46**, 505.

Faust, R. G. and Shearin, S. J. (1974). *Nature* **248**, 60.

Fettiplace, R., Andrews, D. M. and Haydon, D. A. (1971). *J. Membr. Biol.* **5**, 277.

Finean, J. B. (1958). *Exptl. Cell Res.* Suppl. 5, 18.

Finean, J. B. (1973). In: *Form and Function of Phospholipids* (ed. by Ansell, G. B., Hawthorne, J. N., and Dawson, R. M. C.), p. 171. Elsevier, Amsterdam—London—New York.

Finkelstein, A. and Mauro, A. (1963). *Biophys. J.* **3**, 215.

Folch-Pi, J. and Stoffyn, P. J. (1972). *Ann. N. Y. Acad. Sci.* **195**, 86.

Fox, C. F., Law, J. H., Tsukagoshi, N., and Wilson, G. (1970). *Proc. Natl. Acad. Sci. USA* **67**, 598.

Franklin, T. J. (1974). In: *Industrial Aspects of Biochemistry* (ed. by Spencer, B.), p. 549. North Holland/American Elsevier, Amsterdam—London—New York.

Frazier, H. S. and Leaf, A. (1963). *J. Gen. Physiol.* **46**, 491.

Freysz, L., Bieth, R., and Mandel, P. (1969). *J. Neurochem.* **16**, 1417.

Frizzel, R. A. and Schultz, S. G. (1970). *J. Gen. Physiol.* **56**, 462.

Fromm, H. J. (1970). *Biochem. Biophys. Res. Commun.* **40**, 692.

Frommhertz, P. (1970). *FEBS Lett.* **11**, 205.

Frye, L. D. and Edidin, M. (1970). *J. Cell. Sci.* **7**, 391.

Fukui, S. and Isobe, K. (1973). *Biochim. Biophys. Acta,* **328**, 114.

Fullmer, C. S. and Wasserman, R. H. (1973). *Biochem. Biophys. Acta* **317**, 172.

Furlong, C. E. and Weiner, J. H. (1970). *Biochem. Biophys. Res. Commun.* **38**, 1076.

Fürth, R. (1956). In: A. Einstein, *Investigations on the Theory of the Brownian Movements* (ed. with notes by Fürth, R.), p. 101. Dover Publications, New York.

Galliard, T. (1973). In: *Form and Function of Phospholipids* (ed. by Ansell, G. B., Hawthorne, J. N., and Dawson, R. M. C.), p. 253. Elsevier, Amsterdam—London—New York.

Garrahan, P. J. and Glynn, I. M. (1967). *J. Physiol.* **192**, 237.

Geck, P. (1971). *Biochim. Biophys. Acta,* **241**, 462.

Getz, G. S. (1972). In: *Membrane Molecular Biology* (ed. by Fox, C. F. and Keith, A.), p. 386. Sinauer Ass., Stamford, Conn.

Ghuysen, J. M. and Shockman, G. D. (1973). In: *Bacterial Membranes and Walls* (ed. by Leive, L.), p. 37. Marcel Dekker, New York.

Glynn, I. M. and Lew, V. L. (1970). *J. Physiol.* **207**, 393.

Goldman, D. E. (1943). *J. Gen. Physiol.* **27**, 37.

Goldstein, D. A. and Solomon, A. K. (1960). *J. Gen. Physiol.* **44**, 11.

Goodwin, T. W. (1973). In: *Lipids and Biomembranes of Eukaryotic Microorganisms* (ed. by Erwin, J. A.), p. 1. Academic Press, New York—London.

Gorter, E. and Grendel, F. (1925). *J. Exptl. Med.* **41**, 439.

Green, D. E. (1974). *Biochim. Biophys. Acta,* **346**, 27.

Green, D. E. and Perdue, J. F. (1966). *Proc. Natl. Acad. Sci. USA* **55**, 1295.

Green, D. E., Haard, N. F., Lenaz, G., and Silman, H. I. (1968). *Proc. Natl. Acad. Sci. USA* **60**, 277.

Green, D. E. and Ji, S. (1972). *Bioenergetics* **3**, 159.

Green, D. E., Ji, S., and Brucker, R. F. (1972). *Bioenergetics* **4**, 527.

Greenfield, N., Davidson, B., and Fasman, G. (1967). *Biochemistry* **6**, 1630.

Greville, G. D. (1969). In: *Current Topics in Bioenergetics* (ed. by Sanadi, D. R.), vol. **3**, p. 1. Academic Press, New York.

Griffith, T. W., Carraway, C., and Leach, F. R. (1971). *Fed. Proc.* **30**, 1115 Abs.

Grisham, C. M. and Barnett, R. E. (1973). *Biochemistry* **12**, 2635.

Gulik-Krzywicki, T., Shechter, E., Luzzati, V., and Faure, M. (1969). *Nature* **223**, 1116.

Gunstone, F. D. (ed.) (1970). *Topics in Lipid Chemistry.* Logos Press, London.

Gutfreund, H. (1972). *Enzymes: Physical Principles.* Wiley — Interscience, London.

Györgyi, S. and Sugár, I. (1974). *Abstr. 9th FEBS Meeting* Budapest, p. 245.

Hardman, K. D. and Ainsworth, C. F. (1973). *Biochemistry* 12, 4442.

Hansch, C. and Fujita, T. (1964). *J. Am. Chem. Soc.* 86, 1616.

Harold, F. M., Baarda, J. R., Baron, C., and Abrams, A. (1969). *J. Biol. Chem.* 244, 2261.

Hartley, G. S. and Crank, J. (1949). *Trans. Faraday Soc.* 45, 801.

Helmkamp, G. M. Jr., Harvey, M. S., Wirtz, K. W. A. and van Deenen, L. L. M. (1974). 9th FEBS Meeting, Budapest, Abstracts of Communications, p. 366.

Haurowitz, F. (1963). *The Chemistry and Function of Proteins.* Academic Press, New York.

Hazelbauer, G. L. and Adler, J. (1971). *Nature New Biol.* 230, 101.

Hechemy, K. and Goldfine, H. (1971). *Biochem. Biophys. Res. Commun.* 42, 245.

Hechter, O. (1965). *Fed. Proc.* 24 (2), S91.

Heinz, E. (ed.) (1972). *Na-Linked Transport of Organic Solutes*, p. 1. Springer Verlag, Berlin.

Heller, J. and Lawrence, M. A. (1970). *Biochemistry* 9, 864.

Henderson, P. (1907). *Z. physik. Chem.* 59, 118.

Henderson, P. (1908). *Z. physik. Chem.* 63, 325.

Henning, U. and Schwartz, U. (1973). In: *Bacterial Membranes and Walls* (ed. by Leive, L.), p. 413. Marcel Dekker, New York.

Heppel, L. A. (1971). In: *Structure and Function of Biological Membranes* (ed. by Rothfield, L. I.), p. 223. Academic Press, New York—London.

Hill, A. V. (1928). *Proc. Roy. Soc. B.* 104, 39.

Himmelspach, K., Westphal, O., and Teichman, B. (1971). *Eur. J. Immunol.* 1, 106.

Hinman, N. D. and Phillips, A. H. (1970). *Science* 170, 1222.

Hladky, S. B. and Haydon, D. A. (1970). *Nature*, 225, 451.

Hochstadt-Ozer, J. and Stadtman, E. R. (1971). *J. Biol. Chem.* 246, 5304.

Hodgkin, A. L. and Horowicz, P. (1959). *J. Physiol.* 148, 127.

Hoffman, P. G. and Tosteson, D. C. (1971). *J. Gen. Physiol.* 58, 438.

Hogg, R. and Englesberg, E. (1969). *J. Bacteriol.* 100, 423.

Hokin, L. E. (1974). *Ann. N. Y. Acad. Sci.* 242, 12.

Holden, J. T. (1968). *J. Theoret. Biol.* 21, 97.

Holzwarth, G. (1972). In: *Membrane Molecular Biology* (ed. by Fox, C. F. and Keith, A.), p. 228. Sinauer Ass., Stamford, Conn.

Horák, J. and Kotyk, A. (1973). *Eur. J. Biochem.* 32, 36.

House, C. R. (1974). *Water Transport in Cells and Tissues.* Edward Arnold (Publishers) Ltd., London.

Hsu, C. C. and Fox, C. F. (1970). *J. Bacter.* 103, 410.

Hubbard, A. L. H. and Cohn, Z. C. (1972). *J. Cell. Biol.* 55, 390.

Hubbell, W. L. and McConnell, H. M. (1971). *J. Am. Chem. Soc.* 93, 314.

Hybl, A. and Dorset, D. (1970). *Biophys. Soc. Abstr.* 49a.

Israelachvili, J. N. and Mitchell, D. J. (1975). *Biochim. Biophys. Acta* 389, 13.

Itzkowitz, M. S. (1967), *J. Chem. Phys.* 46, 3048.

Ivanov, V. T., Laine, I. A. Abdulaev, N. D., Senyavina, L. B., Popov, E. M., Ovchinnikov, Yu. A. and Shemyakin, M. M. (1969). *Biochem. Biophys. Res. Commun.* 34, 803.

Jacobs, M. H. (1952). In: *Modern Trends in Physiology and Biochemistry* (ed. by E. S. G. Barron), p. 149. Academic Press, New York.

Jacobs, M. H. (1967). *Diffusion Processes*. Springer Verlag, Berlin—Heidelberg—New York (reprint from Ergebnisse der Biologie, Zwölfter Band, 1935).

Jacobs, M. H., Glassman, H. N. and Parpart, A. K. (1950). *J. Exptl. Zool.* **113**, 277.

Jaffe, L. F. (1974). *J. Theoret. Biol.* **48**, 11.

Jain, M. K., Strickholm, A., and Cordes, E. H. (1969). *Nature* **222**, 871.

Jain, M. K., White, F. P., Strickholm, A., Williams, E., and Cordes, E. H. (1972). *J. Membrane Biol.* **8**, 363.

Jakoby, W. D. (ed.) (1971). *Methods in Enzymology*, vol. XXII. Academic Press, New York—London.

Jakovcic, S., Getz, G. S., Rabinowitz, M., Jakob, H., and Swift, H. (1971). *J. Cell. Biol.* **48**, 490.

Janáček, K., Rybová, R., and Slavíková, M. (1972). *Biochim. Biophys. Acta* **288**, 221.

Ji, T. H. and Urry, D. W. (1969). *Biochem. Biophys. Res. Commun.* **34**, 404.

Johnson, A. R. and Davenport, J. B. (1971). *Biochemistry and Methodology of Lipids*. Wiley and Sons, New York.

Johnson, J. A. and Wilson, T. A. (1967). *J. Theoret. Biol.* **17**, 304.

Johnson, W. C., Silhavy, T. J. and Boos, W. (1975). *Appl. Miccobiol.* **29**, 405.

Jones, T. H. D. and Kennedy, E. P. (1968). *Fed. Proc.* **27**, 644.

Jost, J. P. and Rickenberg, H. V. (1971). *Annu. Rev. Biochem.* **40**, 741.

Jost, P., Wagoner, A. S., and Griffith, O. H. (1971). In: *Structure and Function of Biological Membranes* (ed. by Rothfield, L. I.), p. 83. Academic Press, New York—London.

Kaback, H. R. (1972). *Biochim. Biophys. Acta* **265**, 367.

Kagawa, Y. and Racker, E. (1971). *J. Biol. Chem.* **246**, 5477.

Kanemasa, Y., Akamatsu, Y., and Nojima, S. (1967). *Biochim. Biophys. Acta* **144**, 382.

Katchalsky, A. (1961). In: *Membrane Transport and Metabolism* (ed. by Kleinzeller, A. and Kotyk, A.), p. 69. Academic Press, New York.

Katchalsky, A. and Curran, P. F. (1965). *Nonequilibrium Thermodynamics in Biophysics*. Harvard University Press, Cambridge, Massachusetts.

Katz, B. (1966). *Nerve, Muscle, and Synapse*. Mc Graw-Hill, New York.

Kauzmann, N. (1959). *Adv. Prot. Chem.* **14**, 1.

Kedem, O. (1961). In: *Membrane Transport and Metabolism* (ed. by Kleinzeller, A. and Kotyk, A.), p. 87. Academic Press, New York.

Kedem, O. (1972). *J. Membrane Biol.* **10**, 213.

Kedem, O. and Katchalsky, A. (1958). *Biochim. Biophys. Acta* **27**, 229.

Kellermann, O. and Szmelcman, S. (1974). *Eur. J. Biochem.* **47**, 139.

Kepner, G. R. and Macey, R. I. (1968). *Biochim. Biophys. Acta* **163**, 188.

Kilbourn, B. T., Dunitz, J. D., Piodo, L. A. R., and Simon, W. (1967). *J. Mol. Biol.* **30**, 559.

King, E. L. and Altman, C. (1956). *J. Phys. Chem.* **60**, 1375.

Kirkwood, J. G. (1954). In: *Ion Transport across Membranes* (ed. by Clark, H. T.), p. 119. Academic Press, New York.

Klenk, H.D. (1973). In: *Biological Membranes* (ed. by Chapman, D. and Wallach, D. F. H.), vol. **2**, p. 145. Academic Press, London—New York.

Koch, A. S. and Fehér, G. (1973). *J. Gen. Vir.* **18**, 319.

Konings, W. N., Barnes, E. M. Jr., and Kaback, H. R. (1971). *J. Biol. Chem.* **246**, 5857.

Korn, E. D. (1968). *J. Gen. Physiol.* **52**, Suppl. 257.

Kornfeld, R. and Kornfeld, S. (1970). *J. Biol. Chem.* **245**, 2536.

Koryta, J. (1973). *Chem. listy* **67**, 897.

Kostyuk, P. G., Krishtal, O. A. and Pidoplichko, V. I. (1972). *J. Physiol.* **226**, 373.

Kotyk, A. (1973). *Biochim. Biophys. Acta* **300**, 183.

Kotyk, A. (1974). In: *Biophysics of Membrane Transport* (ed. by Miękisz, S. and Gomułkiewicz, J.), vol. **1**, p. 49. Inst. Plant Biol. Biophys. Agric. Academy, Wrocław.

Kotyk, A. and Řihová, L. (1972). *Biochim. Biophys. Acta* **288**, 380.

Kotyk, A. and Janáček, K. (1975). *Cell Membrane Transport: Principles and Techniques.* Plenum Press, New York.

Kroger, A. and Klingenberg, M. (1970). *Vitamins Hormones* **28**, 533.

Kubišta, V. (1974). *Vesmír* **53**, 230.

Kulaev, I. S. (1975). *Rev. Physiol. Biochem. Pharmacol.* **73**, 131.

Kundig, W., Ghosh, S., and Roseman, S. (1964). *Proc. Natl. Acad. Sci. USA* **52**, 1067.

Kundig, W. and Roseman, S. (1971). *J. Biol. Chem.* **246**, 1407.

Kuzuya, H., Bromwell, K. and Guroff, G. (1971). *J. Biol. Chem.* **246**, 6371.

Kyte, J. (1971). *J. Biol. Chem.* **246**, 4157.

Ladbrooke, B. D., Jenkinson, T. J., Kamat, V. B., and Chapman, D. (1968). *Biochim. Biophys. Acta* **164**, 101.

Laico, M. T., Ruoslahti, E. I., Papermaster, D. S., and Dreyer, W. J. (1970). *Proc. Natl. Acad. Sci. USA* **67**, 120.

Langmuir, I. (1933). *J. Chem. Phys.* **1**, 756.

Lapetina, E. G. and Michell, R. H. (1973). *FEBS Lett.* **31**, 1.

Law, J. H. and Snyder, W. R. (1972). In: *Membrane Molecular Biology* (ed. by Fox, C. L. and Keith, A. D.), p. 3. Sinauer Ass., Stamford, Conn.

LeFevre, P. G. (1973). *J. Membrane Biol.* **11**, 1.

LeFevre, P. G. (1975). *Ann. N. Y. Acad. Sci.* **264**, 398.

LeFevre, P. G. and McGinnis, G. F. (1960). *J. Gen. Physiol.* **44**, 87.

Lester, R. L., Smith, S. W., Wells, G. B., Rees, D. C. and Angus, W. W. (1974). *J. Biol. Chem.* **249**, 3388.

Lieb, W. R. and Stein, W. D. (1969). *Nature* **224**, 240.

Lieb, W. R. and Stein, W. D. (1970). *Biophys. J.* **10**, 585.

Lieb, W. R. and Stein, W. D. (1971). *J. Theoret. Biol.* **30**, 219.

Lieb, W. R. and Stein, W. D. (1974). *Biochim. Biophys. Acta* **373**, 178.

Linden, C. D., Wright, K. L., McConnell, H. M., and Fox, C. F. (1973). *Proc. Natl. Acad. Sci. USA* **70**, 2271.

Ling, G. N. and Gerard, R. W. (1949). *J. Cell. Comp. Physiol.* **34**, 382.

London, Y., Demel, R. A., van Kessel, W. S. M. G., Zahler, P., and van Deenen, L. L. M. (1974). *Biochim. Biophys. Acta* **332**, 69.

Longley, R. P., Rose, A. H., and Knights, B. A. (1968). *Biochem. J.* **108**, 401.

Lowenstein, J. M. (ed.) (1969). *Methods in Enzymology*, vol. XIV. Academic Press, New York—London.

Lucy, J. A. and Glauert, A. (1964). *J. Mol. Biol.* **8**, 727.

Lutz, M. and Breton, J. (1973). *Biochem. Biophys. Res. Commun.* **53**, 413.

Luzzati, V. (1968). In *Biological Membranes* (ed. by Chapman, D.), p. 71. Academic Press, New York.

Machtiger, N. A. and Fox, C. F. (1973). *Annu. Rev. Biochem.* **42**, 575.

MacInnes, D. A. (1961). *The Principles of Electrochemistry.* Dover Publications, New York.

MacRobbie, E. A. C. (1965). *Biochim. Biophys. Acta* **94**, 64.

Markham, R., Frey, S., and Hills, G. J. (1963). *Virology* **20**, 88.

Markin, V. S. and Chizmadjev, Yu. A. (1974). *Indutsirovannyi ionnyi transport.* Izd. Nauka, Moscow.

Martonosi, A. (1968). *J. Biol. Chem.* **243**, 71.

Matile, P. and Wiemken, A. (1967). *Arch. Mikrobiol.* **56**, 148.

McElhaney, R. and Tourtellotte, M. E. (1969). *Science* **164**, 433.

McElhaney, R. N., de Gier, J., and van Deenen, L. L. M. (1970). *Biochim. Biophys. Acta* **219**, 245.

Medweczky, N. and Rosenberg, H. (1969). *Biochim. Biophys. Acta* **192**, 369.

Medweczky, N. and Rosenberg, H. (1970). *Biochim. Biophys. Acta* **211**, 158.

Meissner, G. (1973). *Biochim. Biophys. Acta* **298**, 906.

Meister, A. (1973). *Science* **180**, 33.

Melancon, M. J. and DeLuca, H. F. (1970). *Biochemistry* **9**, 1658.

Miller, C. (1974). *Biochim. Biophys. Acta* **339**, 71.

Miller, D. M. (1965). *Biophys. J.* **5**, 417.

Mitchell, P. (1966). *Biol. Rev.* **41**, 445.

Mitchell, P. (1973). *Bioenergetics* **4**, 265.

Moffitt, W. and Moscowitz, A. (1959). *J. Chem. Phys.* **30**, 648.

Montal, M. and Mueller, P. (1972). *Proc. Natl. Acad. Sci. USA* **69**, 3561.

Morowitz, H. J. and Terry, T. (1969). *Biochim. Biophys. Acta* **183**, 276.

Mueller, P., Rudin, D. O., Ti Tien, H., and Wescott, W. C. (1962). *Nature* **194**, 979.

Mueller, P., Rudin, D. O., Ti Tien, H., and Wescott, W. C. (1964). In *Recent Progress in Surface Science* (ed. by Daneilli, J. F., Pankhurst, K. G. A., and Riddiford, A. C.), vol. **1**, p. 379. Academic Press, New York—London.

Mueller, P. and Rudin, D. O. (1968). *J. Theoret. Biol.* **18**, 222.

Mueller, P. and Rudin, D. O. (1969). *Curr. Topics Bioenergetics* **3**, 157.

Mullis, K. B., Pollack, J. R., and Neilands, J. B. (1971). *Biochemistry* **10**, 4894.

Naftalin, J. (1970). *Biochim. Biophys. Acta* **211**, 65.

Nägeli, K. W. and Cramer, K. (1855). *Pflanzenphysiologische Untersuchungen.* F. Schultess, Zürich.

Nakamura, K., Ostrovsky, D. N., Miyazawa, T., and Mizushima, S. (1974). *Biochim. Biophys. Acta* **332**, 329.

Nakao, M., Nakao, T., Hara, Y., Nagai, F., Yagasaki, S., Koi, M., Makagawa, A. and Kawai, K. (1974). *Ann. N. Y. Acad. Sci.* **242**, 24.

Neal, J. L. (1972). *J. Theoret. Biol.* **35**, 113.

Nikaido, H. (1973). In: *Bacterial Membranes and Walls* (ed. by Leive, L.), p. 131. M. Dekker, New York.

Nishimune, T. and Hayashi, R. (1973). *Biochim. Biophys. Acta* **328**, 124.

Njus, D., Sulzman, F. M., and Hastings, J. W. (1974). *Nature* **248**, 116.

Nossal, N. G. and Heppel, L. (1966). *J. Biol. Chem.* **241**, 3055.

Nurminen, T. and Suomalainen, H. (1971). *Biochem. J.* **125**, 963.

O'Brien, J. S. (1967). *J. Theoret. Biol.* **15**, 307.

O'Brien, I. G. and Gibson, F. (1970). *Biochim. Biophys. Acta* **215**, 393.

Ohta, H., Matsumoto, J., Kagano, K., Fujita, M., and Nakao, M. (1971). *Biochem. Biophys. Res. Commun.* **42**, 1127.

Okaya, Y. (1964). *Acta Crystallogr.* **17**, 1276.

Opekarová, M., Kotyk, A., Horák, J. and Kholodenko, V. (1975). *Eur. J. Biochim.* **59**, 373.

Osborn, M. J. (1969). *Annu. Rev. Biochem.* **38**, 501.

Oster, G. F., Perelson, A. S., and Katchalsky, A. (1973). *Quart. Rev. Biophys.* **6**, 1.

Ostwald, T. J. and MacLennan, D. H. (1974). *J. Biol. Chem.* **249**, 974.

Overbeek, J. T. G. (1952). In: *Colloid Science* (ed. by Kruyt, H. R.), vol. I, p. 115. Elsevier Publ. Co., Amsterdam.

Overath, P., Schairer, H.U., Hill, F. F., and Lamnek-Hirsch, I. (1971*a*). In: *The Dynamic Structure of Cell Membranes* (ed. by Wallach, D. F. H. and Fischer, H.), p. 149. Springer-Verlag, New York.

Overath, P., Hill, F. F., and Lamnek-Hirsch, I. (1971*b*). *Nature New Biol.* **234**, 264.

Overath, P. and Träuble, H. (1973). *Biochemistry* **12**, 2625.

Pardee, A. B. (1967). *Science* **156**, 1627.

Patterson, P. H. and Lennarz, W. J. (1970). *Biochem. Biophys. Res. Commun.* **40**, 408.

Pavlasová, E. and Stejskalová, E. (1972). *Folia Microbiol.* **17**, 471.

Payne, J. W. and Gilvarg, C. (1971). *Adv. Enzym.* **35**, 187.

Penniston, J. T., Beckett, L., Bentley, D. L. and Hansch, C. (1969). *Mol. Pharmacol.* **5**, 333.

Phillips, D. R. and Morrison, M. (1971). *FEBS Lett.* **18**, 95.

Phillips, S. K. and Cramer, W. A. (1973). *Biochemistry* **12**, 1170.

Pitot, H. C., Sladek, N., Raglaud, W., Murray, R. K., Moyer, G., Soling, H. D., and Jost, J. P. (1969). In: *Microsomes and Drug Oxidation* (ed. by Gillette, J. R.), p. 59. Academic Press, New York.

Planck, M. (1890). *Ann. Physik* **40**, 561.

Polissar, M. J. (1954). In: *The Kinetic Basis of Molecular Biology* (by Johnson, F. H., Eyring, H., and Polissar, M. J.), p. 515. John Wiley and Sons, New York.

Pollack, J. R. and Neilands, J. B. (1970). *Biochem. Biophys. Res. Commun.* **38**, 989.

Post, R. L. and Orcutt, B. (1973). In: *Organization of Energy-transducing Membranes* (ed. by Nakao, M. and Pocker, L.), p. 25. Tokyo University Press, Tokyo.

Pressman, B. C. (1970). In: *Membranes of Mitochondria and Chloroplasts* (ed. by Racker, E.), p. 213. Van Nostrrand-Reinhold, New York.

Pressman, B. C. and de Guzman, N. T. (1974). *Ann. N. Y. Acad. Sci.* **227**, 380.

Racker, E. (1970). In: *Membranes of Mitochondria and Chloroplasts* (ed. by Racker, E.), p. 127. Van Nostrand-Reinhold, New York.

Racker, E., Horstman, L. L., Kling, D., and Fessenden-Raden, J. M. (1969). *J. Biol. Chem.* **244**, 6668. Ratledge, C. and Marshall, B. J. (1972). *Biochim. Biophys. Acta* **279**, 58.

Ratledge, C. and Marshall, B. J. (1972). *Biochim. Biophys. Acta* **279**, 58.

Razin, S. (1972). *Biochim. Biophys. Acta* **265**, 241.

Redwood. W. R., Müldner, H., and Thompson, T. E. (1969). *Proc. Natl. Acad. Sci. USA* **64**, 989.

Regen, D. M. and Morgan, H. E. (1964). *Biochim. Biophys. Acta* **79**, 151.

Regen, D. M. and Tarpley, H. L. (1974). *Biochim. Biophys. Acta* **339**, 218.

Renkin, E. M. (1954). *J. Gen. Physiol.* **38**, 225.

Repke, K. R. H. and Schön, R. (1973). *Acta Biol. Med. Germ.* **31**, 19 K.

Reusch, V. M. Jr. and Burger, M. M. (1973). *Biochim. Biophys. Acta* **300**, 79.

Richardson, S. H., Hultin, H. O., and Green, D. E. (1963). *Proc. Natl. Acad. Sci. USA* **50**, 821.

Robertson, J. D. (1959), *Biochem. Soc. Symp.* **16**, 3.

Robinson, J. D. (1970). *Arch. Biochem. Biophys.* **139**, 17.

Romeo, D., Girard, A., and Rothfield, L. (1970a). *J. Mol. Biol.* **53**, 475.

Romeo, D., Girard, A., and Rothfield, L. (1970b). *J. Mol. Biol.* **53**, 491.

Rorive, G., Nielsen, R., and Kleinzeller, A. (1972). *Biochim. Biophys. Acta* **266**, 376.

Rosen, B. P. (1971). *J. Biol. Chem.* **246**, 3653.

Rosen, B. P. (1973). *J. Biol. Chem.* **248**, 1211.

Rosen, B. P. and Vasington, F. D. (1971). *J. Biol. Chem.* **246**, 5351.

Rosenberg, T. and Wilbrandt, W. (1957). *J. Gen. Physiol.* **41**, 289.

Rosenberg, T. and Wilbrandt, W. (1963). *J. Theoret. Biol.* **5**, 288.

Rosenheck, K. and Doty, P. (1961). *Proc. Natl. Acad. Sci. USA* **47**, 1775.

Rotman, B. and Papermaster, B. W. (1966). *Proc. Natl. Acad. Sci. USA* **55**, 134.

Rouser, G., Nelson, G. J., Fleischer, S., and Simon, G. (1968). In: *Biological Membranes* (ed. by Chapman, D.), p. 5. Academic Press, London—New York.

Sackmann, E. and Träuble, H. (1972). *J. Am. Chem. Soc.* **94**, 4482.

Sackmann, E., Träuble, H., Galla, K.J., and Overath, P. (1973). *Biochemistry* **12**, 5360.

Sarkadi, B. and Schubert, A. (**1972**). *Acta Biochim. Biophys. Acad. Sci. Hung.* **7**, 367.

Schleif, R. (1969). *J. Mol. Biol.* **46**, 185.

Schlögl, R. (1964). *Stofftransport durch Membranen.* Dr. D. Steinkopff-Verlag, Darmstadt.

Schlögl, R. (1969). *Quart. Rev. Biophys.* **2**, 305.

Schnaitman, C. A. (1969). *Proc. Natl. Acad. Sci. USA* **63**, 412.

Schnaitman, C. A. (1970). *J. Bacter.* **104**, 890.

Schnebli, H. P. and Abrams, A. (1970). *J. Biol. Chem.* **245**, 1115.

Schultz, S. G. and Curran, P. F. (1970). *Physiol. Rev.* **50**, 637.

Schwencke, J. S., Farias, G. and Rojas, M. (1971). *Eur. J. Biochem.* **21**, 137.

Seelig, J. (1970). *J. Am. Chem. Soc.* **92**, 3881.

Seelig, J., Axel, F. and Limacher, H. (1973). *Ann. N. Y. Acad. Sci.* **222**, 588.

Shamoo, A. E. (1974). *Ann. N. Y. Acad. Sci.* **242**, 389.

Shporer, M. and Civan, M. M. (1972). *Biophys. J.* **12**, 114.

Sigler, K. and Janáček, K. (1971). *Biochim. Biophys. Acta* **241**, 528.

Singer, S. J. and Nicolson, G. L. (1972). *Science* **175**, 720.

Siñeriz, F., Farías, R. N., and Trucco, R. E. (1973). *FEBS Lett.* **32**, 30.

Sjöstrand, F. S. (1963). *J. Ultrastruct. Res.* **9**, 561.

Sjöstrand, F. S. (1968). In: *Regulatory Functions of Biological Membranes* (ed. by Järnefelt, J.), p. 11. Elsevier Publ. Co., Amsterdam.

Skou, J. C. (1972). *Bioenergetics* **4**, 203.

Slater, E. C. (1958). *Rev. Pure Appl. Chem.* **8**, 221.

Slayman, C. (1973). In: *Current Topics in Membranes and Transport* (ed. by Bronner, F. and Kleinzeller, A.), vol. **4**, p. 1. Academic Press, New York—London.

Sonnenberg, M. (1971). *Proc. Natl. Acad. Sci. USA* **68**, 1051.

Spanner, D. C. (1964). *Introduction to Thermodynamics.* Academic Press, London—New York.

Steck, T. L. (1974). *J. Cell Biol.* **62**, 11.

Steck, T. L. and Wallach, D. F. H. (1970). *Methods Cancer Res.* **5**, 93.

Steck, T. L. and Fox, C. F. (1972). In: *Membrane Molecular Biology* (ed. by Fox, C. F. and Keith, A. D.), p. 27. Sinauer Ass., Stamford, Conn.

Stein, W. D. (1967). *The Movement of Molecules across Cell Membranes.* Academic Press, New York—London.

Stein, W. D. and Danielli, J. F. (1956). *Disc. Faraday Soc.* **21**, 238.

Stein, W. D., Lieb, W. R., Karlish, S. J. D., and Eilam, Y. (1973). *Proc. Natl. Acad. Sci. USA* **70**, 275.

Shipley, G. G., Leslie, R. B., and Chapman, D. (1969). *Nature* **222**, 561.

Storelli, C., Vögeli, H., and Semenza, G. (1972). *FEBS Lett.* **24**, 287.

Suomalainen, H., Nurminen, T., and Oura, E. (1973). In: *Progress in Industrial Microbiology* (ed. by Hockenhull, B. J. D.), vol. **13**, p. 109. Churchill Livingstone, Edinburgh—London.

Szubinska, B. (1971). *J. Cell Biol.* **49**, 747.

Takacs, B. J. and Holt, S. C. (1971). *Biochim. Biophys. Acta* **233**, 278.

Taylor, R. T., Norrell, S. A., and Hanna, M. L. (1972). *Arch. Biochem. Biophys.* **148**, 366.

Teorell, T. (1953). In: *Progress in Biophysics and Biophysical Chemistry* (ed. by Butler, J. A. V. and Randall, J. T.), vol. **3**, p. 305. Academic Press, New York.

Thomas, L. (1973). *Biochim. Biophys. Acta* **291**, 454.

Thomas, R. C. (1972). *Physiol. Rev.* **52**, 563.

Tien, H. T., Carbone, S., and Dawidowicz, E. A. (1966). *Nature* **212**, 718.

Tomasz, A. (1973). In: *Bacterial Membranes and Walls* (ed. by Leive, L.), p. 321. M. Dekker, New York.

Tourtellotte, M. E. (1972). In: *Membrane Molecular Biology* (ed. by Fox, C. F. and Keith, A. D.), p. 439. Sinauer Ass., Stamford, Conn.

Träuble, H. (1971). *J. Membrane Biol.* **4**, 193.

Träuble, H. and Haynes, D. H. (1971). *Chem. Phys. Lipids* **7**, 324.

Träuble, H. and Overath, P. (1973). *Biochim. Biophys. Acta* **307**, 491.

Träuble, H. and Eibl, H. (1974). *Proc. Natl. Acad. Sci. USA* **71**, 214.

Tsukagoshi, N., Fielding, P., and Fox, C. F. (1971). *Biochem. Biophys. Res. Commun.* **44**, 497.

Tsukagoshi, N. and Fox, C. F. (1973a), *Biochemistry* **12**, 2816.

Tsukagoshi, N. and Fox, C. F. (1973b). *Biochemistry* **12**, 2822.

Tsukagoshi, N., Tamura, G., and Arima, K. (1970a). *Biochim. Biophys. Acta* **196**, 204.

Tsukagoshi, N., Tamura, G., and Arima, K. (1970b). *Biochim. Biophys. Acta* **196**, 211.

Uesugi, S., Dulak, N., Dixon, J. F., Hexum, T. D., Dahl, J. L., Perdue, J. F., and and Hokin, L. E. (1971). *J. Biol. Chem.* **246**, 531.

Ussing, H. H. (1949). *Acta Physiol. Scand.* **19**, 43.

Ussing, H. H. (1960). In: *The Alkali Metal Ions in Biology.* Handbuch der experimentellen Pharmakologie, vol. **13**, p. 10. Springer-Verlag, Berlin—Göttingen—Heidelberg.

Ussing, H. H. and Zerahn, K. (1951). *Acta Physiol. Scand.* **23**, 110.

Vandenheuvel, F. A. (1963). *J. Am. Oil Chem. Soc.* **40**, 455.

Vanderkooi, G. and Capaldi, R. A. (1972). *Ann. N. Y. Acad. Sci.*, **195**, 135.

Vanderkooi, G. and Green, D. E. (1970). *Proc. Natl. Acad. Sci. USA* **66**, 615.

Vanderkooi, G. and Sundaralingam, M. (1970). *Proc. Natl. Acad. Sci. USA* **67**, 233.

van Zupthen, H., Demel, R. A., Norman, A. W., and van Deenen, L. L. M. (1971). *Biochim. Biophys. Acta* **241**, 310.

Verkeleeij, A. J., Zwaal, R. F. A., Roelofsen, B., Comfurius, P., Kastelijn, D., and van Deenen, L. L. M. (1973). *Biochim. Biophys. Acta* **323**, 178.

Verwey, E. J. W. and Overbeek, J. T. C. (1948). *Theory of the Stability of Lyophobic Colloids.* Elsevier Publ. Co., New York.

Vidaver, G. A. and Shepherd, S. L. (1968). *J. Biol. Chem.* **243**, 6140.

Vieira, F. L., Caplan, S. R., and Essig, A. (1972). *J. Gen. Physiol.* **59**, 60.

Vincent, J. M. and Skoulios, A. E. (1966). *Acta crystallogr.* **20**, 432.

von Mohl, H. (1851). *Grundzüge der Anatomie und Physiologie der vegetabilischen Zelle.* Braunschweig.

von Szyszkowski, B. (1908). *Z. physik. Chem.* **64**, 385.

Wallach, D. F. H. (1967). In: *The Specificity of Cell Surfaces* (ed. by Davis, B. D. and Warren, L.), p. 129. Prentice-Hall, Englewood Cliffs, N. J.

Wallach, D. F. H. (1969). *J. Gen. Physiol.* **54**, 3s.

Wallach, D. F. H. (1971). In: *The Dynamic Structure of Cell Membranes* (ed. by Wallach, D. F. H. and Fischer, H.), p. 181. Springer-Verlag, New York.

Wallach, D. F. H. (1972). *The Plasma Membrane.* Springer-Verlag, New York—Heidelberg—Berlin.

Wallach, D. F. H. and Gordon, A. S. (1968). In: *Regulatory Functions of Biological Membranes* (ed. by Järnefelt, J.), p. 11. Elsevier Publ. Co., Amsterdam.

Wallach, D. F. H. and Lin, P. S. (1973). *Biochim. Biophys. Acta* **300**, 211.

Warren, G. B., Toon, P. A., Birdsall, N. J. M., Lee, A. G. and Metcalfe, J. C. (1974). *Proc. Natl. Acad. Sci. USA* **71**, 622.

Weiner, J. H., Berger, E. A., Hamilton, M. N., and Heppel, L. A. (1970). *Fed. Proc.* **29**, 341.

Weiner, J. H. and Heppel, L. A. (1971). *J. Biol. Chem.* **246**, 6933.

Wiebelhaus, V. D., Sung, C. P., Helander, H. F., Shah, G., Blum, A. L. and Sachs, G. (1971). *Biochim. Biophys. Acta* **241**, 49.

Wiemken, A. and Nurse, P. (1973). *Proc. IIIrd Internat. Special. Symp. Yeasts*, Part II, p. 331. Print OY, Helsinki.

Wilbrandt, W. (1954). *Symp. Soc. Expt. Biol.* **8**, 136.

Wiley, W. R. (1970). *J. Bacter.* **103**, 656.

Willis, R. C. and Furlong, C. E. (1974). *Fed. Proc.* **33**, Abstr. no. 963.

Wilson, G. and Fox, C. F. (1971a). *J. Mol. Biol.* **55**, 49.

Wilson, G. and Fox, C. F. (1971b). *Biochem. Biophys. Res. Commun.* **44**, 503.

Winne, D. (1973). *Biochim. Biophys. Acta* **298**, 27.

Winzler, R. J. (1969). In: *Red Cell Membrane Structure and Function* (ed. by Jamieson, G. A. and Greenwalt, T. J.), p. 157. Lippincott, Philadelphia.

Wirtz, K. W. A. and Zilversmit, D. B. (1968). *J. Biol. Chem.* **243**, 3596.

Wirtz, K. W. A., Kamp, H. H. and van Deenen, L. L. M. (1972). *Biochim. Biophys. Acta* **274**, 606.

Wolfe, L. S. (1964). *Can. J. Biochem.* **42**, 971.

Wong, J. T. F. and Hanes, J. (1962). *Can. J. Biochem. Physiol.* **40**, 763.

Yamashita, S. and Racker, E. (1969). *J. Biol. Chem.* **244**, 1220.

Yang, J. T. (1967). In: *Poly-α-amino Acids* (ed. by Fasman, G.), p. 239. M. Dekker, New York.

Yariv, J. J., Kalb, Katchalski, E., Goldman, R., and Thomas, E. W. (1969). *FEBS Lett.* **5**, 173.

Young, I. G., Cox, C. B. and Gibson, F. (1967). *Biochim. Biophys. Acta* **141**, 319.

Zadunaisky, J. A., Candia, O. A., and Chiardini, D. J. (1963). *J. Gen. Physiol.* **47**, 393.

Zahler, P. (1969). In: *Modern Problems of Blood Preservation*, p. 1. G. Fischer-Verlag, Stuttgart.

SUBJECT INDEX